A Guide to Remote Sensing

To Joan

A Guide to

Remote Sensing

Interpreting Images of the Earth

S. A. Drury
Lecturer in Earth Sciences
The Open University

Oxford New York Tokyo
OXFORD UNIVERSITY PRESS
1990

Oxford University Press, Walton Street, Oxford OX2 6DP

Oxford New York Toronto
Delhi Bombay Calcutta Madras Karachi
Petaling Jaya Singapore Hong Kong Tokyo
Nairobi Dar es Salaam Cape Town
Melbourne Auckland

and associated companies in
Berlin Ibadan

Oxford is a trade mark of Oxford University Press

Published in the United States
by Oxford University Press, New York

British Library Cataloguing in Publication Data

Drury, S. A. (Stephen A., 1946–)
A Guide to Remote Sensing: Interpreting Images of the Earth
1. Remote Sensing
I. Title
621.36′78
ISBN 0-19-854494-4
ISBN 0-19-854495-2 (pbk.)

Library of Congress Cataloging-in-Publication Data

Drury, S. A. (Stephen A.), 1946–
A Guide to Remote Sensing: Interpreting Images of the Earth
Includes bibliographical references.
1. Remote sensing. I. Title
G70.4.D78 1990 621.36′78–dc20 89-37597 CIP
ISBN 0-19-854494-4
ISBN 0-19-854495-2 (pbk.)

Typeset by Joshua Associates Ltd, Oxford
Printed in Hong Kong

Preface

Remote sensing is not a new subject, but it is one that has undergone an explosive growth over the last 20 years, in terms of investment, research, and design of new techniques. Although a routine tool in military and economic intelligence—indeed it is the single most important tool for verification of arms-control treaties—apart from in a few civilian applications, it is yet to come of age. Most activity outside the intelligence community still revolves around testing more and more sophisticated technologies on a variety of scientific problems, often covering small, relatively well-known test areas. Yet the very nature of the way information is gathered in remote sensing means that there is a rapidly growing stockpile of hitherto unused data.

This book sets out to demonstrate that there is a huge range of intrinsically practical uses to which these data can be directed. Taking advantage of the opportunities depends on a much more widely spread understanding of what remote sensing is all about, the problems that it can address, and the costs and efficiency involved. Rather than addressing the professional scientific community, which is well catered for anyway, my target is extremely broad. On the one hand are those individuals in government, in national surveys of one kind or another, those working for aid and development agencies, and managers in commercial concerns, who wield considerable power in directing and managing the collection of information appropriate to their responsibilities. Given a background briefing in remote sensing, beyond a somewhat hazy understanding that such a thing exists, they can ensure growth in useful practical applications. Then there is the future generation of users of remotely sensed information, for whom there is minimal provision of basic educational material. If at all, they have to rely on either coffee-table glossies or advanced textbooks in individual subjects. Finally, and most numerous, is that ill-defined concept of the interested layperson, who wishes to deepen and broaden his or her understanding of the environment that they inhabit.

Because it aims at familiarization, much complexity relating to basic scientific principles, technology, and the scientific areas which remote sensing can serve has been glossed over, or not mentioned at all. But that does not mean that some demand is not made on the reader to come to grips with the rudiments of an extremely sophisticated subject. I have not hesitated to use and define a fair amount of remote-sensing jargon—any reader going further will otherwise be overwhelmed—and an easy reference to definitions and acronyms is contained in a glossary. Because the book aims to draw readers into *using* remote sensing, it is as comprehensive as I can make it, and that means that it is copiously illustrated with images, the *sine qua non* of the subject. Because there are so many, often three or four to a page, and because each is discussed from the standpoint of its content and application, this is not a book for the casual browser, but aims to be read. That being said, the nature of the subject has allowed me to assemble some of the most aesthetically compelling and intellectually stimulating pictures of every aspect of the planet that we inhabit. In large measure this is thanks to the generosity of a large number of colleagues who loaned their best images to the project.

I did not set out to write a book, but began with the zealous intention of setting up a familiarization course in applied remote sensing, based on the distance-teaching methods that we have developed at the Open University over the last two decades. It was intended to be a 'world course', a grand design for a grand subject. Such grandiose ambitions demand grandiose resources. It took no more than a few months of visiting international agencies with the wherewithal to stake such a project—I shall not mention them here for the sake of politeness—to discover that there is a certain resistance to 'world courses' in anything. Most grandiose agencies have a vested interest in running their own courses in remote sensing, along quite conventional and controllable lines, so they were none too keen to fund an upstart competitor.

It was at this dismal stage that Geoff Holister of UNESCO stepped in with sound advice; effectively, 'Do what is possible, the rest might come later'. On and off for three years Geoff has kept the pot boiling, and his support and encouragement are important reasons why I started the book, and finished it. Finding a publisher willing to commit to the level of colour demanded by the subject was no easy matter. My editor at the Oxford University Press hardly batted an eyelid. I hope that I have not misled him or my readers.

Milton Keynes S. A. D.
December 1989

Cover illustration

Specially enhanced SPOT image of the Aravalli Hills, Rajasthan, India. Reproduced by permission of SPOT-IMAGE, Toulouse, France.

Contents

Introduction

There is no doubt that our planet increasingly faces problems, and that the scale and complexity of the problems are increasing too. Many are generally put down to overgrowth in population and finite natural resources, whereas the majority have an economic and political source. The post-war disintegration of the great empires of France, Britain, Holland, and Portugal, and the struggle for self-determination by the emergent nations has seen a growth in aspirations to the material standards of their former masters. These justifiable ambitions have become channelled with the economic survival of the former imperial powers and other developed countries. As the world economy enters a recessionary phase, we can expect a worsening of many of the problems, including military conflict. The areas most prone to disaster are the emergent, once colonial nations, and it is there that global economic decline could be manifested in its most savage form.

That we are increasingly aware of the magnitude and impact in human or environmental terms of most of the problems is due to the rapid improvement in telecommunications over

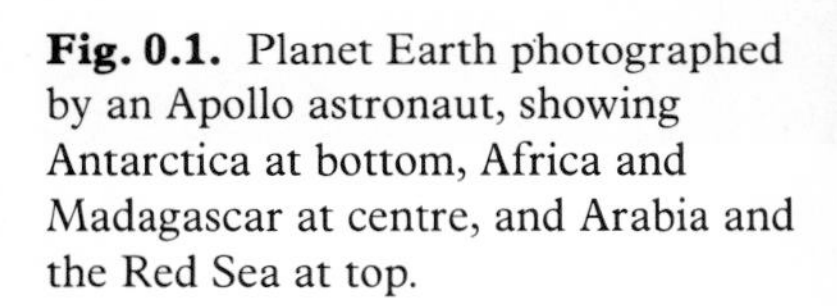

Fig. 0.1. Planet Earth photographed by an Apollo astronaut, showing Antarctica at bottom, Africa and Madagascar at centre, and Arabia and the Red Sea at top.

Fig. 0.2. The Himalaya photographed by a member of the Space Shuttle crew. The plains of northern India are at right, the High Himalaya at centre and the Tibetan Plateau at left.

the last three decades. Thanks to scientific and socio-economic research there is a growing understanding of what the problems and their root causes are. Technological advances are beginning to offer opportunities to soften or even eradicate many of them. The research has shown where the areas of greatest risk occur, and it is becoming clearer that many issues are related and may be served by common solutions. Together with more widely accessible television, the results of this research have been passed on to the population at large through a wealth of educational broadcasts, which are by no means the exclusive privilege of the developed world. Knowing that a problem exists, and even knowing its cause, is of little comfort to those afflicted, particularly when remedies are apparent. Solutions must somehow be implemented, but with the assurance that they do not bring on further, and perhaps worse problems.

A simple case is the emergency provision of food aid to a famine-struck area. Frequently this is accompanied by a sudden drop in local food prices that severely affects indigenous agriculturalists and pastoralists. Another example is major irrigation, seemingly an obvious solution to famine, but one which often results in deposition of salts in the soil, eventually rendering it infertile. Huge reservoirs carry many benefits of water supply, hydropower, and fisheries, but increase the dangers of water-borne disease, displace previous occupants of the site, and may even trigger earthquakes through loading of the Earth's crust.

Major problems are geographic in nature; they involve vast areas of land and sea, climatic systems, regional distribution of water, vegetation, soil and rocks, lines of physical communication and trade, political, ethnic, and language boundaries. The bulk of knowledge about the problems, their causes, and possible solutions comes mainly from investigations of tiny and often isolated examples. A change for the better demands that the issues are addressed on the scale at which they really present themselves. Remote sensing does not directly hold forth solutions, but supplies a publically accessible pool of information that matches the interconnectedness, diversity, and scale of global problems in a way that has never been possible before the last two decades of the twentieth century. As we shall see, it maximizes diversity of information and area of coverage at minimal cost, and allows the interpreter to address several problems far more rapidly than by any other method.

Remote sensing developed from military intelligence applications, starting with photographs of the disposition of Confederate trenches from balloons during the American Civil War. It has since extended into means of discriminating camouflage from natural surfaces, night-time surveillance by radar, and detecting the presence of individual troops and disturbed ground above weapons caches. Economic and scientific pressures, particularly in the United States, forced many of the techniques into the public domain. Now it is safe to say that the basic technology is available to all, the lid of security covering only methods capable of the finest resolution from the greatest possible distance. The secrecy generally covers what the military intelligence experts are looking at and the detail available to them, rather than how they are able to maintain surveillance. It is an interesting reflection on the dominance of economic interests that a large volume of military remote sensing is now dependent on civilian satellites, such as the US Landsat and French SPOT.

In this book, I have focused on remotely sensed images of the Earth, because they convey much of the information that we need very efficiently, and in a form that anyone can learn to understand with a minimum of effort. It starts with a review of global problems, though by no means all the problems are addressed. Chapter 2 shows how natural processes involving electromagnetic radiation—light, infrared, and microwaves—can be exploited to acquire knowledge from a distance. Chapter 3 deals with how the Earth's natural features appear using remote sensing. This leads on to examples of how images of the Earth have served a comprehensive range of applications in Chapter 4. The final Chapter examines some of the practicalities involved in using remotely sensed information, such as price, availability, and advanced training, as well as some political issues. Please note that north is towards the top of the satellite images, except where specified.

1 The problems

1.1 Water

The greatest problem with water is that there is often either too little or too much of it for human requirements. The first provision of any human society is a safe, dependable water supply. The second is the siting of habitation away from areas prone to inundation. All development agencies recognize that for the bulk of the world's population the priority of a water supply suitable for drinking, watering stock, and irrigation is still awaiting adequate solution. Population growth inevitably makes the problem more acute. Children are most at risk, easily avoidable water-borne disease and dehydration still accounting for a high proportion of world infant mortality and morbidity. Without the guarantee of regular and sufficient crops that stable water supply helps provide, or with the threat of floods destroying viable crops, the basic economy of a region cannot break the poverty barrier to full development.

Fig. 1.1. During drought the poor of Rajasthan, India, in common with millions worldwide, are forced to turn to heavily polluted bodies of standing water for both drinking and washing.

Drought and its attendant famine increasingly provide the most searing images of human misery and desperation. The limit for existence by dependence on rainfall alone for crops and livestock is between 400 and 100 mm precipitation per year, the semi-arid deserts. Below 100 mm, in truly arid regions, drought is endemic and survival depends entirely on supplies from beneath the ground. Outside the semi-arid regions, an annual shortfall of a few hundred millimetres of rain is not a life-threatening problem, provided agricultural productivity is not overwhelmed by the sheer density of population, as it can be in the Indian subcontinent for instance. In arid to semi-arid areas it is potentially disastrous. For millions in the sub-Saharan area of northern Africa—the Sahel—survival is, at present, entirely dependent on adequate and timely rainfall.

As well as causing food shortage, drought cripples the economy of subsistence farmers. If they survive, they are often made destitute and must flee their former homeland. Gravitation to cities, the immediate source of food aid, reduces the productivity of rural land when adequate rainfall does return, and pressures urban society to breaking point. Despite famine, population in the Sahel continues to grow, and the vicious circle of drought, poverty, and migration now seems to most outside observers and to many of its victims to be an unstoppable force, with a life of its own. This need not be so, as numerous examples demonstrate. In semi-arid areas of North and South America, Australia, Arabia, the Mediterranean fringe of Africa, and southern Africa agricultural economies in low-rainfall areas support large populations and maintain a thriving export trade. Survival and agricultural development there is buffered from drought by exploiting ground-water, and by managing rainfall through dams and water transfer.

Fig. 1.2. Economically crippled by a three year drought, these Rajasthani farmers have abandoned their land and are seeking a means of survival in India's rapidly growing cities.

Water is an ephemeral part of the environment, continually on the move. Its ultimate source and eventual resting place is the sea, but without extremely costly desalination plants seawater is of little comfort. Moreover, most drought-prone areas are far from the coast. Depending on the more or less fixed climatic zones and wind belts, water vapour evaporated from the oceans falls as rain or snow on the land, a major control being the cooling effect of air being forced to rise over mountains, as well as the collision of air masses of different temperatures and densities.

At high elevations or high latitudes some precipitation enters long-term storage as glaciers and ice caps. For the habitable parts of the planet, precipitation is in the form of rain. This suffers three main fates. It may flow at the surface, augmented seasonally in places by snow melt. Some of this run-off can enter temporary storage in lakes, but most reaches the sea within a year or less. A proportion of rainfall soaks into the ground, where it may enter the entirely hidden regime of ground-water. This too forms a temporary store, but residence times can be as long as tens of thousands of years, although they are usually in tens of years before the water emerges in springs. Thirdly, a proportion of rainfall returns to the atmosphere as vapour before reaching storage. It may evaporate directly from surface water, by the heating of damp soil or through the intervention of vegetation that takes up water from below the surface, uses it in metabolism, and then exhales or transpires it in the form of vapour. The relative amounts of precipitation apportioned to run-off, ground-water storage, evaporation, and transpiration depend on many factors. These include topographic relief, surface temperature, rock and soil type, and vegetation cover.

Like most natural cycles, that of water involves all the attributes of the Earth's surface environment. Understanding this cycle requires inputs from climatology and meteorology, geomorphology and hydrology (the study of surface water), geology and hydrogeology (the study of water's behaviour below surface), and knowledge of the distribution of natural vegetation. Exploiting it means focusing on those parts of the cycle involving storage—lakes, and ground-water—and intervening in the form of dams, wells, aqueducts, and pipelines. This intervention does not necessarily involve high technology and huge capital. Hand-built dams, wells, and aqueducts date back over 4000 years, even in extremely arid areas of west Asia and the Indian subcontinent. Since the bulk of obvious sites for this simple kind of development have long been exploited, what is needed is information pointing the way to the less obvious.

Fig. 1.3. One solution to drought is the damming of large rivers and canalization of water for irrigation. In Karnataka, the State government and local people have constructed thousands of kilometres of canals by hand.

Fig. 1.4. Highland agriculture benefits from assured rainfall, but faces disruption by soil erosion. In the Palni Hills of south India, this is combated by terraced plots of land.

Remote sensing has an important role to play in this, since it can provide information on landforms, drainage systems, near-surface geology, and vegetation. It has another role too, perhaps more urgent while drought remains such a massive threat. It can monitor cloud cover—a fair sign of the distribution of rainfall—atmospheric moisture, and the season-by-season progress of natural vegetation on a continuous basis, as a means of giving early warning of the onset of drought over the vast areas that are prone to it.

1.2 Food

Agricultural food supply is the product of a web linking soil, water, sunlight, seed stock, fertilizers, technique, human effort, economy, and planning. The dominant world economy of the twentieth century has culminated in North America and Europe having vast surpluses when globally there is gross undersupply. That the surpluses are locked in permanent storage and huge stocks of food are destroyed or denatured to maintain prices is a savage indictment of that economy and its management. Post-war developments in seedstock, fertilizers, and agricultural methods, could, given the political will and economic resources, remove the unequal distribution of food in a very short time. However, it is not merely a question of throwing money, equipment, and knowledge at the problem of agricultural underdevelopment.

A large agricultural bureaucracy in the developed world, with a system of compulsory crop and livestock returns, is able to inventory production on an annual basis. Despite their well publicized failings, such as those connected with the EEC Common Agricultural Policy, such statistics are able to highlight what is being produced in different areas, how they match the demands of the market and how efficiency varies from area to area. They could and should form the basis for rational agricultural planning globally. For the bulk of the Earth's surface area, nothing is known in detail, except for the produce that passes through ports for export. Not only are the volume, type, and efficiency of agricultural production vaguely known, but even the location of existing agricultural areas. Destructive practices (such as slash and burn cropping), being undocumented go unchecked. Areas of temporary local surplus or crop failure go unnoticed. Planning and central direction of farming in an information vacuum is futile.

Likewise, compulsory reporting of crop diseases and pest infestation in developed economies enables precautionary measures to be implemented swiftly. Where communications and infrastructure are poor, often in areas with far greater natural threats from diseases and pests, such problems can go unnoticed and unchecked for long periods. Then it is often too late for remedies.

Where agricultural practices have been changed as part of development programmes, they need to be monitored to check if the hoped-for results are forthcoming and to suggest modification. To put such changes into effect, or to open up new areas to agriculture requires more than just well-meaning notions of what ought to improve the situation in a particular area. What may have been successful in one location is not guaranteed to run smoothly in another. As well as the obvious need for adequate water supplies to support the venture, much needs to be known about the type of soils involved, for many crops are only suited to a limited range of soils, and some soils quickly lose water and fertilizers through their rapidly draining structure. Soil itself is not as durable as it might appear. Under

Fig. 1.5. The scourge of many semi-arid areas is a combination of overgrazing, especially by sheep and goats, and fuel gathering. In the Aravalli Hills of Rajasthan hundreds of square kilometres have been stripped bare of all vegetation since 1980.

Fig. 1.6. Despite frequent air raids by Ethiopian forces, new agricultural developments, based on flood irrigation and tube wells, are growing rapidly in the arid mountains of Eritrea.

natural conditions it is balanced and has stabilized with surface run-off, slope angle, and the binding effect of vegetation roots. The same land may sustain primitive agriculture in small fields without endangering the soil. However, highly technological methods, involving land levelling, deep ploughing, and larger fields can break the balance, resulting in erosion of the soil by surface water or wind destroying the agricultural potential rapidly and permanently. All the variables relating to soil suitability and erosion need to be known in advance of commitment to change.

One of the contributory factors in the spiral of famine and poverty in the Sahel and many other marginal areas is destruction of the vegetation that helps to bind soil through overgrazing by herds, and the encroachment of desert. As drought and famine force once nomadic herders to remain around assured water supplies and feeding stations, so the pressure on local grazing increases. Overgrazing is not an exclusively Sahelian issue, however, but is common to all the great savannahs and steppes of Asia, Africa, and South America simply through population growth among traditional pastoralists. A more closely managed use of rangeland is required, and once again greater and more timely knowledge of availability of browse is needed to reverse the situation. Nomadic herders have traditional migration routes and seasonal pasturelands, yet climatic fluctuations change the conditions from area to area. They require reliable and timely direction to fertile areas.

Probably the greatest untapped resource of protein resides in the oceans. Conventional fishing centres on a limited range of marketable species on traditional fishing grounds, many of which are near to exhaustion. Much of the sustainable biomass of the oceans remains a largely unknown quantity. The economic and physical risks involved foster neither a change in cropping patterns nor an increase in fishing, especially in Third World maritime areas. Knowing that a resource exists and where it is likely to be on a timely and reliable basis could form the impetus for a growth in fisheries.

Modernizing, developing, and managing food resources require inputs of knowledge from climatology, meteorology, soil science, ecology, hydrology, oceanography, and economics. As we shall see, remote sensing forms a source of much of the information needed for implementing more efficient agricultural methods, management of food resources, and development of new agricultural areas. Since it can detect and quantify the aquatic plants the form the start of the marine food chain, it has important implications for mariculture too.

Fig. 1.7. Many coastal people in the less-developed world subsist by fishing. Remote sensing can help locate shoals of many food species, as described in Chapter 4.

1.3 Natural disasters

By definition, disasters appear suddenly and with little warning. They are short-lived, extreme events bringing death, injury, and destruction of buildings and communications. They often have little, if any, connection with economic or political systems, affecting rich and poor alike. Their aftermath can be equally as damaging as their intrinsic physical effects, through disruption of sanitation and water supplies, destruction of housing and hospitals, and breakdown of transport for food, temporary shelter, and emergency services.

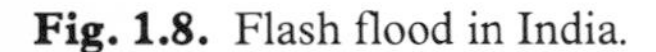
Fig. 1.8. Flash flood in India.

Any natural phenomenon acting to extremes can constitute a disaster if it occurs in an area of human habitation. If there is no warning of the disaster, the physical threat to life and problems in its aftermath are all the greater. Primary causes are earthquakes, volcanic eruptions, intense rainfall, and high winds. Frequently the worst effects stem from secondary phenomena. Landslides accompany heavy rain and earthquakes. Mudflows may result from lavas melting snow and ice high on the flanks of a volcano. Hurricanes can conspire with tides to force inundation of low-lying coastal areas, with much the same results as tsunamis generated by sub-sea earthquakes.

There are several types of knowledge that are needed in relation to disasters; knowledge concerning their cause, means of early warning, their physical effects, and the areas of immediate and long-term need for aid that they create. Analysing the cause of disasters is crucial for long-term planning. From this can stem an assessment of future risk and identification of other sites prone to the same phenomenon. Measures can then be taken to avoid concentrating population or siting potentially dangerous installations, such as nuclear power stations and dams, in areas at risk. At the very least, contingency plans for evacuation and emergency services can be drawn up and practised in areas identified as risky. Scrutinizing records for signs of impending events relating to the period immediately before disaster has struck can help design means of prediction and warning. To a large extent, the rarity and unique nature of most disasters provides only a scanty database on which to build causative models and scenarios for future events. For some considerable time disasters will come unannounced and the main thrust will focus on immediate and

Fig. 1.9. Cyclonic storm in the Pacific, photographed by a Space Shuttle astronaut.

long-term relief. However, because disasters manifest natural processes on a large scale, information provided by remote sensing is a most apporpriate input to analysis of actual events and investigation of other areas of potential risk.

The greatest problem for disaster relief is one of physical communications. The disruption of telecommunications often means that viable routes are unknown, as too are the locations of greatest need for assistance. Planning emergency aid is not dissimilar to a battle—intelligence of terrain is vital. Restoration of damage in the aftermath requires some assessment of priorities, itself based on an inventory of the area affected. In either case, speed is of the essence. In this regard, only remote sensing can provide the information required in time.

Fig. 1.10. Volcano Colima in Mexico during a glowing avalanche or *nuee ardente* eruption in 1983. The eruption has moved rapidly down the flank of the volcano and has entered a forested area. (Courtesy of R. S. Thorpe, The Open University.)

1.4 Physical communications

Transport of goods, people, and in cases of disaster, relief supplies, requires clearly defined routes combining safety with convenient linkages of populated areas. Those of us in developed countries take all forms of physical communications for granted, they are there and being continually upgraded. There is little need for major new routes, except to relieve the strain of increasing traffic. Elsewhere, development is dependent to a large degree on establishing strategically aligned new routes to a standard suitable for modern vehicles. In some regions it is not even known with any precision where large numbers of semi-nomadic peoples are located. During the 1984–85 famine in north-east Africa this was a great hindrance to distribution of relief supplies, let alone the physical problem of transport in areas lacking all but the most rudimentary tracks.

Planning a new road or rail link requires a considerable range of information. The objective is to minimize length whilst maximizing safety and stability. Costs must be kept to a minimum, by avoiding high-cost bridges and tunnels but balancing this with the expense of increased length through detours or greater transportation distances for suitable construction materials. Another important factor is ensuring that gradients are not excessive, and that the costs of cuttings and embankments to reduce them do not rise too far. Stability demands knowledge of the bearing capacity of rocks and soils that are crossed,

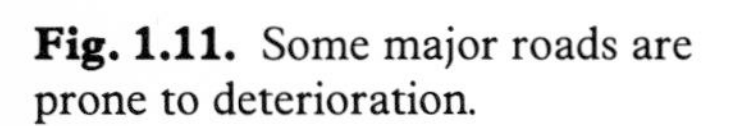

Fig. 1.11. Some major roads are prone to deterioration.

Fig. 1.12. Silting of rivers and rapid changes in the position of sandbanks creates great problems for commercial navigation.

the locations of potential hazards from earthquakes, landslips, and floods, and the potential sources of suitable aggregate. In spite of all these conditions, the route must serve as many centres of population or sites of economic importance as possible.

Water-borne transport is a rather less complex issue, but has its own difficulties and problems. Where is the best site for a port? It must be protected from storms yet have easy access to deep water without the chances of rapid silting. The near-shore bathymetry must be known in detail. River traffic is dependent on a sound knowledge of the year-to-year changes in the courses of channels and the location of sandbanks that characterize most navigable waterways. Marine shipping suffers hardly at all from route restriction, but is exposed to considerable hazard from wind, storm waves, fog, and ice.

Fig. 1.13. Shifting sand forms a major threat to highway development in arid and semi-arid terrains. Areas prone to this must be identified well before construction begins.

Inputs to planning new communications and improving safety on existing routes are from social and economic geography, geomorphology, geology, oceanography, meteorology, and engineering. Perhaps more than any other set of problems, those associated with transport are most appropriately addressed by remotely sensed information, simply because they involve such large areas of the Earth's surface.

1.5 Physical resources

A simple piece of arithmetic highlights the principal issue relating to commodities won from the Earth. Take the average yearly consumption per capita in the developed world of energy or of any metal, fertilizer, or building material, and multiply it by the total world population. In every case the product exceeds current annual production by mines or oilfields by a staggering amount. Beyond mere survival, that is what development is really about. Despite the notions of well-heeled pundits concerning depletion of finite resources and environmental effects of mining and energy use, some three billion disadvantaged people have the right to expect the same material standards as those prevailing in the advanced economies. The opportunities presented by modern science and technology demand it. Development means nothing more nor less than finding and winning the shortfall in supplies of physical resources and sustaining them.

Historically, the majority of existing mines and oilfields were almost literally tripped over by explorationists with quite rudimentary skills. They were visible at the surface and extraordinarily rich. As these easily found sources become depleted, the search switches to other areas for similarly easy deposits—the economics of industry demands that this approach will continue until all of them are exhausted. Exhaustion comes first in the developed countries and exploration increases in the less fortunate areas, but it is dominated by the multinational corporations to whom the eventual profits flow. The commodities feed the developed economies and only trickle back to their source in the form

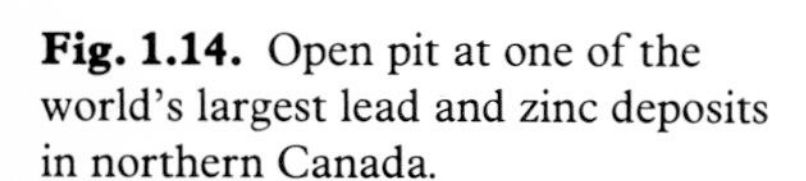

Fig. 1.14. Open pit at one of the world's largest lead and zinc deposits in northern Canada.

of expensive manufactured goods. Development means the finding, extracting, and processing of physical resources to finished goods by the peoples of the areas in which they are located. Only through that route is security and self-sufficiency guaranteed.

The discovery of physical resources concentrates mainly on the fact that they consist of unusual concentrations of the materials that are found in common rocks. The concentrations are the results of geological processes acting in an extreme way. Part of the exploration seeks to detect anomalies, in the chemical composition of rocks due to concentration of metals, in their physical properties due to the compounds in which metals are found, or in their structure which may have been conducive to the deposition of metallic compounds or the migration and trapping of hydrocarbon liquids and gases. This strategy employs a large range of instruments aimed at monitoring rock properties that are critical to the presence or absence of deposits. The other main thrust is in analysing the geological relationships and evolution of known deposits, to understand the processes involved in their formation. These notions, or models, set in the context of general geological knowledge of an area, can then be used in a predictive way to suggest particular kinds of rock or structure in which such processes may have operated. Equally important, they can filter out those settings where such processes are unlikely to have occurred, or if they did occur were unable to cause the necessary concentration. This strategy allows the instrument-based exploration methods to be deployed efficiently, rather than in a very costly blanket fashion.

As the 'trip-over' deposits are found and worked out, so it becomes increasingly difficult to replace the supplies that they represented. Less easy targets buried at depth, in a form more like common rocks, located in remote areas or with very subtle geological controls, force increased costs on exploration as more sophisticated methods and analysis are required. Such deposits are more financially risky than bonanzas, and capitalist market forces, such as fluctuations in commodity prices, have a more rapid effect on their viability. Cutting down the element of financial risk therefore plays a major role in exploration for the kinds of deposits upon which development will depend. Put very simply, this means

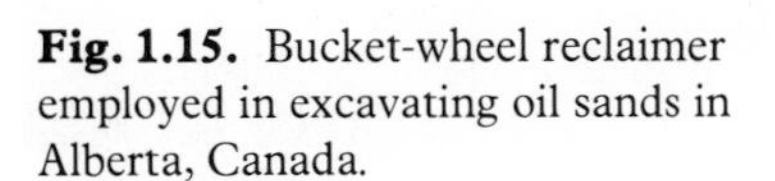

Fig. 1.15. Bucket-wheel reclaimer employed in excavating oil sands in Alberta, Canada.

cheaply and quickly reducing the area over which expensive exploration methods must be deployed to small parcels of ground in which chances of a find are high. Where the geological features of the crust are well mapped, this is a relatively simple matter of applying the vast and growing body of knowledge about the ways in which resources were formed during the Earth's evolution to the available geological information of the area.

There is a catch. The best geological maps are of areas with a long history of resource development. More than 70 per cent of the Earth's continental surface—mainly in the Third World—is known geologically only in a sketchy, and often outdated way. This is despite the attentions of thousands of geologists for more than a century. To complete the geological map of the Earth, by conventional methods, to the standards required for exploration will take centuries. A more timely completion needs either a huge increase in the number of well-financed geologists or the development of rapid, comprehensive, and cheap methods of mapping the rocks. This fundamental geological input can, within limits, be supplied through remote sensing.

Finding deposits is not the culmination of the process. They have to be carefully assessed in relation to economic factors, and means sought for bringing in the heavy equipment needed for exploitation. During production, waste must be disposed of as safely as possible and without covering possible extensions to the deposit. A water supply will probably be needed. All these considerations can be served in some way by remote sensing.

1.6 Natural environments

Humanity, its economy, and all the attendant activities are inseparable from the natural world. We are dependent on it, even as far as the ideas in our heads. Parts of the environment change imperceptibly, such as the landscape in which we live and the rocks upon which it is built. Other parts change very rapidly indeed, due entirely to natural processes, such as earthquakes, volcanic eruptions, hurricanes, floods, and landslides. There are some natural changes that are not noticeable on a day-to-day basis, but which are nevertheless measureable, such as the advance and retreat of glaciers, the meandering of rivers, the

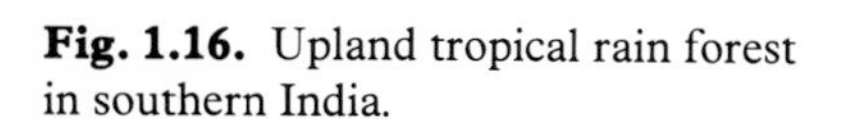

Fig. 1.16. Upland tropical rain forest in southern India.

Fig. 1.17. The Red Sea Hills, Eritrea.

encroachment of deserts on fertile land, and climatic fluctuation in temperature and rainfall. Some of these moderately rapid natural changes stem from others that are at first sight unconnected. A good example is the effect of smoke from forest fires and ash from volcanoes, which block the sun's energy and result in a temporary drop in atmospheric temperature. As humanity grows in size and intervenes more and more in natural processes, there are changes that are set in motion by us alone. The interconnectedness of things, however, means that human-induced changes are seized on and magnified by nature itself, to be passed through the chain of natural events to have far-reaching and sometimes unexpected effects.

Several of these chain reactions are well publicized. Sometimes their trigger seems perfectly safe, albeit unnatural or crude. The use of non-toxic fluorocarbon propellant in aerosols is connected to the destruction of ozone in the upper atmosphere. This removes the shield protecting life from excess ultraviolet radiation from the sun. In Antarctica it has already begun to happen. Overgrazing in the savannah lands and steppes, or repeated ploughing, removes the protection of soil from the action of wind. Desert conditions spread as the soil is eroded and the vegetation mat is destroyed. Deforestation in the tropical rain forests does not merely render the soil beneath infertile, but the transpiration of water vapour back to the climatic system is reduced. This may be connected to far-distant reductions in rainfall. These forests are also the main sites where oxygen is generated through photosynthesis. The harmless gas, carbon dioxide, has the strange property of

Fig. 1.18. The high desert of California.

absorbing thermal energy that the Earth would otherwise lose by radiation. An increase in its proportion in the atmosphere through burning of carbon-based fuels, which is indeed a well-documented fact, could lead to a small rise in the global average temperature. A matter of a degree Celsius or so would alter the climate with some effect on weather, but also with knock-on effects on the melting or polar ice-caps and the global rise of sea level. More than this, such a change feeds back to climate by increasing the polar regions' ability to absorb solar energy, much of which is reflected back to space by the white snow and ice cover. This 'greenhouse effect' is then a self-sustaining mechanism.

The clearest examples of human-induced change in the natural environment result from obviously unsavoury activities. The very nature of many industries forms a potential threat to plants, animals, and humanity through the release of waste products. These range from the downright dangerous, in the shape of radioactive isotopes and toxic metals and compounds, through fertilizers washed from fields into drainage systems, to untreated sewage. Among the toxic wastes, the most widely pervasive is sulphur dioxide gas released by the burning of fuels containing trace amounts of metallic sulphides. This is spread by winds and eventually falls as acid rain, with severe effects on many kinds of tree. Acid water in the soil dissolves naturally occurring but generally stabilized toxic metals, which then wreak havoc on fisheries when the mineralized water enters rivers. Fertilizers entering watercourses do what they are designed to do; they encourage the growth of aquatic plants, generally microscopic ones. The death and decay of such organic 'blooms' consumes dissolved oxygen, as does raw sewage, effectively killing a river or lake as regards its animal life. The ultimate human-induced change, with both toxic and non-toxic intermediaries, is of course nuclear warfare, whose effects hardly need summary here.

The natural world is the source of all our ideas, most obviously those involved in science and technology. Our curiosity demands that we continually learn more about the components, workings, and interconnectedness of the environment. The physical and economic safety of human society require that natural changes are detected, documented, and analysed at the earliest possible moment. It is not at all sentimental to state that we have a responsibility to ensure that changes caused by human intervention are recognized, understood, and rectified.

Fig. 1.19. Hot, acid, oil-charged water is the main by-product of working the oil-sand deposits of Alberta, Canada.

For most of history, natural sciences have been confined to the laboratory or, at best, to areas of the surface that could be encompassed by teams deployed on the ground. Perhaps because of this, the network of links between natural things and processes, which are only now becoming apparent, were hidden in much the same way as individual trees hide the true nature of the forest. Remote sensing steps back from the documentation of the individual to reveal the wealth of variety and interrelation on a grand scale. In this respect it is ideal for the study of change and the wider relations between many components of the environment. It lends itself wonderfully to maps and inventories. So it acts as a database to link the little, disconnected parcels of existing understanding, which result from traditional means of study, in all-sided concepts that are no longer static but more truly reflect the dynamics of nature.

1.7 Security

The least natural facet of human society is physical conflict—warfare. As we shall see, remote sensing has an enormous potential for ensuring human survival, development and harmony with the rest of the natural world. It is ironic, then, that all the techniques that are deployed in remote sensing have their source in helping ensure the destruction of life and wealth as efficiently and cheaply as possible.

The earliest practical exercise in remote sensing was photography of the disposition of forces during trench warfare in the American Civil War. Aerial photography became a dependable tool in reconnaissance for artillery bombardment and monitoring damage in the charnel house of World War I. Infrared photography to discriminate camouflage from vegetation, and radar to allow all-weather detection of aircraft and shipping we owe to the second World War. The post-war period has seen a rapid increase in sophistication and coverage of all conceivable attributes of human and natural phenomena, and, of course, the carrying of surveillance devices on spacecraft. There is very little that can escape detection, or so the main users of military remote sensing think, as we shall show.

Military use of remote sensing falls nicely into several distinct categories. Military planning requires intelligence, primarily to ensure that forces can be deployed in unfamiliar

Fig. 1.20. The aftermath of battle and overgrazing at Afabet, Eritrea, in 1988. (Courtesy of S. Berhe.)

terrain, as efficiently and safely as possible, to inflict maximum damage on opposing forces. The lie of the land has always been of crucial tactical importance, revealing lines of advance and retreat, laying trajectories for artillery and knowing secure bases and sources of water and food supply ahead of operations. This needs accurate topographic maps, and knowledge of habitations, cover, and the nature of the terrain relative to the means of transport. In the era of Cruise missiles and 'smart' bombs, topographic detail and the location of targets are needed with an accuracy of metres. An added bonus is advance knowledge of the disposition of enemy forces, in the shape of manpower, equipment, and defensive structures, in both quantitative and qualitative terms. In the case of counter-insurgency warfare, detail must strive to pinpoint individual guerrillas, arms caches, and anti-personnel devices. At sea the presence and routes of surface and submarine vessels must be monitored if possible.

Increasingly, military action is governed by a warped sense of the market place. Missiles cost money. As well as getting more 'bang for their buck', the warriors require evidence of the cost–benefit of their hardware. Voyeurism reaches its highest pitch in collecting the statistics of destruction after action, by remote means, once again using the best attribute of remote sensing, the detection of change.

With remote sensing, intelligence can expand beyond tactical issues into the strategic realm, where conflict is not an immediate issue. The potential wealth in resources and state of military readiness of any country can be monitored. Foreknowledge of impending problems, such as the failure of the Soviet wheat crop, can allow economic measures of destabilization to be readied in advance. Further into the area of clandestine economic activities, detailed knowledge of potential sites of oil or mineral wealth in undeveloped areas can given multinational conglomerates the 'edge' in negotiating favourable terms for exploration.

Given the intensity of cover and great detail of modern military remote sensing, it is an expensive, though extraordinarily useful tool. Dependence on it for action, with even higher economic cost, is beginning to foster countermeasures, exploiting the very scientific knowledge that guides the technology itself. This is not new. Nazi bombers were coaxed into bombarding areas of open farmland in southern Britain by the construction of mock airfields and factories. A similar, but far more sophisticated and possibly much cheaper tactic today could tie up entire fleets of surveillance satellites and hundreds of interpreters, in the sure knowledge that high-cost ground or air forces would be committed to no avail, while the real offensive could be threaded through the blanket of spoof. This is an approach that lends itself more to the guerrilla than to the conventional army.

The bulk of remote-sensing technology has been transferred into the public domain, largely because of pressures from potential industrial users. Only the most advanced forms of detector, with the capacity to resolve detail to less than a metre from orbital distances, remain secret. The main secret now is not how they are looking but the object of their attention. Consequently it comes as no surprise to learn that the main purchasers of products from civilian remote sensing are the very originators of its technology.

2 Getting the information

Until a matter of 30 to 40 years ago, capturing images of our surroundings was restricted to only a tiny proportion of the spectrum of electromagnetic radiation in which the environment is bathed—the range to which our eyes are sensitive. It is now possible to exploit the full range of this spectrum. In order to understand the wealth of information that new techniques provide requires that at least the rudiments of how radiation and matter interact are understood. This means delving into some aspects of the physics involved, but because the bulk of interpretation hinges on displaying the new information in a visible form, it is vital to know something of the powers and limitations of human vision itself.

2.1 Human perception

Knowledge of our surroundings comes from information picked up by the five senses, of which sight is by far the most important. As well as giving the most varied information, the visual appearance of things is the only source of knowledge at a great distance, without the aid of complex instruments. Even when we probe the interior of the Earth, using the vibrations from earthquakes, or the invisible depths of space from the radio waves emitted by distant galaxies, the results from instruments must be rendered in some visible form before we can make anything of them other than in an abstract way. Since most remote sensing, especially that covered in this book, centres on the acquisition, display, and interpretation of images, it is vitally important that we understand at least the rudiments of how we see.

The human eye has a lot in common with a camera. The *lens* and *cornea* focus an image on a curved photosensitive surface called the *retina*, coated with about 130 million individual detectors (Fig. 2.1). The area of the retina is about the same as that of a 35 mm frame. The human lens has a focal length of about 17 mm compared with the normal lens on a camera of 50 mm. This gives us a *field of view* of about 180° compared with 50° in a camera. Such a short focal length gives a fish-eye effect in a camera, but the curvature of the retina compensates for this. A wide-angle lens has another great advantage. The range in focus, or the *depth of field* of a lens is inversely proportional to its *focal length*, and ours is nine times better than a camera, for a particular *aperture* or *f-stop*. The eye's f-stop is governed by the diameter of the iris, which changes to give a range from f/8 to f/2. The optimum for gathering light and best depth of field is f/4. Curiously, with normal light this is only achieved by being in the right mood and with heightened interest—more interesting still is that this optimum is best achieved by visual and, better still, physical contact with an attractive partner!

The detectors, or *receptors*, on the retina are unevenly distributed over its surface (Fig. 2.2). They are made up of *rods* sensitive to brightness only, which are 20 times more abundant than the *cones*, which are sensitive to colour. Both rods and cones form an integral part of the brain, to which they are directly linked by the *optic nerve*. Cones are concentrated at the very back of the eye, where there are no rods, around the colour-sensitive *fovea*. Both types of receptor fall off in number with increasing *visual angle*. Rods

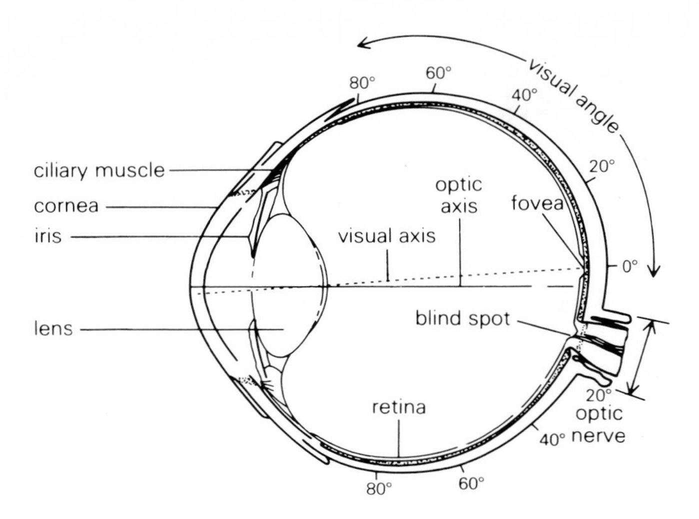

Fig. 2.1. A cross-section of a human eye shows the main elements involved in the focusing of an image on the retina. The chambers in front of and behind the lens contain fluids with different refractive index. They perform most of the refraction in the eye. Muscles which change the shape and focal length of the lens are not shown.

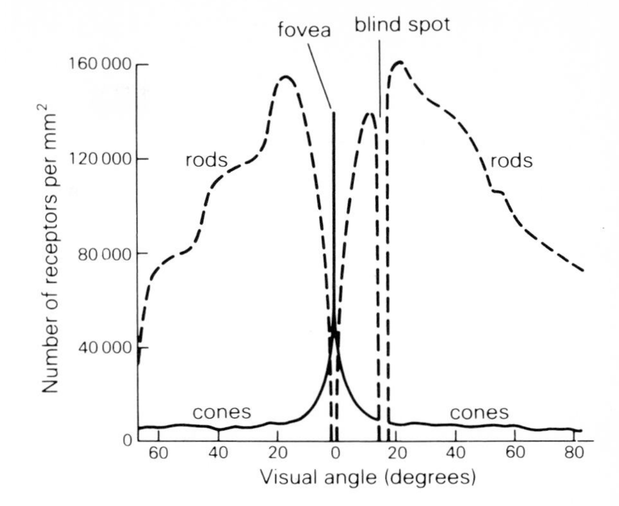

Fig. 2.2. A plot of the density of receptors in the retina shows their uneven distribution, from a maximum around the fovea to a minimum at the periphery of vision.

and cones function by light energy bleaching pigments, which triggers a nerve signal to the *visual cortex* in the brain. Because the regeneration of pigment takes time, overstimulation by looking at a bright light temporarily puts receptors out of action, illustrated, for example, by the black spot seen after the sun has been accidentally looked at.

There are three types of cone with peaks of sensitivity at blue, green, and red (actually orange) wavelengths. Below a certain level of illumination they do not function, rods alone giving *night vision*. The rods are sensitive across the whole range of the visible spectrum, with a peak sensitivity in the blue. An interesting fact is that day vision peaks in the green, the dominant colour emitted by the sun, and the blue sensitivity of night vision (Fig. 2.3) corresponds to the main colour in light from the sky, as we shall see in Section 2.2.

The different receptors, retinal nerves, and neural cells in the visual cortex are grouped in several different types of neural circuit, which perform different functions, some to sense brightness, others colour, lines and boundaries, directions and angles, shapes and three-dimensional vision. All these circuits are linked to short- and long-term memory segments in the brain. Our visual abilities are almost wholly learned (Fig. 2.4), rather than instinctive,

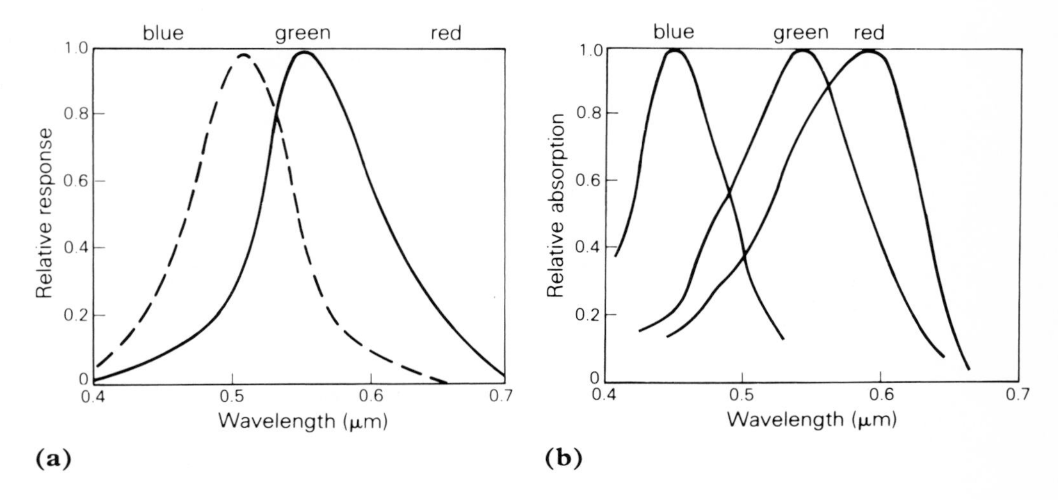

Fig. 2.3. (a) The overall sensitivity of the eye peaks at different wavelengths for night and day vision. (b) The spectral absorption curves for the three different kinds of cone show that two peak at blue and green but that the cone normally associated with red light actually peaks at orange.

unlike those of many other animals. For example a geologist will be better able to detect and interpret features relating to rock structures in a scene than a lay person, simply because he or she has learned the association of particular shapes and patterns that relate to geological features. By the same token, the geologist who has seen more rocks will be proportionally more perceptive. Since short-term memory is involved, everyone begins to see more and more in a scene as time goes by—they get their 'eye in'. Both the expert and the beginner, however, have learned in the framework of the visible range of light and normal horizontal viewing in familiar surroundings. Blue skin, red grass, and green meat disorient us, as do views from unfamiliar angles, such as vertically downwards from an aircraft or satellite, or where we have no clues to scale and depth as in fog or a snowstorm. This *visual dissonance* is frequently encountered in remote sensing, partly by the preponderance of views looking vertically downwards, partly by the absence of clues to scale and depth, but more by the rendition of invisible parts of the spectrum in the visible range. The most common example, as we shall see, is the red appearance of vegetation when images extend detection into the infrared part of the spectrum, but there are many more. For each part of the spectrum and different combinations of these parts in colour images, there is much to learn. There are a few important fundamental principles, however.

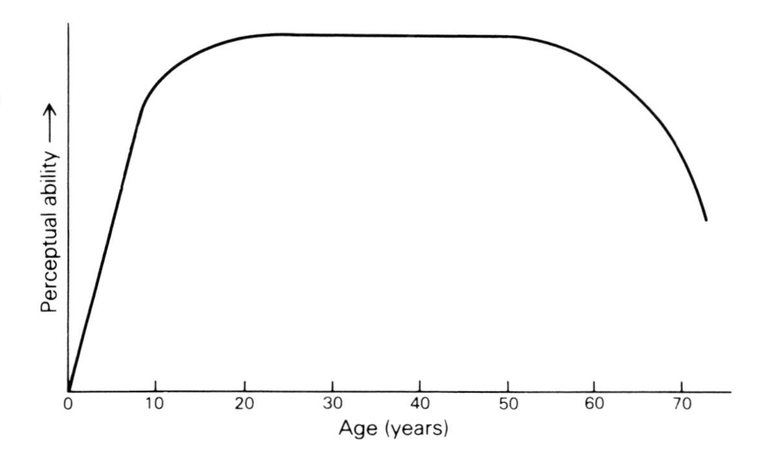

Fig. 2.4. This schematic graph shows the relationship between perceptual ability and age. Most rapid development happens between birth and the age of eight. The ability of adults can be significantly changed by training and new experience.

Because the number of receptors of both kinds is finite, there is a limit to our ability to perceive small objects. In fact, although an image on the retina is broken up or dissected into a mosaic, the scanning motion of the eye and repeated signals to the brain smooth out the mosaic effect. Reality is a continuum, but every image is *dissected* in some way—the brush strokes of a painting, the grain of a photograph, and the *raster* of a television picture. As we shall see later, many remotely sensed images are dissected in a very regular way.

Part of what we see depends on our ability to distinguish spatial detail—our *spatial acuity*. This varies according to our attention. There is a difference between detecting the presence of something unusual and recognizing what it is. How long we look also governs the information that we can extract. A small bright object on a dark background is easier to detect than the opposite, a good example being light hairs on a dark carpet. In this respect *contrast* is very important; black on white is easier to spot than dark grey on light grey. Other factors are the level of illumination and the light-capturing ability of the eye, itself governed by the pupil diameter, which in turn depends on illumination and on our subjective mood. Because the red and blue parts of the spectrum are most affected by aberration in the eye's optical system, yellow and green objects are more finely discriminated. Whatever the conditions, there is a general rule: fine features are seen much

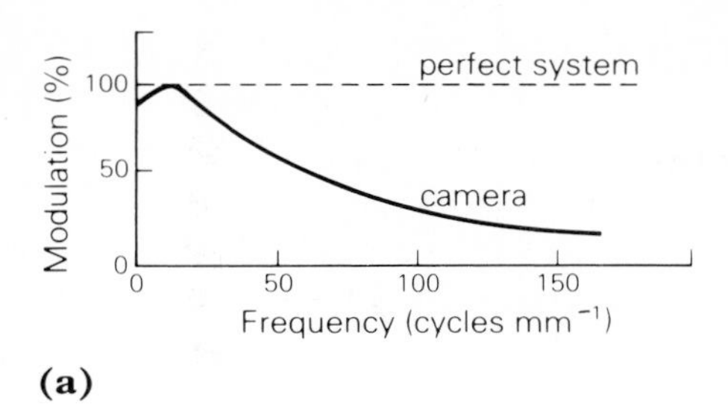

(a)

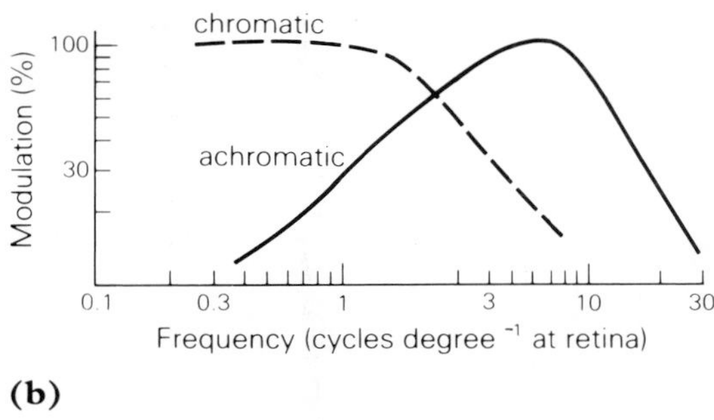

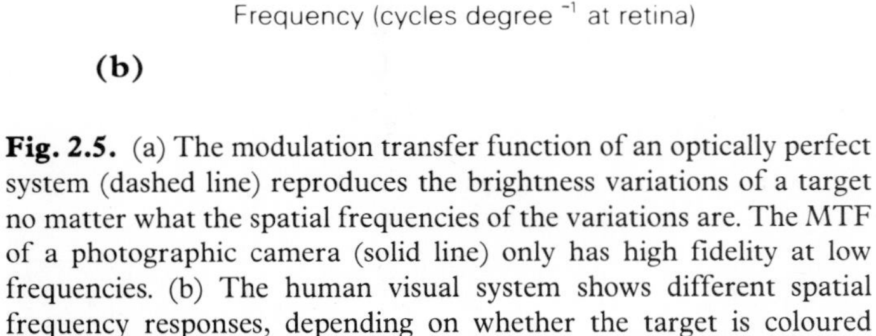

(b)

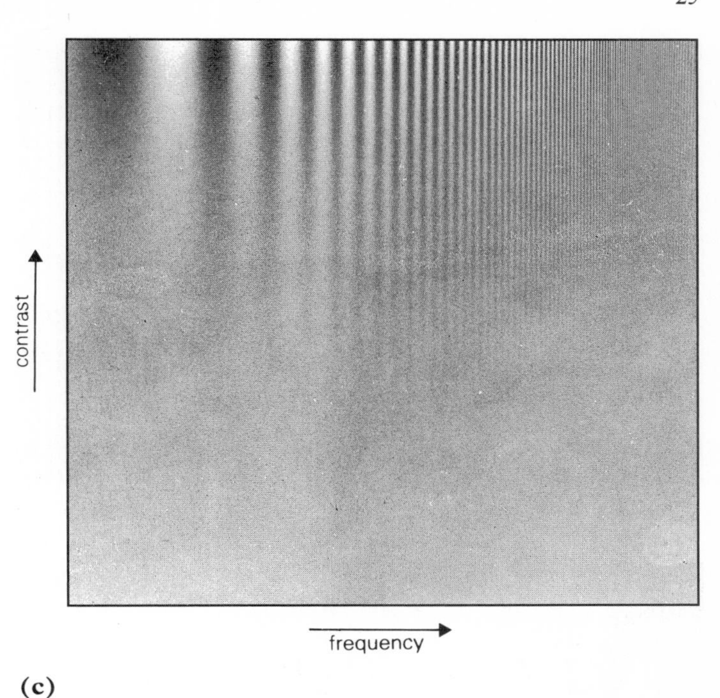

(c)

Fig. 2.5. (a) The modulation transfer function of an optically perfect system (dashed line) reproduces the brightness variations of a target no matter what the spatial frequencies of the variations are. The MTF of a photographic camera (solid line) only has high fidelity at low frequencies. (b) The human visual system shows different spatial frequency responses, depending on whether the target is coloured (chromatic) or black and white (achromatic). Note that the human MTFs are shown relative to frequency on the retina's surface, unlike (a). The test target (c) has been produced so that frequency increases from left to right and contrast increases upwards. At low frequencies the variations are only seen at high contrast, whereas poorer contrast variations become more easily seen as the frequency increases, and then they become less easy to see.

more clearly in a black and white image than in colour, simply because there are far more rods than cones. Detection of broad, subtle features relies more on colour vision (Fig. 2.5).

The full power of sight is not always needed in order to recognize something. Spatial information is only part of the interpretive process: shape, patterns, textures, relative scale, and context also play important roles. This is why a caricature of a well-known face in a familiar setting, based on only a few lines, is as easily recognized as a detailed photograph. Many images contain *redundant information* through which the interpreter has to sift (Fig. 2.6). Simplifying through interpretation is extremely important in passing on an expert's view of the world to others less well-trained or skilled.

The eye is amazingly efficient at detecting light, down to levels of about 10 *photons*, but very poor at discriminating different levels of brightness. Normally we can only distinguish about 30 brightness levels (Fig. 2.7). However, in a colour image hundreds of thousands of hues are easily distinguished, as anyone who has tried to match paint on a scratched automobile will know. But how do we perceive colour?

The cones are sensitive to red, green, and blue light. The British physicist, Thomas Young (1773–1829), showed that whereas white light is broken down into a continuous spectrum by a glass prism it can be reconstituted by mixing just red, green, and blue light of equal brightness. Moreover, all possible colours that we can see can be reproduced by combining just red, green, and blue in different proportions (Fig. 2.8). There is far more to colour and colour vision than this, but Young's *additive theory of colour* is all that is necessary here.

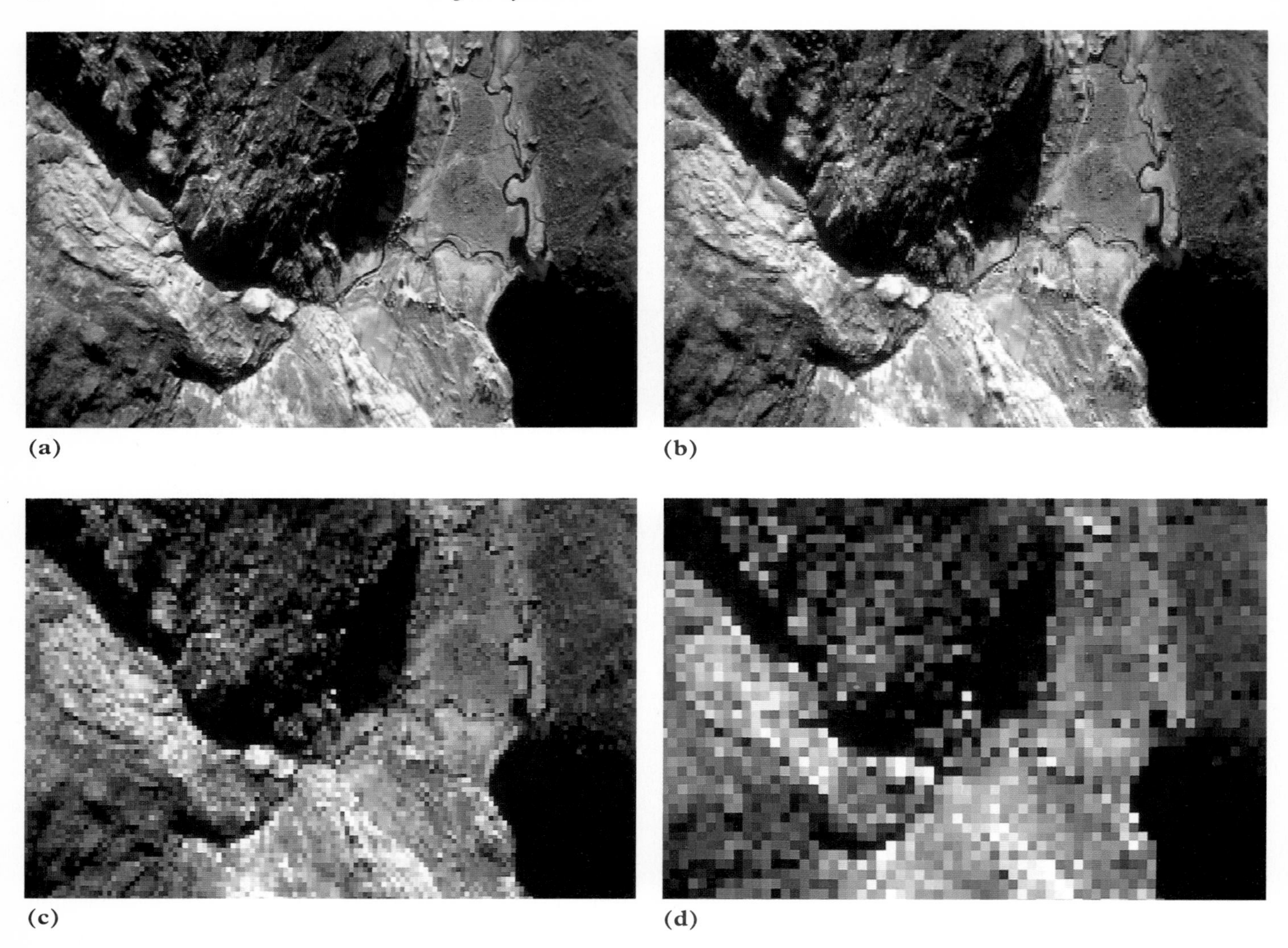

Fig. 2.6. This series of images shows the appearance of the same data after dissection to component picture elements which increase in size on the ground from (a) 10 m, through (b) 20 m, (c) 40 m to (d) 80 m. For best effect it should be examined at normal reading distance, at arms length and from about 3 metres.

Fig. 2.7. In this series of photographs the same image is shown with 128, 64, 32, 16, 8, and 4 grey levels to show the futility of trying to perceive more than 20 to 30 grey tones in a black and white image.

Perceiving *depth* is a vital attribute of everyday life. For anyone with normal binocular sight, focusing an object on both retinas allows us to visualize its position in 3-D. How this is achieved is demonstrated by holding up a pencil and blinking one eye then the other. The pencil seems to shift relative to the more distant background; the further the pencil the less the apparent shift. This *parallax* shift associated with each eye's image allows the brain to reconstruct the 3-D position, which works up to distances of about half a kilometre. Beyond that we rely on what we have learned—the size of familiar objects and decreasing contrast with distance due to haze. The absence of haze from pictures in space or on the Moon explains their unreal appearance to us. Most curious of all is our unconscious use of shadows, normally based on our knowledge of the sun's position in the sky. Without shadows, on a foggy day or in a blizzard, we easily become disoriented. In an image including shadows we seem innately to assume illumination from the top left corner of the image, a fact long exploited by artists and cartographers, and which leads to illusions of inverted topography when the actual illumination is from another direction.

Fig. 2.8. The 'paint box' of colours has been produced by combining the three additive primary colours in the proportions shown as red, green, and blue.

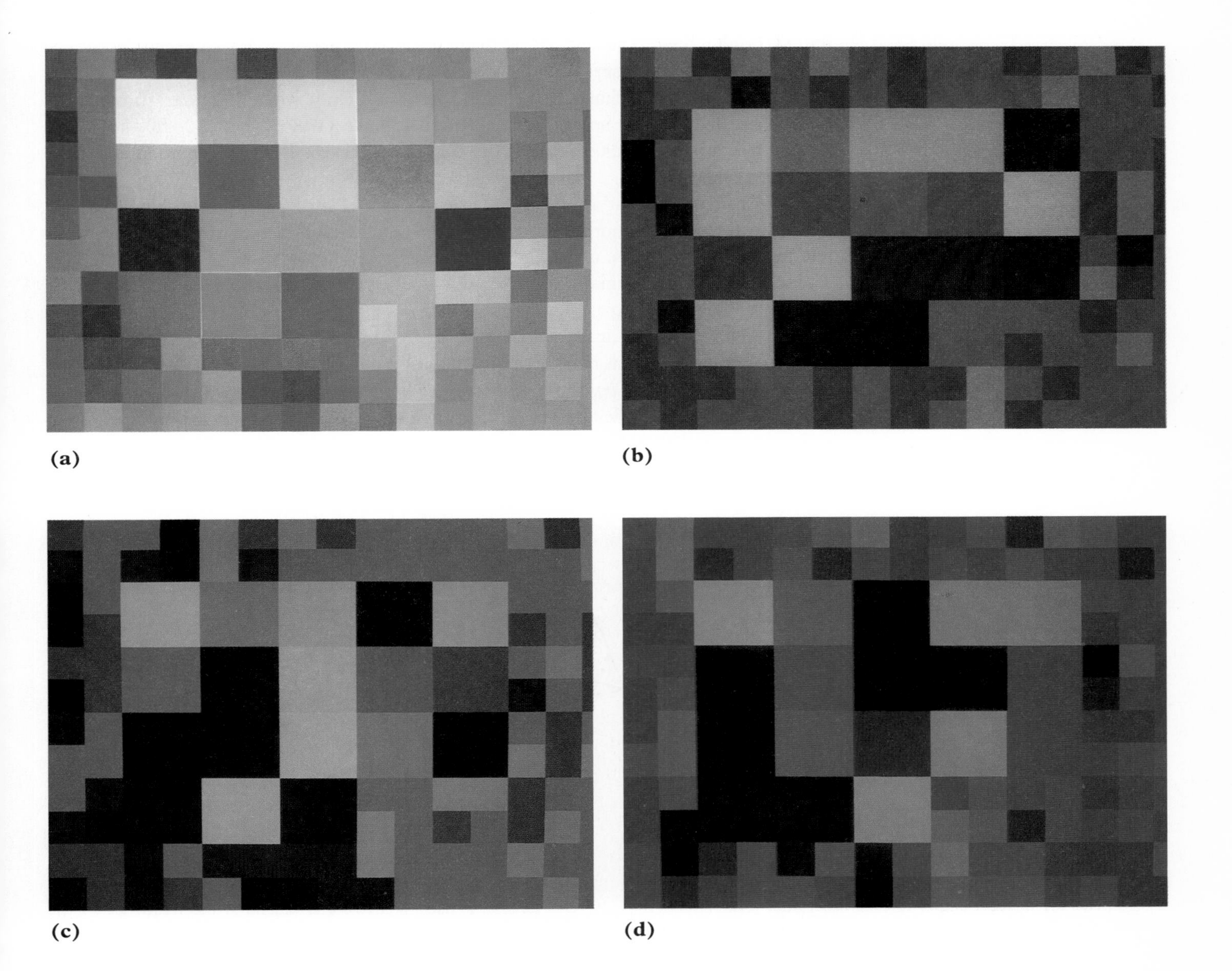

Finally, although the science of human vision is well developed, one aspect is very poorly understood—aesthetics. It is of great importance in remote sensing, because the most subjectively pleasing image is always the most easily interpreted. For instance, our ability to discriminate differences in colour decreases from blues through reds to greens. In general all people associate reds and yellows with warmth, while blues and greens suggest cold. Polls indicate a preference for blue, red, green, violet, orange, and yellow in decreasing order. We all know about colours that harmonize well and those that clash: yellow and green, violet and red, yellow and blue go well together, purple and orange, red and green, pink and turquoise are awful combinations to most people. Getting the best from images used to be the field of the graphic artist. In the case of many kinds of remotely sensed image it is the artful use of the image-processing computer.

2.2 Radiation and matter

Sight relies partly on the way things reflect, absorb, give off, and allow light to be transmitted through them. A second part is due to the information-processing capacity of the eye and brain, discussed earlier. Equally as important, light and the information that it carries travels without being changed through a vacuum. Only when it passes through the atmosphere does it suffer any distortion or change. Since it travels about 300 000 km in one second, it is a nearly instant means of capturing and transmitting information about the Earth in its near vicinity.

The information about an object conveyed to us by light is of two main kinds. First, it relates to the object's morphology from the way it is illuminated and shadowed by its relationship to the source of the light. This gives information on *size*, *shape*, and *texture*. The second kind of information is due mainly to the way light is reflected and absorbed by the object. This shows up as the object's brightness and colour.

Light behaves in two inseparable ways, as rapidly moving and indivisible packets of energy, or photons, and as regular fluctuations of electrical and magnetic fields. This is the *wave–particle duality of light* and all similar radiation. Each photon has a unique pair of electrical and magnetic fields vibrating at right angles to each other and to the direction in which they travel (Fig. 2.9). This vibration is most easily expressed as either the distance

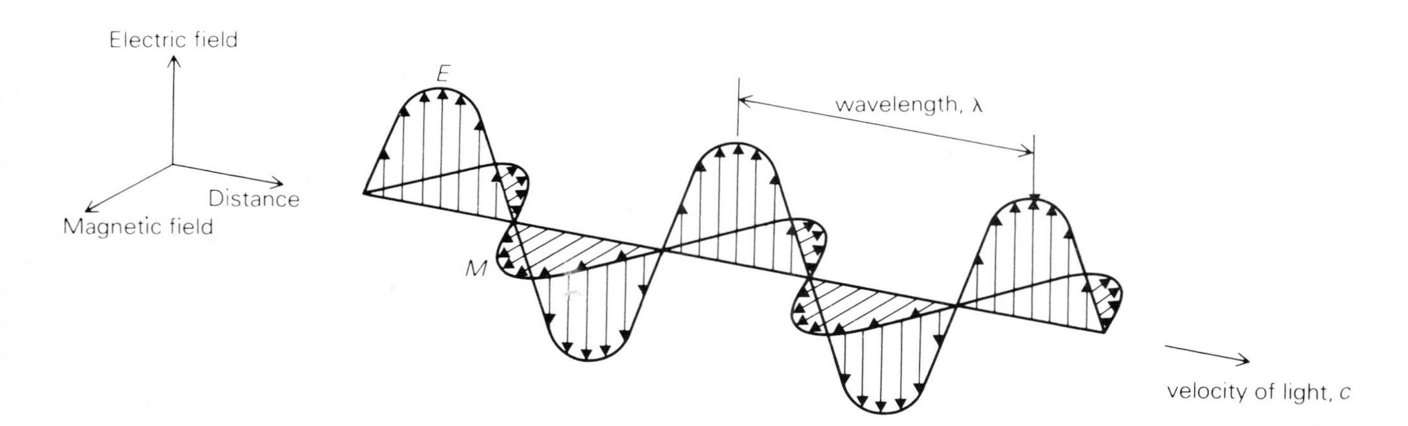

Fig. 2.9. Electromagnetic radiation is most easily represented by waves in electrical and magnetic fields travelling at the speed of light. These fields are at right angles to each other and to the direction of propagation. They correspond to regular fluctuations in the fields and are described by sine functions. The distance occupied by a complete cycle—from peak to peak—is the wavelength, and the number of cycles passing a fixed point in one second is the frequency of the radiation.

between peaks in the electromagnetic fluctuations—as a wavelength—or as the number of waves that pass a point in one second—its frequency. The wavelength and frequency associated with a photon are related to its energy, so that the shorter the wavelength or the higher the frequency, the greater the energy of the photon. This is the gist of *Planck's Law*, and it means that light carries energy. The shorter the wavelength the fewer the photons that are required to carry a particular amount of energy.

White light from the sun consists of a continuous range of wavelengths in which photons corresponding to any wavelength occur. The continuum of colours in a rainbow spectrum ranges through wavelengths from 400 (indigo) to 700 nanometres (red). The colours of objects that we see, when they are lit by the sun, result from parts of this continuous range of wavelengths being absorbed more than others. A black material absorbs all wavelengths completely, grey means equal but not complete absorption of all, and white means that all are reflected equally strongly. A red material absorbs shorter wavelength light, while one that is blue takes in energy from longer wavelengths. Green, the usual colour of plant leaves, results from greater absorption of wavelengths both shorter and longer than that corresponding to what we perceive as green. Other colours that do not appear in the rainbow spectrum result from more complex absorptions, purples from green absorption, browns from blue absorption, and pastel shades from relatively weak absorptions at a range of visible wavelengths. How we translate the results of these physical processes into what we see as colours was covered in the previous section.

Light is not the only kind of *electromagnetic radiation* in nature. In fact it occupies only a tiny range of the wavelengths that can be detected, from these comparable with the dimensions of atoms, gamma rays, through X-rays, ultraviolet, visible light, infrared, microwaves, radio waves to wavelengths longer than a hundred kilometres, whose role in nature we do not understand at all well (Fig. 2.10). In one way or another, all electromagnetic radiation interacts with matter, potentially allowing some information about its constituents to be conveyed to the distant observer. There is one important restriction, however. This is the atmosphere, through which all radiation has to pass on its way to the observer.

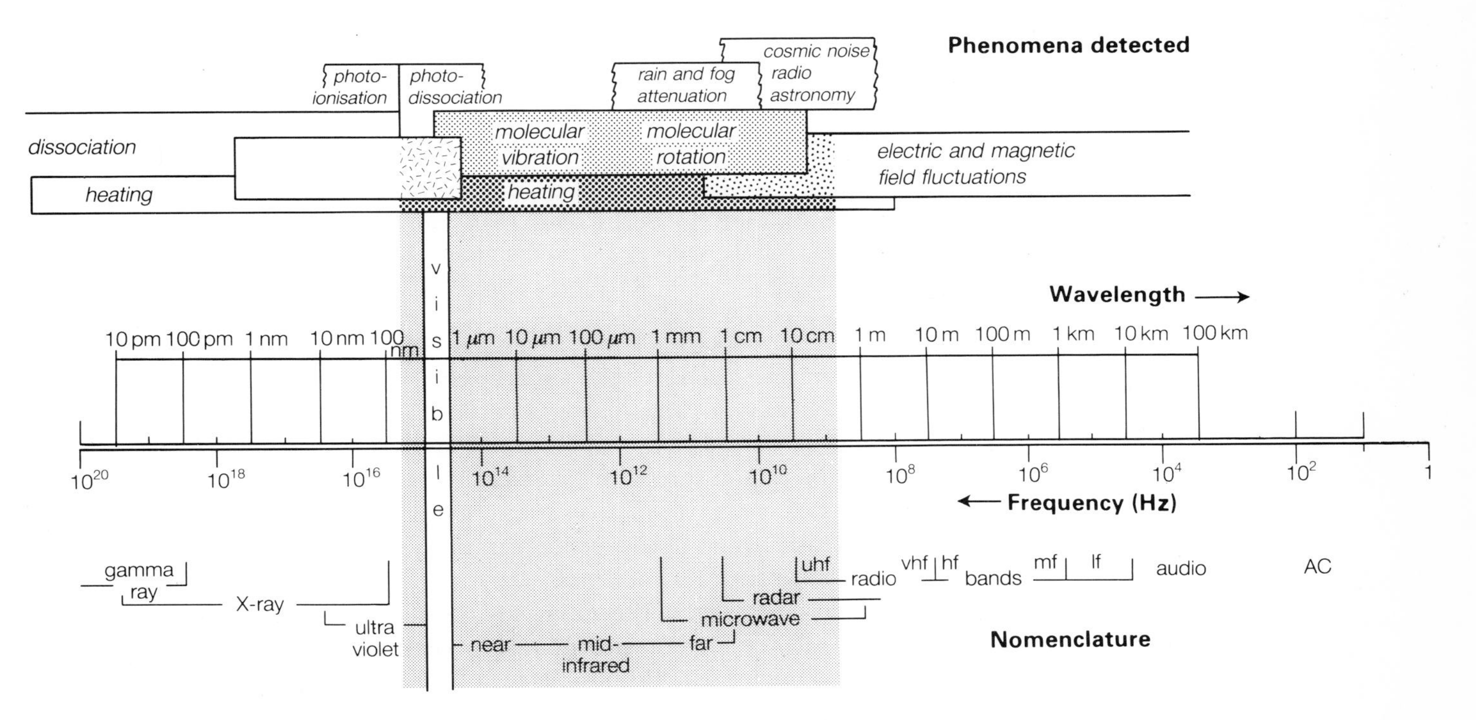

Fig. 2.10. This summary of that part of the EM spectrum which is routinely detected by scientists shows the relationship between wavelength and frequency, the phenomena which are involved in generation and interaction of EMR, and the nomenclature for different parts of the spectrum. Those portions covered by this book are highlighted, together with the processes relevant to geological remote sensing. The narrow visible band is useful as a reference. The wavelength and frequency scales are logarithmic.

Three gases in the *atmosphere*—oxygen, carbon dioxide, and water vapour—all absorb different ranges of wavelengths, more or less completely. This means that only some regions of the continuum of electromagnetic radiation can be used for remote sensing of materials at the Earth's surface. These are known as *atmospheric windows*. Other atmospheric interactions further restrict what is possible. Clouds halt all radiation except that with wavelengths in the microwave region. The physical presence of molecules of oxygen and nitrogen, giant molecules of water vapour and dust result in a phenomenon known as *scattering*, which produces haze, mainly in the visible range. This is most easily visualized from the blue appearance of distant mountains, but also includes the blue colour of the sky and the reds of dawn and sunset. In all cases it begins to mask the true appearance of an object in the absence of an atmosphere. Other gases such as ozone and various pollutants themselves have absorption effects at specific wavelengths, enabling them to be detected and measured (Fig. 2.11).

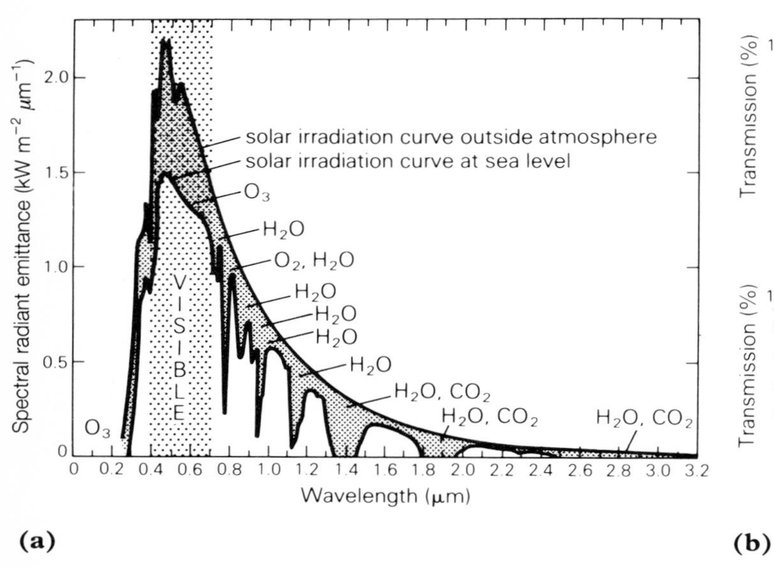

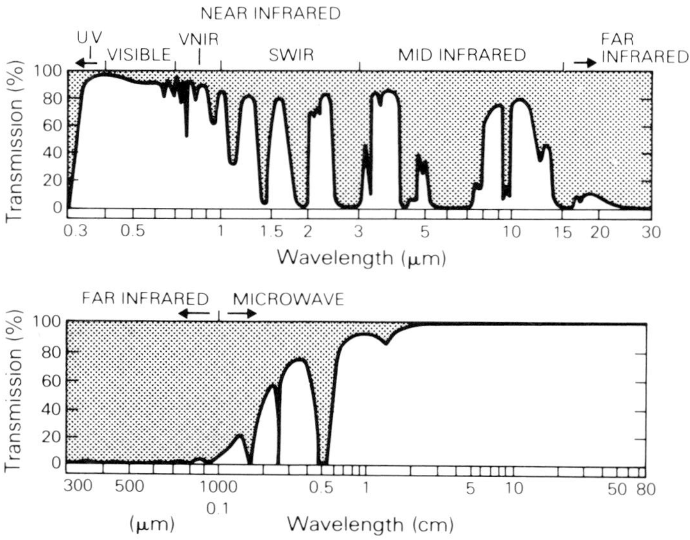

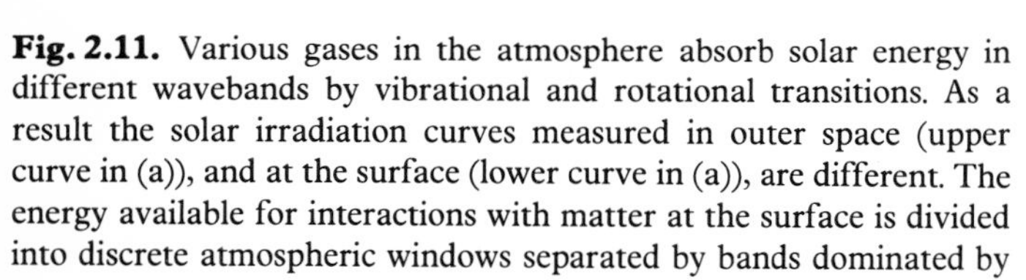

Fig. 2.11. Various gases in the atmosphere absorb solar energy in different wavebands by vibrational and rotational transitions. As a result the solar irradiation curves measured in outer space (upper curve in (a)), and at the surface (lower curve in (a)), are different. The energy available for interactions with matter at the surface is divided into discrete atmospheric windows separated by bands dominated by atmospheric absorption—grey. In (b) the main atmospheric windows throughout the whole of the useful part of the EM spectrum are shown on a logarithmic scale, in terms of the percentage transmitted through the atmosphere. These two graphs, together with the spectral properties of natural materials, form the basis for designing remote-sensing systems.

Most remote sensing depends on the sun as an energy source, in two main contexts. The first examines the ways in which materials reflect and absorb visible and near-visible infrared radiation coming directly from the sun. Part of this range, which carries the bulk of energy from the sun to the Earth's surface, is absorbed and helps heat up the near-surface layer. The re-emittance of this stored solar radiation forms the basis of a second kind of remote sensing. A third relies on the artificial production of electromagnetic energy and its interaction with natural materials.

The most important means of generating electromagnetic radiation in nature is through the vibration of atoms and molecules induced by heat energy, causing the electrical and magnetic fields associated with them to fluctuate. Just how much energy is emitted in the

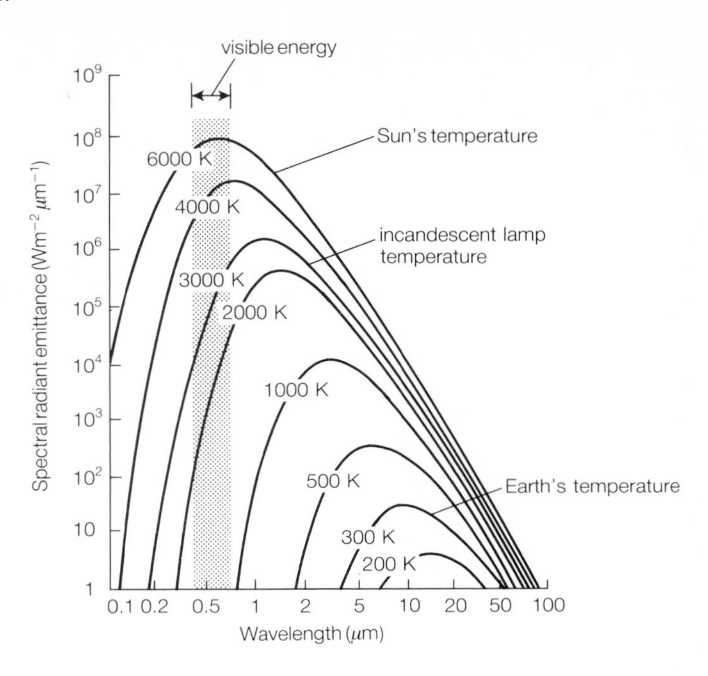

Fig. 2.12. This family of curves on logarithmic axes expresses how energy emitted by a square metre of a perfect emitter, or blackbody, at different temperatures varies with wavelength, and how the wavelength of maximum emittance and the range of wavelengths emitted change with absolute temperature. The area under each curve represents the total energy emitted at each temperature. Both the Stefan–Boltzmann and Wein Laws control the shapes.

form of radiation as a result is proportional to the fourth power of the material's temperature. The range of wavelengths and the wavelength at which maximum energy—the greatest number of photons—is emitted is inversely proportional to the temperature. So, as temperature increases, total energy emitted rises very rapidly and the wavelength carrying most energy decreases. This is expressed by the *Stefan–Boltzmann* and *Wien displacement laws* (Fig. 2.12).

The sun's surface has a temperature of about 6000 °C. It emits radiation according to these fundamental laws with a peak in the visible spectrum around the wavelength of green light. The energy tails off to both shorter and longer wavelengths. The sun also emits gamma rays and radiowaves because of other processes. The mean surface temperature of the Earth (30 °C) allows it to emit energy too, peaking at much longer wavelengths (around 10 μm). However it emits no energy shorter than about 4 μm and very little longer than 50 μm. Interestingly, the peak coincides with one of the atmospheric windows, which means that we can monitor terrestrial temperature from a distance. Volcanic lavas, being as hot as 1100 °C, can emit some visible radiation.

The Earth's *surface temperature*, apart from in volcanic areas, is mainly controlled by energy coming from the sun. The solar heating effect is interesting, since it is not due to long-wavelength radiation, which forms a very small proportion of solar radiation, but to the absorption of radiant energy near the peak of its emission in the visible and near infrared. This absorbed energy is converted mainly to heat, which the Earth re-emits at mid-infrared wavelengths. How much energy is absorbed depends on the nature of the surface. A light surface reflects much of the sun's energy back to space and it stays relatively cool. Such a surface is said to have a high *albedo*. A dark surface is due to much stronger absorption that results in a greater heating effect. The temperature that is reached is a balance between absorbed incoming energy and that emitted, and is controlled by the thermal properties of the surface material. These are *density, specific heat capacity*, and the ability to transfer energy deeper within itself—its *thermal conductivity*. The temperature is highest just after the hottest part of the day and falls as the sun goes down and at night. How quickly temperature falls is also governed by thermal properties acting together and expressed as the material's *thermal inertia* (Fig. 2.13).

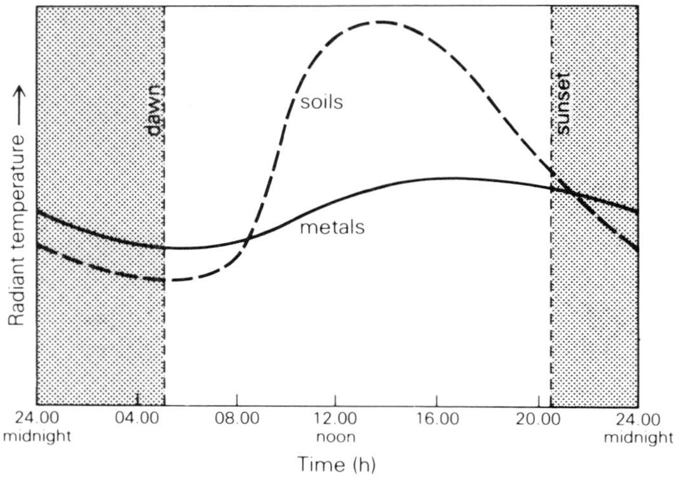

Fig. 2.13. Materials with high thermal inertia, such as metals, show little range in temperature from day to night since they heat and cool slowly. Those like soils with low thermal inertia are prone to rapid heating and cooling, and so they reach high daytime and low night temperatures.

This simple relationship between solar energy, thermal properties, and surface temperature is complicated by a number of other effects. Wind causes a chilling effect both by the movement away of warmed air at the surface and by the evaporation of moisture. Living organisms further modify the processes by their own metabolism. Warm-blooded animals have temperatures generally above that of the surface and so emit more energy in the mid-infrared, especially at night. *Vegetation* contains water and does not heat up as quickly as its surroundings by day, because of the high specific heat capacity of water. Consequently it has a relatively cool signature, which is assisted by its transpiration of water vapour. Its high heat capacity allows it to retain heat at night and so it cools down more slowly than soils and rocks, and appears relatively warm at night. Dead plants contain air and make good insulators, so that they also appear cool by day and warm at night.

2.2.1 Reflection and absorption

It is possible to measure accurately how materials reflect and absorb radiation according to the wavelength in the visible and near infrared regions. This gives a *spectral reflectance curve*, and all materials are intrinsically unique spectrally.

The most familiar and pleasing feature of our natural surroundings is living green vegetation. The visible region of a spectral curve for a green leaf clearly demonstrates why it looks green. Although the whole of the visible spectrum is one of net absorption, to fuel photosynthesis more energy is absorbed in the blue and red parts than in the green. The green peak is due to the photosynthetic pigment *chlorophyll*. Some plants have other pigmentation, appearing red, brown, and blue-green. What is really unique to living vegetation is not colour as we perceive it, but how plants deal with radiation of longer wavelength than red light—the *very-near infrared (VNIR)*. Plant cells are very efficient at preventing VNIR entering their structure. If it did enter entirely then temperature would rise, chlorophyll would break down, and photosynthesis would stop. The plant would die. This self-defence mechanism against overheating is due to the structure of the water-rich cells in leaves. They cause any VNIR wavelengths to be efficiently reflected out of the leaf before damage is done (Fig. 2.14).

The characteristic high reflectance of infrared radiation by plants was the reason for the military development of infrared-sensitive film to distinguish between real plants and *camouflage*. As it happens, many different species of vegetation are distinguishable using this part of the spectrum and so the military breakthrough now finds its most frequent use

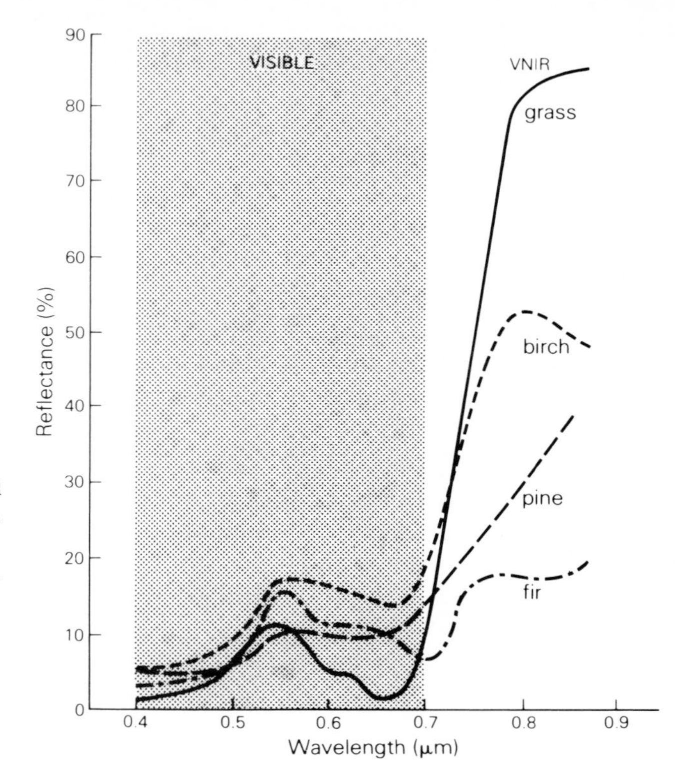

Fig. 2.14. All the different attributes of chlorophyll content, leaf shape, area, and number, together with overall plant structure contribute to the spectral reflectance properties of a plant species. Whereas all four plants shown have rather similar properties in the visible spectrum, they are clearly distinguished by their near-infrared reflectance.

in the civilian remote sensing of natural vegetation patterns and the monitoring of crops. Since the spectral curves of plants undergo quite marked changes when they are diseased or short of water, it is also possible to monitor remotely the health of natural and agricultural vegetation.

Spectrally a soil or rock is very much different from a plant. Their spectral reflectance rises steadily with wavelength to a peak and then falls off once more, without having a sudden jump at the red/infrared boundary. The broad features in the visible range control our perception of their colour. However, it is the fine detail in the spectral reflectance curves of rocks and soils that give clues to the different minerals that they contain—the basis for subdivision and classification. The most common colouring agent is iron, bound in various minerals. Nearly all *iron minerals* break down during weathering to produce oxides and hydroxides of iron. They are coloured in various shades of reds, oranges, and browns. Complex interactions between radiation and both the iron atoms and the molecular structure of iron minerals produce absorption features in the VNIR whose wavelength, breadth, and depth characterize the different mineral species (Fig. 2.15).

Some other minerals also produce clearly identifiable features in spectral reflectance curves, most notably those containing *hydroxyl* (OH^-) ions, and *carbonates* and *sulphates*, due to the distortion of molecular bonds. Their most distinctive features are in the 2.0 to 2.4 μm range, or the *short-wave infrared (SWIR)*. The hydroxylated minerals, mainly *clays*, are particularly interesting because the intricacies of their SWIR spectral curves enable them to be uniquely identified (Fig. 2.16).

With rocks and soils it is important to notice that the most readily detected minerals in the reflected range are not those that make up the bulk of crystalline rocks—quartz, feldspar and iron, and magnesium-rich silicates—but those produced by alteration, which dominate weathered surfaces and are common in sedimentary rocks. Radiation in the reflected range cannot penetrate more than a few micrometres into the surface rind of

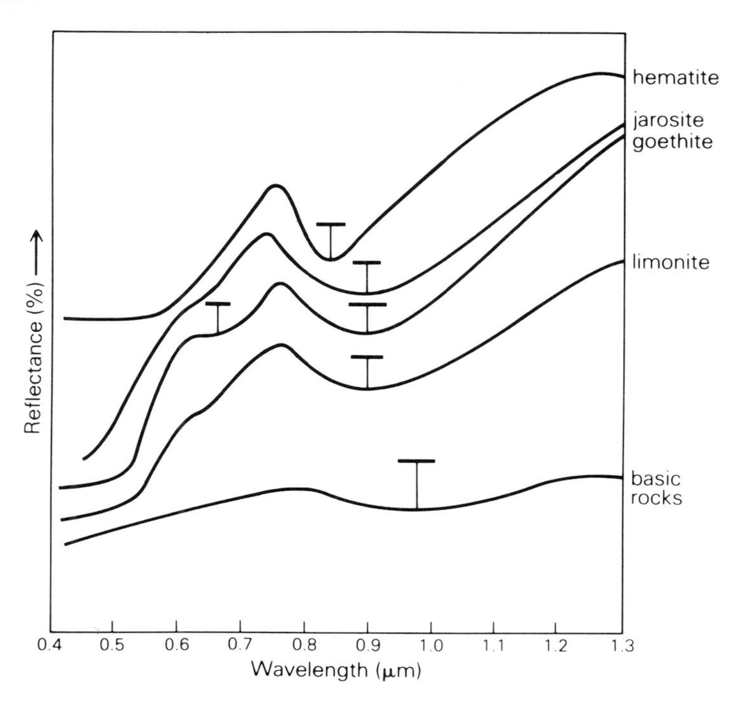

Fig. 2.15. Iron oxides and hydroxides display absorption features in their reflectance spectra in the VNIR—T-shaped symbols. These are due to energy at specific wavelengths being absorbed to accomplish adjustments in their molecular structure. The general decrease in reflectance in green and blue, which imparts the reddish colour of these compounds is connected to the way in which atoms are bonded in the minerals' molecules. The spectra are offset for clarity.

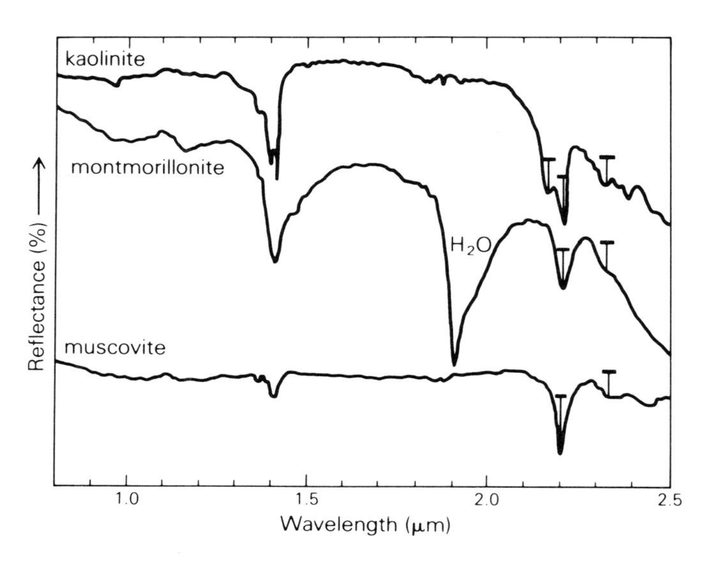

Fig. 2.16. Energy that is absorbed to accomplish distortion of Al–OH and Mg–OH bonds in clay minerals and micas produces distinctive absorption features in their reflectance spectra—T-shaped symbols. Together with other features of the spectra, they form a potentially powerful means of discriminating these minerals, which are important products of hydrothermal and sedimentary processes. The spectra are offset for clarity.

weathered material. So we are reliant for geological mapping on *alteration products* rather than on the bulk mineralogy and chemistry of rocks and soils. Moreover, geological materials are rarely exposed bare of vegetation at the surface, except in deserts or recently glaciated areas. This means that it is more difficult to map directly different classes of rock or soil than vegetation, and geological interpretation is dependent on an understanding of the relationship of rocks, soils and the plants that they support, and the small-scale textures in the landscape that weathering and erosion of rocks produce.

One property of water stands out very clearly. The top few centimetres of a body of water almost totally absorb all infrared radiation. Even in the visible range absorption by water is high. Relatively little energy is transmitted, reflected from the bottom, and re-emerges from clear water, the amount that does being inversely proportional to the wavelength. So red penetrates only a few metres while green and blue can reach depths of tens of metres giving the gradation from green to blue as water depth increases over a white base, as in a swimming pool. These shades can be used to give a crude measure of water depth, as we shall see in Section 3.2.

Other materials suspended in water can change its spectral properties. High sediment load increases visible reflectivity, really turbid rivers often being indistinguishable from the soils over which they flow. Water colour also changes with the content of organic dyes, such as those produced by decaying vegetation. Much the most important effect on water's spectral properties is that of minute floating plants or phytoplankton, the first step in the marine food chain. They do impart a degree of reflectance in the visible region, enabling them to be detected and their proportions estimated.

Water that has seeped into soil imparts some of its spectral properties to those of the soil itself. Soil moisture causes a general decrease in reflectance, particularly noticeable around the 1.4 and 1.9 μm wavelengths where distortion of the bonds between hydrogen and oxygen in water is induced by these wavelengths, so causing absorption of energy.

2.2.2 Thermal emission

The simplest role of *thermal remote sensing*, mainly in the mid-infrared but extending into the microwave region, where a little energy is emitted by materials above absolute zero, is the monitoring of surface temperature. This can be used qualitatively over land as another means of discriminating materials by their *emissivity*. To use it quantitatively means assuming perfect emissive properties and uniform reflectance, which is only justifiable for bodies of water. They approximate perfect *blackbody emitters* so that their temperature can be calculated from the Stefan–Boltzmann law. Such monitoring of sea-water temperature is very important because of its links with currents, marine-biological productivity, and controls over climate.

As well as being able to monitor the variations in temperature of the Earth's surface, which hold clues to the materials of which it is comprised, there are two other possibilities dependent on thermal emission.

The first involves the use of emission data from both day and night for the same area. After allowing for the effect on heating and cooling of varying surface reflecting properties, the temperature difference is a good measure of the *thermal inertia* of materials (Fig. 2.13). This is a very interesting property, especially for rocks. For a start, the energy emitted by the surface represents several centimetres beneath the surface, not just the tiny veneer that is involved in the shorter-wavelength, reflected region. Secondly, several important kinds of rock that have similar reflecting properties are strongly contrasted in the way they emit thermal infrared (Fig. 2.17). Thermal remote sensing of this kind is thus the only way to discriminate them easily.

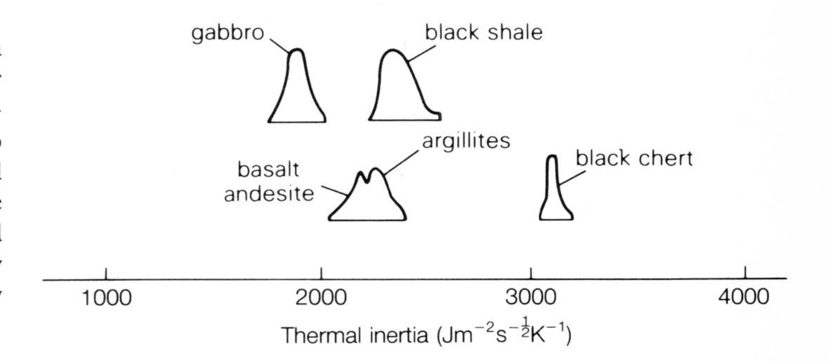

Fig. 2.17. Remote sensing in the visible and NIR part of the spectrum fails to discriminate between rocks with low albedos—they all appear dark. In thermal images, uniform low albedos offer distinct advantages. First, more solar energy is absorbed, allowing the surfaces to emit more energy. Secondly, the variations in temperature produced by this roughly equal absorption, and the different rates at which the different rocks cool at night relate to differences in the thermal properties of the rocks. The apparently similar black rocks are clearly separated from each other in the thermal region, best expressed by their differences in thermal inertia.

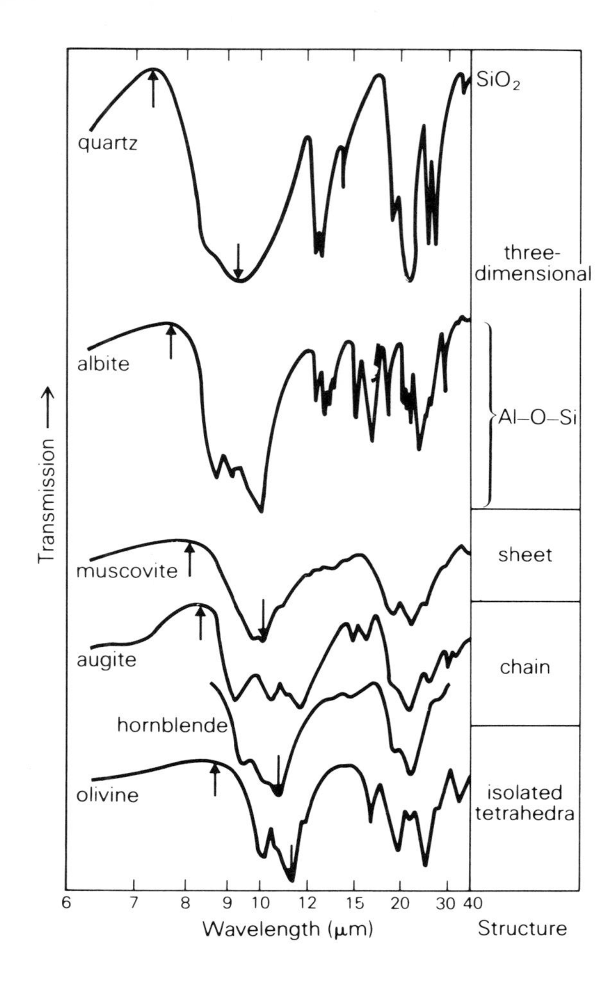

Fig. 2.18. Because of differences in the structure of silicates, each group in this family of minerals deviates from the thermal emittance curve of a perfect blackbody in a different way. In particular, the wavelength region over which emission is low shifts position, depending on the groups' molecular structure. The spectra are offset for clarity.

The second exploitable feature in the thermal range is also important for geology. No mineral is a perfect emitter of energy following the Stefan–Boltzmann law. Instead, there are regions of the thermal spectrum where much less energy is emitted than others. It is used up instead by causing important bonds in the molecular structure of minerals to become distorted. Many minerals display these poor emission zones at different wavelengths. In particular, the silicates, of which there are many kinds and from which the bulk of crystalline rocks are made up, show considerable variation (Fig. 2.18). As with the absorption features in the reflected range, this means that emitted spectral features can be used to detect rock-forming minerals and classify the rocks of which they are a part.

2.2.3 Artificial radiation

Although it is impossible to illuminate very much of the Earth's surface using a flashgun, there are so-called *active remote-sensing* methods that rely on the same principle. They are deployed for two reasons, either because they enable unusual phenomena to be stimulated or because they are simple and use little energy.

As well as the familiar minerals from museum displays and various dyes, many organic materials *fluoresce* when exposed to ultraviolet radiation. Fluorescent behaviour is simply the absorption of short-wavelength radiation and its re-emission at longer wavelength. The most common examples absorb ultraviolet and re-emit in the visible range. *Laser* technology now allows high-intensity beams of radiation of a single wavelength to be focused on the surface with little spreading of the beam. Since most ultraviolet of solar origin is absorbed by ozone in the upper atmosphere, *ultraviolet lasers* are the only reliable means of exploiting fluorescence for remote-sensing purposes. Clearly, the method is costly but potential is high. Another use of lasers in remote sensing is *lidar* (*l*ight-*i*ntensity *d*etection *a*nd *r*anging). Lasers emitting different wavelengths can be used to stimulate fluorescence from various atmospheric compounds, including clouds, to estimate their concentration and to sound the levels at which they occur. In ranging mode they can be used to measure the elevation of the Earth's surface to within a few centimetres, enabling the growth and ablation of ice masses to be monitored. When directed at reflectors on the ground, space-borne lasers can determine position with such great precision that lidars may be able to measure the tiny relative movements associated with sea-floor spreading and continental drift. By using the tiny shift in wavelength of a laser due to the relative motion of material from which it is reflected it is also possible for lidars to measure wind velocity and direction.

Radar (*ra*dio *d*etection *a*nd *r*anging) is the most common active method of remote sensing, exploiting the low energy needed to generate radiation at long wavelength. Most radar is operated in the 1–50 cm region, and is generated using antennae and alternating currents in them controlled by complex electronic systems. As well as its low cost, radar possesses the advantage of being able to penetrate the atmosphere, even when cloudy, with no loss of signal, except for very short-wave radar which can be affected by rainfall. It is an all-weather, day-or-night method.

Unlike reflected visible and infrared radiation, or infrared that is emitted, radar interacts with the macroscopic features of surface materials, rather than on the molecular level. The electrical properties of a material govern how much energy from a radar beam may penetrate into it. Good conductors of electricity absorb little if any radar energy, and since most rocks and soils, and vegetation of course, contain sufficient water to be electrically conductive they are opaque to radar and reflect it efficiently. However, there are some areas on Earth that have so little rainfall that sand and gravel are dry enough for some radar energy to be absorbed, transmitted, reflected from deeper zones, and re-emitted. Although this is rare and only depths up to 6 m can be penetrated, in deserts radar has the unique property of allowing subsurface information to be gathered from a distance (Figs 3.35 and 4.6(b)).

Understanding how radar interacts with surfaces that it cannot penetrate depends on the simple example of the metallic *corner reflectors* that are commonly placed at the mast-heads of small boats. These consist of three plates with larger dimensions than the normal range of radar wavelengths at right angles to each other. If a radar beam enters this device from any angle it is reflected back and forth by the surfaces and eventually re-emerges exactly along its original path. The reflector appears as a bright signal on a radar screen whereas the small boat does not. Natural surfaces, whether they are the complex canopies of plants or assemblages of rock and mineral fragments, approximate 'families' of corner reflectors. Provided the dimensions of this *surface roughness* are greater than about one-eighth of the radar wavelength, a proportion of the energy falling on the surface returns to the source due to the phenomenon of *back-scattering*. If the surface is smoother than this it acts like a mirror to radar (Fig. 2.19). As we shall see, radar images depend on microwave energy being aimed at the surface from the side, so smooth surfaces reflect all energy away from the source and thus appear dark, while rough surfaces are much brighter. Since surface roughness is often dependent on the physical properties of rock or soil, and is a major variable from one species of plant to another, radar remote sensing is yet another means of discriminating types of surface.

The sideways-looking nature of imaging radar also means that a surface which is inclined towards the source is more likely to scatter energy back to the source than one

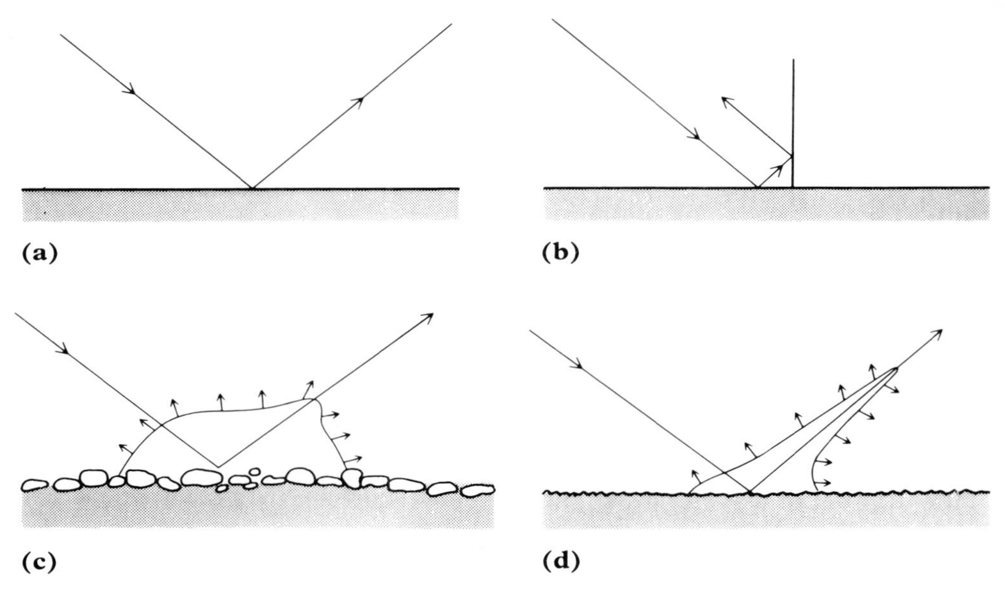

Fig. 2.19. How radar energy is scattered from the surface depends on its angle of incidence, but more importantly on the roughness of the surface. Four possibilities are shown: (a) for a perfectly smooth surface, which acts as a mirror to the radar; (b) for a corner reflector, where the energy is reflected back along its original path; (c) for a rough surface, where complex reflections at the irregular surface result in the energy being reflected in all directions, so that a proportion can return to the radar antenna, and (d) for a smooth surface, where the bulk of the energy is reflected away, only a tiny proportion being diffusely scattered.

facing away. Consequently radar is extremely efficient at displaying subtle topographic attributes of the surface, compared with methods exploiting reflected and emitted natural radiation.

The atmospheric windows in the spectrum of electromagnetic wavelengths form the framework within which different remote-sensing methods have been devised. The different ways in which surface materials interact with radiation in the reflected, emitted, and microwave windows are the basis for the strategies for deploying the different methods and interpreting their results.

2.3 Detecting radiation

As outlined earlier, information about our surroundings is best expressed visually as some kind of image. Remote sensing is dominated by image-acquiring systems, and the results from those that do not directly produce images are most easily understood if they are presented in image form. That may be some form of graph, such as the proportion of ozone at different depths in the atmosphere, but more usually it is a representation of how some property varies in the two dimensions that define position on the Earth's surface, as in a map.

Reality is a continuous variation of properties in space and also time, which we perceive visually as an image that is very finely dissected by the distribution of rods and cones on the retina. With light we perceive only about 30 steps in the continuous variation of tone of a black and white scene, though we can distinguish more than a million different colours. What we see is directly related, albeit in a complex way, to the energies and wavelengths that we can detect. We see things as an analogue of how they really are. A photograph is also an *analogue image*, where the number of photosensitive grains that become fixed by exposure bear some numerical relationship to the energy variation in a real scene. Because of the grain structure of film, photographs are also dissected.

Until relatively recently, most remotely sensed images were acquired by photographic techniques. New data-gathering methods have developed rapidly for two main reasons. First, planetary monitoring by unmanned satellites means that it is impossible to retrieve film, except by costly return and re-entry manoeuvres. Systems that transmit data by microwave telemetry are needed. Secondly, photographic methods cannot operate in many useful parts of the electromagnetic spectrum. This has required the development of new kinds of instrument based on electronic systems. Some of them, like television cameras, collect and transmit images which correspond to a 'snap shot' made up of electronic signals stripped from regularly arranged elements on photosensitive plates or *vidicons*, one for each primary colour. They consist of representations of the energy reflected from or emitted by a scene in the form of modulated radio signals corresponding to the regular grid of detector elements. The signals are continuous and produce dissected analogue images in a way similar to human vision and photographs. Video methods, however, are limited to much the same spectral range as photographs.

Other systems also gather data as a sequence of small rectangular parcels of a scene arranged in a geometrically regular grid. The sequence of data is reconstructed later to form a picture. The parcels are known as picture elements, or *pixels*. The intensities for each individual pixel are first expressed in analogue form as electrical potentials produced in a radiation detector by photons. However, precise analogue measurements, like 1.763 millivolts, are cumbersome to transmit back to the user. It is more efficient to rescale these numbers and transmit them as integers in a limited range. Transmission, and later manipulation and display using computers, is easiest when numbers are in binary form, usually spanning the range of eight binary bits or one byte from 0 to 255. Images in this form are known as *digital images*. Whereas photographs can only be digitized after their capture by such a digital imaging device, a video image may be transmitted or recorded in either analogue or digital form.

We are most used to handling visual information about our surroundings when we look straight ahead towards the horizon, or latterly, looking obliquely down from an aircraft port. In doing this, perspective has an important role in our innate interpretation. However, perspective plays havoc when we wish to record our observations in the form of a map, because the scale on an oblique image varies. Although it is quite easy to record oblique views of the Earth, most remote sensing depends on the view vertically downwards, which, with suitable compensation for the geometry of the sensing system, corresponds to a map. As we shall explain later, only radar images are routinely produced by a sideways-looking system.

2.3.1 Photographs

Photographs of the Earth's surface are routinely taken looking vertically downwards using very precisely machined cameras with superb optics. These are called metric cameras, and to avoid even the slightest blur due to motion of the platform, the film is slowly drawn across the focal plane during exposure, at a rate that matches the ground speed. The film itself has a much larger format than that used by photographers, usually about 20 cm wide. To enable a stereoptic view of the surface, exploiting 3-D human vision, frames are exposed in sequence as the platform passes over at such a rate that each overlaps with the next. These overlapping vertical photographs each contain radially distributed distortions due to the height of objects and their position relative to the central part of the lens system

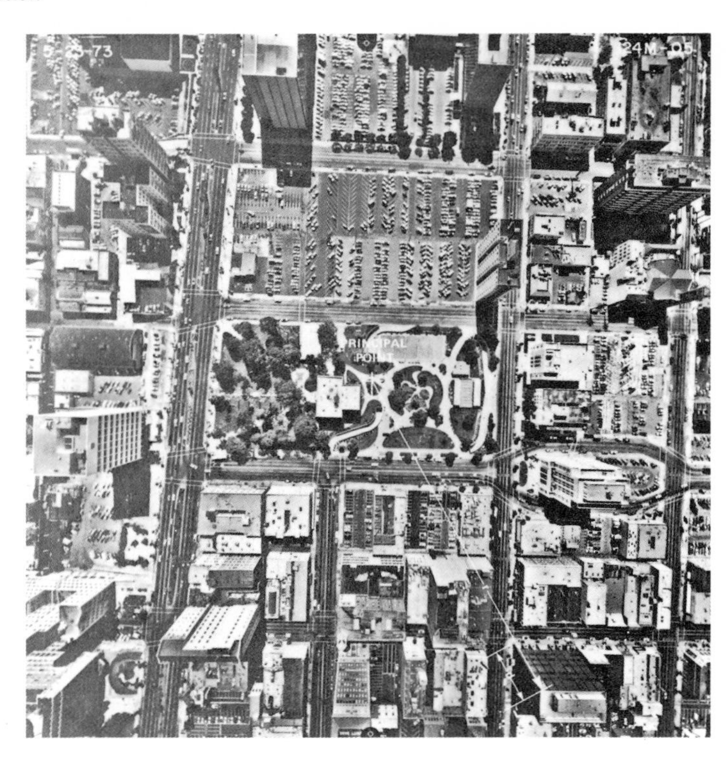

Fig. 2.20. Vertical aerial photographs of urban areas dominated by high-rise buildings show clearly the radial displacement and the increase in scale associated with increased elevation of the surface. This results in the buildings appearing to lean away from the centre of the photograph and to taper downwards. Width is 500 m.

(Fig. 2.20). The distortion due to one object changes from frame to frame, because of the changed viewing position. By viewing one image with one eye and the adjacent one with the other, a 3-D or *stereoscopic* mental model is perceived helping recognition of topographic features and, with suitable instruments, measurement of their relative elevations.

Black and white film comprises minute grains of silver chloride set in an emulsion. The most familiar type of film is sensitive to all visible radiation—so-called *panchromatic film*. When a photon strikes one of the grains an electron is freed which converts a silver ion to an atom of metallic silver. A chain reaction converts the entire grain to silver. The coarser the grain the more chance of a photon striking it and the more sensitive the film is to light—it is said to be a fast film—and vice versa for slow film. Developing dissolves the remaining silver chloride, leaving the silver on the negative film as black grains. Printing reverses this process.

Natural-colour film relies on the blue, green, and red components of light bleaching grains of yellow, magenta, and cyan dyes respectively in different layers of the emulsion. Projecting white light through the developed film, the remaining dye filters the blue, green, and red components. So for a blue object recorded on film, the red and green components are absorbed completely by cyan and magenta dye, the yellow layer, having been bleached away, allows the blue to pass through, and so on.

Because of the atmospheric haze associated with the blue and ultraviolet, to which all film is sensitive, a yellow filter is used with panchromatic film to give a *minus-blue image*, and an ultraviolet filter is used with colour film to reduce some haze. Much of the blue haze remains, accounting for the blue tinge on all natural-colour photographs from space.

To allow the gross distinction between plants and inorganic matter provided by the VNIR part of the spectrum to be exploited, *infrared film* has its range of sensitivity extended into this region, to about 1000 nanometres. A deep red filter is used with black and white infrared film to remove visible wavelengths. With *colour-infrared film*, a yellow

(a)

(b)

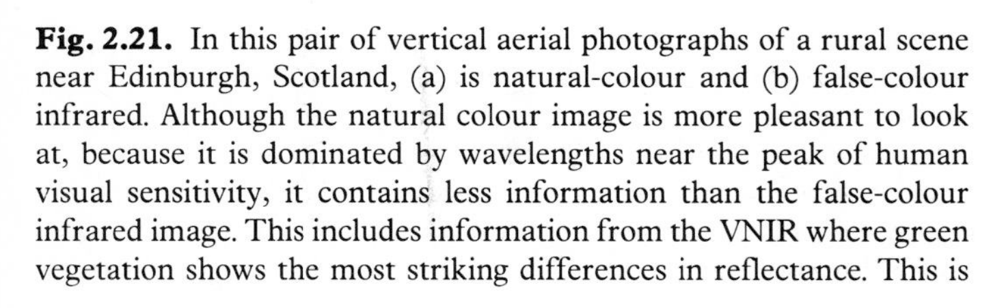

Fig. 2.21. In this pair of vertical aerial photographs of a rural scene near Edinburgh, Scotland, (a) is natural-colour and (b) false-colour infrared. Although the natural colour image is more pleasant to look at, because it is dominated by wavelengths near the peak of human visual sensitivity, it contains less information than the false-colour infrared image. This includes information from the VNIR where green vegetation shows the most striking differences in reflectance. This is particularly obvious in the curved band of woodland near top centre, where the dark conifers on the left are sharply distinguished from a band of bright-red broad-leaved species fringing the wood on the right. On the natural colour image both types of tree have a similar green shade. Among the fields, image (b) shows much more variations due to different crop types at different stages of growth. Width is 2 km, north to right.

filter removes blue, so that the yellow dye layer is bleached by green, magenta by red, and cyan by VNIR. This produces a false-colour photograph, where vegetation with high VNIR reflectivity appears in shades of red (Fig. 2.21). Red soil, absorbing green and reflecting red and VNIR to approximate the same degree, appears yellow, and most other inorganic materials approximating their natural colours. This *false-colour* rendition is discussed extensively in later sections.

The actual *resolution* of film depends on the quality and size of the camera lens, and on the grain size and speed of the film. As explained earlier, what the eye can resolve depends on the scale of the photographic print, whether it is black and white or in colour, and on the contrast. Contrast, in turn, varies according to the variation in reflectance of the surface, on the way the film was developed and on the type of paper used—soft or hard, equivalent to slow or fast film.

2.3.2 Electromechanical methods

Photography and the use of vidicon cameras are the only means of capturing images as 'snap shots', but are limited to the visible and VNIR range. All other systems rely on building up pictures line by line and pixel by pixel. Tedious as this might seem, it offers several great advantages. First, it allows the full range of wavelengths transmitted by the atmosphere to be used by choosing from a number of detectors, each of which has a range of wavelengths to which it is sensitive. Second, freed from the fixed optics of cameras, many detectors can be deployed in one instrument. Third, the spectrum can be split into many wavebands with a great flexibility in their widths. Detectors of this kind can record in either analogue or digital mode. The latter are most useful since the data can be processed, enhanced, and analysed by digital computers, of which more later.

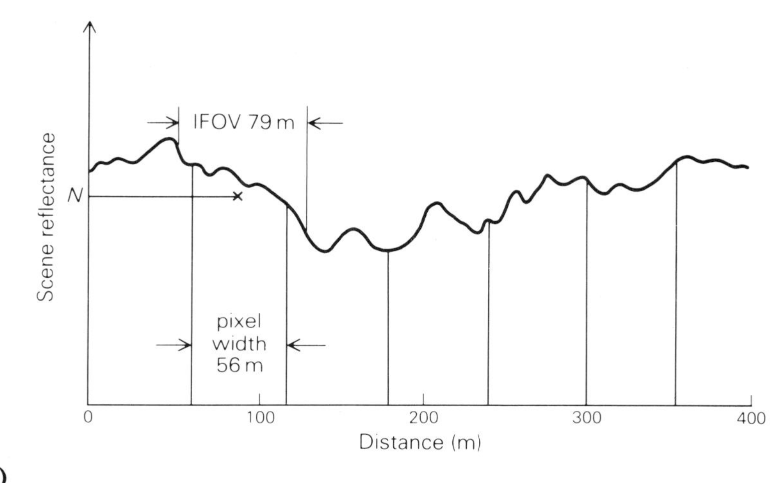

(a)

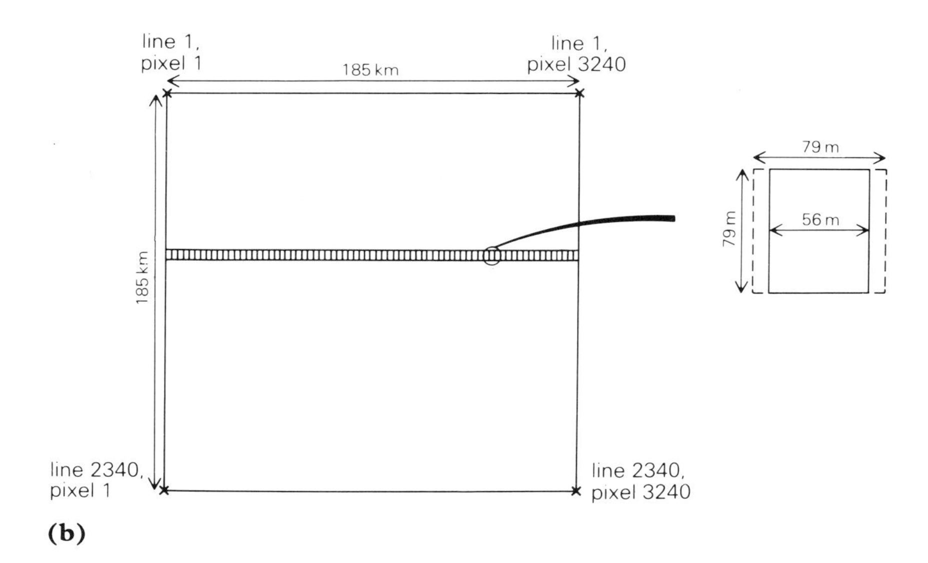

(b)

Fig. 2.22. (a) The irregular graph represents the actual variation of energy reflected by the surface in one waveband along a scan line, in this case one sampled by the Landsat Multispectral Scanner (MSS). The scanner samples this record every 56 m along a scan line to produce a string of pixels. The actual resolution of the MSS is 79 m, so the value assigned to a pixel is the average reflectance over this distance, not that from the pixel itself. The value is converted to a digital number from a signal measured in millivolts. (b) The MSS mirror sweeps across an area 240 m across, the detected radiation being directed onto a bank of six sensors for each of four bands. Consequently, the resulting six lines of pixels produced by each scan are 80 m wide. The raw data from the Landsat MSS is in the form of a square matrix 2340 lines deep, each line containing 3240 pixels. This gives a ground cover of 185 by 185 km. Because the Earth rotates beneath the platform during the 25 seconds taken to produce a whole scene, the square matrix represents a parallelogram on the ground, whose actual shape is determined by the latitude of the scene—the rotational velocity of the Earth decreases towards the poles.

The first of these systems to be devised was the *line-scanner*. This consists of a rotating mirror which collects radiation from the surface as it sweeps across it. The radiation is split into a spectrum by a diffraction grating, chosen bands being directed by various means onto appropriate detectors. These record the intensity of radiation as a continuous signal for each sweep. The signal along each sweep is divided electronically into segments. The average response in each segment is recorded as an analogue on film, via a cathode-ray display, or converted to a digital number for magnetic recording or direct transmission and later conversion to an image. The final image is made up of a sequence of *scan lines*, each of which consists of rectangular pixels (Fig. 2.22). Line scanners are routinely used for both the reflected and emitted infrared parts of the spectrum. Because of the space limitations inherent in such a mechanically complex system relatively few detectors can be deployed. Consequently, the data cover relatively broad spectral wavebands, precluding detailed spectral analysis of the surface. However the bands are generally designed to highlight the most common spectral features that are expected.

A simpler method, in which there are no moving parts, consists of linear arrays of thousands of tiny detectors, each array recording a different waveband. Radiation from the surface is directed onto the arrays through a system of lenses, diffraction gratings, and slits. The detectors, called *charge-coupled devices* or *CCDs*, are very rapidly charged by the radiation in proportion to its intensity, and discharged through a measuring and recording system. In essence, CCDs are light-sensitive capacitors. An image is built up as a series of lines, made up of pixels corresponding to each CCD in the array, as the system is swept over the surface by the motion of the platform. Hence the name *pushbroom system*. Because of its simplicity, a pushbroom system can be pointed in any direction. Moreover, since each detector is exposed to radiation for hundreds or thousands of times longer than those in a line-scan system, a pushbroom can be used either for much narrower wavebands, in which the energy available for recording is proportionally lower, or for much finer spatial resolution than a line scanner. CCDs cannot record in the thermally emitted infrared part of the spectrum. Pushbrooms are deployed in either a broad-band mode for general monitoring, or with many narrow bands to investigate the detailed spectral properties of the surface.

The orderly array of lines and pixels produced by both line-scanners and pushbrooms, and the direction wavebands of radiation from the same portion of the surface on to this array is extremely useful. It means that image data from all the wavebands are exactly registered to one another. Consequently the spectral response from each packet of surface or pixel can be investigated very precisely, or represented in combinations of three as the red, green, and blue intensities of a colour image made up of millions of these pixels, in rather the same way as a mosaic picture made of tiles. Not only natural-colour and false-colour images, similar to conventional photographs, can be displayed, but colour images combining any three wavebands from all parts of the spectrum as red, green, and blue primary colours can be displayed. This allows spectral differences at the surface to be exploited and displayed extremely efficiently.

2.3.3 Passive microwave imaging

Thermal phenomena involving temperatures around the average for the Earth's surface emit radiation beyond the infrared into the microwave region. Although the energies involved are very low, thermally induced microwaves can be detected using fairly simple antenna systems like small radio-telescopes, rather than the sophisticated supercooled solid-state detectors that must be used in the thermal infrared. It is a cheap option requiring relatively little energy, and is thereby attractive for deployment on satellites. Some systems use a fixed antenna to build up profiles of emitted microwave energy, but it is more useful to scan the surface using a movable antenna. This produces *passive microwave images*. The resolution of such a system is expressed as a solid angle, given in steradians as approximately the wavelength divided by the antenna diameter. Since there is a physical limit to the size of a portable antenna, and microwave wavelengths are in millimetres to centimetres, this angle is never small. Consequently at orbital altitudes the spatial resolution is very coarse compared with other imaging systems. Passive-microwave remote sensing is therefore only useful for gross overviews of atmospheric and oceanographic features from orbit. Airborne systems can have geological applications, but the better resolving power of thermal-infrared imagery is generally preferred.

2.3.4 Active microwave or radar imaging

Artifical microwave energy used in radar remote sensing spans the 1–30 cm range. It is usually in the form of a coherent single wavelength, like a laser, generated electronically. Deployed in its most simple form, radar energy is directed vertically downwards as a pulse for a few microseconds. The time taken for the pulse reflected by the surface to travel back to the antenna is a very accurate measure of the average distance from the platform to the patch of surface illuminated by the pulse. This is the basis of *radar altimetry* from orbit, most usefully analysed over the oceans. As we shall see in Fig. 3.15, given enough orbits a low-resolution image of the average elevation of the ocean surface can be produced, from which extremely interesting deductions can be made about bathymetry and the gravitational field.

To produce high-resolution radar images a fundamentally different strategy from all other remote-sensing methods is employed. Radar pulses emitted by an antenna are directed downwards and to the side, so illuminating a narrow strip perpendicular to the flight path of the platform (Fig. 2.23(a)). The motion of the platform is used to build up a continuous image by careful synchronization of subsequent pulses with platform velocity. As described earlier, part of the sideways-directed energy is reflected away from the antenna and some is scattered back to it, depending on the attitude of the surface and its roughness. The time at which the back-scattered pulse is received is proportional to the distance that it has travelled, and the energy at that time records the interaction with the surface at the corresponding distance (Fig. 2.23(b)). At its most simple, a radar image is built up by the modulation of the brightness along a succession of lines on a cathode-ray tube or on film by the energy–time history of successive pulses (Fig. 2.23(c)). This is corrected for the geometry of the system—the angle of the radar beam relative to the surface decreases with distance to the side—to produce an image with a constant scale. Another consequence of the angular geometry is that the resolution perpendicular to the track of the platform, *range resolution*, falls off in inverse proportion to the cosine of the incidence angle. The resolution parallel to the track, *azimuth resolution*, is inversely proportional to the length or aperture of the antenna.

The main problem with this simple *real-aperture radar imagery* is that the beam spreads out with distance. This means that resolution parallel to the platform track—azimuth resolution—decreases dramatically with range. Because azimuth resolution is inversely proportional to antenna aperture, above about 10 km real aperture radar is virtually useless; to maintain acceptable resolution the antenna must be huge. This is where the coherent nature of radar proves extremely useful. At any instant only the illuminated surface directly to the side of the platform is stationary relative to the platform. Those parts ahead of and behind are moving towards and away from the platform (Fig. 2.24). As a result the radar pulse scattered back from them suffers a *Doppler shift*, so increasing and decreasing radar frequency, respectively. The frequency of the returned radar energy is combined with a reference frequency equal to that emitted. The shifted returns interfere with the reference, and so code information from ahead, aft, and to the side for every pulse. This effectively allows a short antenna to mimic one up to several hundred metres long, hence the name *synthetic-aperture radar* or SAR. The Doppler history of each pulse is

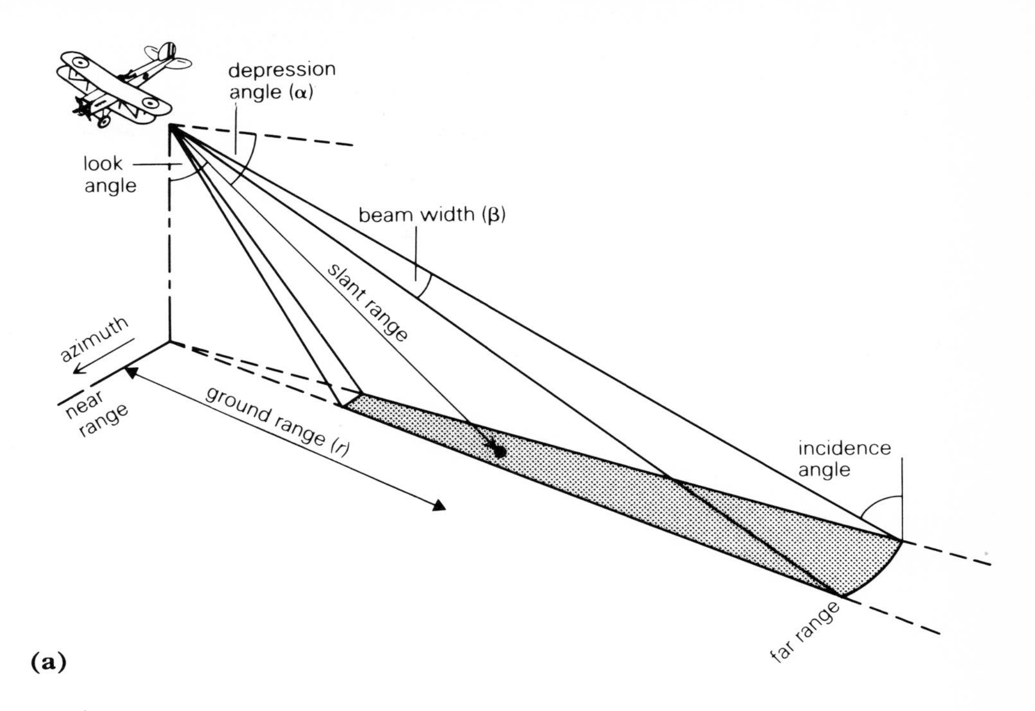

(a)

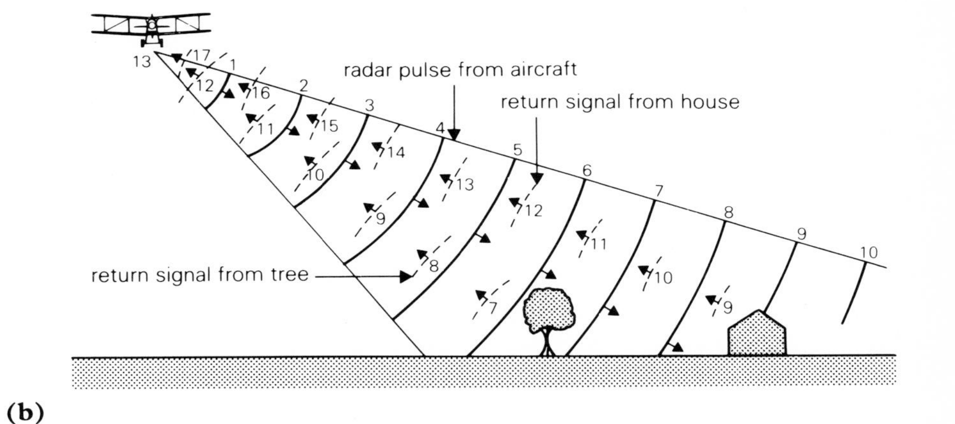

(b)

high energy
output pulse

return from house

return from tree

Signal

0 2 4 6 8 10 12 14 16 18

Time (number of pulses)

(c)

Fig. 2.23. (a) The geometry involved in sideways-looking radar governs many of the intrinsic properties of radar images. The *depression angle* is the angle between horizontal and a radar ray path. The *look angle*, sometimes quoted instead, is simply the angle between vertical and a ray path. The *incidence angle* is the angle between an incident radar ray and a line at right angles to the surface. The further away from the antenna the surface is, the smaller the incidence angle. The *beam width* determines how the illumination of the surface spreads out with distance to the side—from near to far range. The *slant range* is the direct distance from the antenna to object, and is related to the true or *ground range* by the depression angle. (b) The solid curves on this cartoon are spherical radar wavefronts emitted by the antenna, and their numbers indicate the time since they were emitted. The dashed curves are radar wave fronts back-scattered from the house and tree. They have the same number convention indicating time. The wave front from the tree which has just been received at the antenna has travelled for 13 time units, that from the house for 17. The graph (c) shows how radar energy returned from the house and tree is recorded at the antenna.

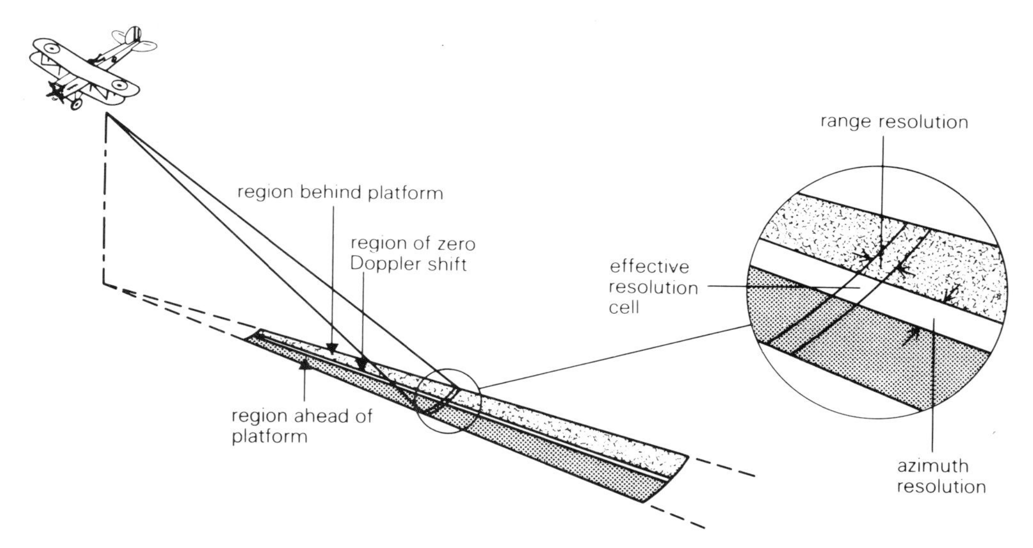

Fig. 2.24. In a synthetic aperture radar system interference occurs between a reference signal and the increased or decreased frequencies of radar waves returning from illuminated areas ahead of and behind the platform. Waves returning from a narrow strip at right angles to the flight path do not suffer a Doppler shift, and no interference occurs. This allows returns at any instant from the in-line strip to be discriminated from those from the rest of the illuminated surface during signal processing. This makes possible a much finer resolution in the azimuth direction than could be achieved with the same antenna used in a real-aperture system. The further down range an object is, the more times it can back-scatter successive radar pulses as the platform flies by. Thus the discrimination of ahead, in-line, and behind positions relative to the moving platform is independent of range. Azimuth resolution is theoretically constant in SAR images. In practice it is degraded with increasing range because the radar energy becomes weaker with distance due to the beam spread. Eventually, energy back-scattered from the furthest range becomes too weak to be detected.

recorded digitally or on film in the form of a radar hologram, requiring a second stage of digital or optical processing to give an image whose azimuth resolution is independent of flying height. Despite the great complexity of the method, SAR is the only means whereby the great advantages of radar imaging can be deployed from orbit.

2.4 Sources of remotely sensed data

Airborne remote-sensing systems are clearly the most flexible sources of data, and provide the most detailed information. However, there are so many aerial survey companies, using both photography and electronic systems, that a comprehensive account would be too lengthy for this book. This section concentrates on publically available data from satellites.

There are three basic types of orbit that cover satellites producing remotely sensed data. All attempt to maintain a nearly constant height so that data are acquired at a uniform scale.

If a satellite is placed in orbit around 41 000 km away and its motion is parallel to that of the Earth's rotation, its velocity matches that of the Earth and it remains above a fixed point on the surface. Such a *geostationary satellite* can monitor almost an entire hemisphere all the time, and can transmit directly to any point on that hemisphere. This is of particular use for climatic observations, but the distance means that resolution is very coarse.

To get a sharper view of the planet means a lower orbit and a narrower field of view. To obtain full global coverage requires the orbit to pass close to the poles and that successive orbits move progressively around the Earth, so that ultimately the fields viewed by the instruments overlap completely. The other important consideration, for passive remote sensing, is that the orbits should cross all parts of the Earth when they are illuminated by the sun at the same time of day. This is so that results can be compared from one area to another. Such *sun-synchronous polar orbits* are limited to a height of between 600 and 950 km. They must be inclined relative to the Equator to ensure correct timing and eventual global coverage (Fig. 2.25). The details of each orbit, together with the width of the swathe covered by the imaging system, determine the rate at which full coverage is repeated. Where remote sensing uses an active method, such as radar, matching orbits to the same time of day is not important. Moreover, both the ascending (northward) and descending (southward) orbits can be used.

The need for eventual re-entry and power considerations have meant that, up to the present, polar-orbiting *manned satellites* are impractical. The orbits of manned craft are limited to about 50° north and south of the Equator, ruling out synchronization with the sun and global coverage. They have rarely orbited higher than about 300 km. Until the building of the NASA/ESA *space station* in the mid- to late-1990s, which will be in sun-synchronous polar orbit, manned platforms are limited to the acquisition of data on an experimental basis.

Since the 1960s a very large number of satellites that deployed remote-sensing instruments have been launched. To list and describe them all would take up too much space here and would be confusing. Only a selection are covered to illustrate the kinds of data that are easily available.

The main controls over the kind and quality of data provided by any platform are partly governed by the optical characteristics of the devices employed on a satellite and by the rate at which data can be transmitted to the ground. For instance, the same system viewing a 60 km by 60 km area from a low orbit could transmit data to give 10 m resolution, but used in geostationary orbit to cover a whole hemisphere would have to be restricted to 5 km resolution using the same data transmission rate. The advantage of low-resolution geostationary systems is that they can provide continuous information, and so they are the work-horses for climatic, atmospheric, and gross oceanographic monitoring. They are often referred to as *metsats*. Currently there are six such platforms stationed above the Equator, covering the Americas and East Pacific (*GOES-E* and -*W*, USA), Africa and Europe (*Meteosat*, European Space Agency), the Indian Ocean (*GOMS*, USSR; *Insat*, India) and the West Pacific (*GMS*, Japan).

The main instrument on all these satellites is some form of imaging device, exemplified by the *Visible Infrared Spin-Scan Radiometer (VISSR)* aboard the GOES satellites. The VISSR is a day and night monitor of visible (900 m pixels, 550–700 nm) and thermal infrared (6.9 km pixels, 10.5–12.6μm) that obtains images of cloud cover, cloud-top, and surface temperature. As the name suggests, this instrument uses the spinning motion of the satellite to scan the surface. At each revolution the optics are tilted slightly so that successive scans cover adjacent strips across the globe. By this means an image of the hemisphere below is built up over a period of about half an hour, giving very detailed coverage of the changes in large-scale weather patterns. The equivalent of VISSR aboard Meteosat is also equipped with a channel aimed at thermal emission by water vapour in the atmosphere in the 5.7–7.1 μm region, whose function is described in Chapter 3. Another system carried by geostationary metsats is non-imaging, and uses filters covering the range 3.94–14.74 μm for atmospheric sounding. This relies on the different transparencies of the atmosphere to radiation of different wavelengths, to build up records of temperature and moisture contents at different levels. On GOES satellites this is called the *VISSR Atmospheric Sounder*, or *VAS*. Figures 3.2, 3.3, and 3.9 give examples of data produced by a geostationary platform.

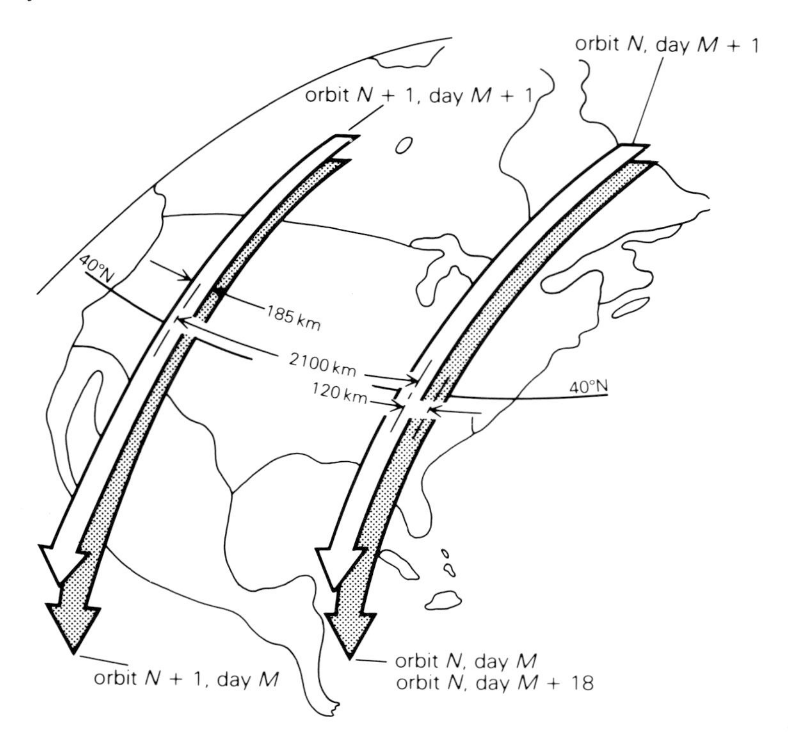

(a)

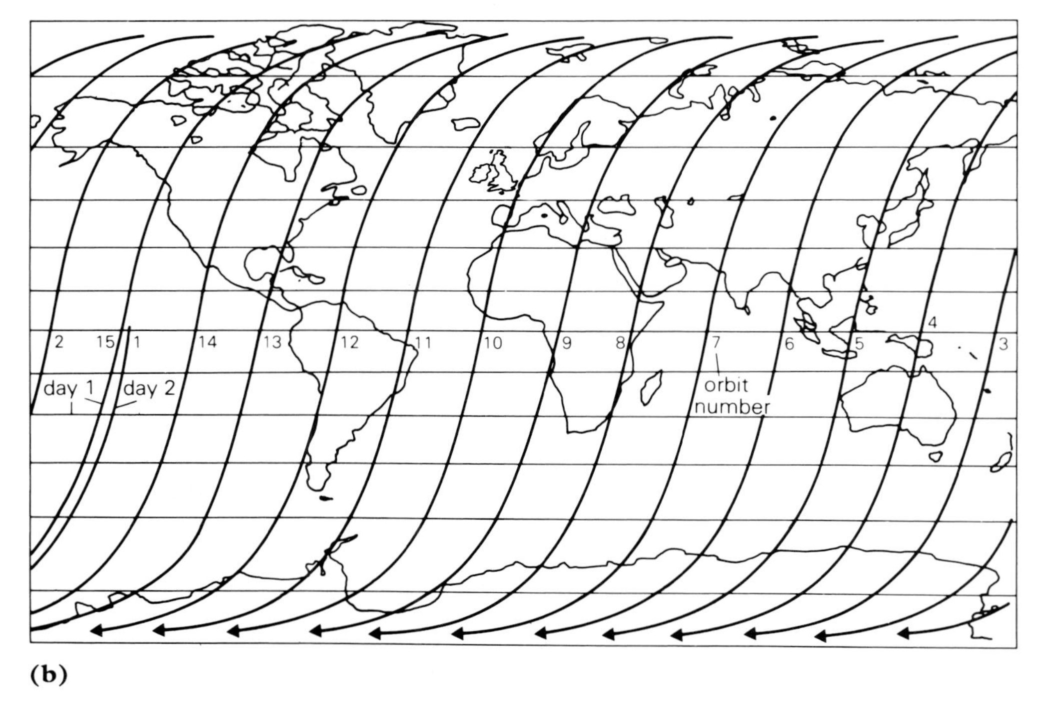

(b)

Fig. 2.25. (a) The diagram shows successive Landsat orbits on two days. On day M, orbit $N + 1$ is shifted 2100 km west of orbit N at this latitude. Orbit N on day $M + 1$ is only 120 km west of the same orbit on the previous day, so the images taken on both days overlap. The higher the latitude the greater the overlap between adjacent scenes. The sequence repeated itself every 18 days for Landsat-1, -2, and -3. This revisit frequency was reduced to 16 days for Landsat-4 and -5. Very much the same principle applies to all sun-synchronous, polar-orbiting systems. In the case of the NOAA AVHRR, the revisit frequency is every 12 hours to give day and night coverage. (b) So that orbits are synchronous with the sun, they are inclined relative to the Equator. On a Mercator projection Landsat orbits plot as sinuous curves. Only the descending—N to S—orbits are shown, the ascending orbits pass on the night side of the Earth.

Geostationary satellites also relay information from ground-based sensors, such as river gauges, earthquake monitors, and unmanned meteorological stations, to data-handling centres. Images are streamed continuously to ground stations so that regular updates of hemispheric conditions can be used by climatologists and meteorologists. The images are distributed to users as low-resolution facsimiles by telephone, sometimes using the GOES satellites as data relays.

More detailed meteorological and oceanographic monitoring is the function of polar orbiting, sun-synchronous satellites with a wide swathe of coverage. Currently there is only one series of satellites that provide publically available data—the *TIROS/NOAA series* operated by the US National Oceanic and Atmospheric Administration (NOAA), of which four are operational or on standby. Another series operated by NASA as the source of data for experimental work ended with the abandonment of *Nimbus*-7 in 1984, but Nimbus data are still used widely for research. Both types are literally bristling with instruments (Fig. 2.26).

Fig. 2.26. Artist's impression of the sun-synchronous, polar orbiting TIROS-N/NOAA satellite, the main work-horse for meteorological and oceanographic remote sensing.

Of the four operational TIROS/NOAA platforms two are 'turned on' and are synchronized to overpass at 07.30 and 19.30, and 14.30 and 02.30 local solar time, giving a six-hourly repetition of all points. The main instrument is a line-scan system called the *Advanced Very High Resolution Radiometer (AVHRR)*, which produces images in five wavebands (visible red, VNIR, 3.55–3.93, 10.3–11.3, and 11.5–12.5 μm thermal infrared). The resolution of the AVHRR allows a pixel size of 1.1 km for points directly beneath the satellite, decreasing in resolution as the scan reaches to either side of the orbit path. The image swathe is 2400 km wide, and because this is achieved by a large scanning angle the edges of the swathe are extremely foreshortened and distorted. AVHRR data can be recorded in 4000 km long strips at full 1.1 km resolution for local area cover (*LAC*), or degraded to 5 km resolution for global mosaics (*GAC*). Because the ground track of successive orbits is shifted west by 821 km, full global cover is possible twice each day. The AVHRR provides most of the images used in many weather forecasts, which are transmitted as facsimiles either directly to the user or by telephone from ground receiving stations. Figures 3.1, 3.8, 3.40, 3.41, 3.53, and 4.7 are examples of various types of AVHRR images.

Other NOAA-series instruments are contained in a package called the *TIROS Operational Vertical Sounder (TOVS)*. The *High Resolution Infrared Spectrometer (HIRS)* records data with a 17.4 km pixel size in 20 non-imaging channels directed at CO, H_2O and N_2O absorption bands (Figs 3.5 and 3.6). The *Stratosphere Sounding Unit (SSU)* computes air temperatures from the surface to 50 km altitude, using 15 μm infrared. The *Microwave Sounding Unit (MSU)* is sensitive to four channels of emitted microwaves, aiming at the water-vapour and oxygen absorption regions around 5–6 mm. Other instruments are the *Solar Back-scatter Ultraviolet Instrument (SBUV)* measuring the ultraviolet spectrum scattered by the atmosphere to estimate ozone content and distribution in the upper atmosphere, and the *Earth Radiation Budget Sensor System (ERBSS)* measuring reflected and re-emitted radiation from 200 nm to 50 μm. In addition, the NOAA satellites act as relays for data from balloons and buoys, and detect signals from distress beacons.

Nimbus-7 also had a daily surveillance role, using eight instruments. The *Scanning Multichannel Microwave Radiometer (SMMR)* is a non-imaging passive-microwave instrument covering five microwave wavelengths aimed at sea-surface roughness and winds, sea-surface temperature, the liquid and gaseous water content of clouds, droplet size of rainfall, soil moisture, and presence of sea ice (Fig. 3.7). The *Stratospheric and Mesospheric Sounder (SAMS)* measures vertical concentrations of water, methane, carbon monoxide, and nitrogen oxides in the upper atmosphere. Two instruments mimic the TIROS/NOAA SBUV and ERBSS (Fig. 3.11). The *Stratospheric Aerosol Measurement Experiment (SAMII)* measures the concentration and optical properties of atmospheric aerosols in three dimensions. The *Temperature-Humidity Infrared Radiometer (THIR)* measures 11 and 6.7 μm thermal infrared by day and night to provide images of cloud cover and depth as well as surface temperature. The *Limb Infrared Monitoring of the Stratosphere Experiment (LIMS)* uses infrared measurements of the Earth's limb (the atmosphere above the satellite's 'horizon') to derive concentrations of ozone, water, and nitrogen oxides together with air temperature in profiles of the upper atmosphere. The most versatile instrument is again an imaging system with an 825 m resolution and 1500 km swathe width, the *Coastal Zone Colour Scanner (CZCS)* which monitors four visible, one VNIR, and one thermal infrared wavebands (Figs 3.16, 3.20, 4.18, and 4.19). This was intended to measure sediment and biological materials in sea-water and its temperature, but like the AVHRR, does provide useful synoptic information over the land surface.

Of all remote-sensing satellites, perhaps the best known by name is the polar-orbiting *Landsat series* aimed at land applications. The first was launched by NASA in 1972, and apart from a brief period in 1984, the Landsat series has given repeated coverage of most parts of the globe up to the present. After a period under the control of the US NOAA, it is now administered by a commercial company called *Eosat*. The current Landsat-5 covers all points at 09.30 local time—the time of minimum cloud cover in northern temperate climates—every 16 days for Landsats 4 and 5, and every 18 days for the three earlier platforms—with an image swathe width of 185 km. The data are divided along each swathe into 185 km segments on a regular grid system. Each scene is assigned a path number, corresponding to the ground track, which is repeated after 251 orbits, and a row number that marks the 185 km lengthwise division of the ground track on receipt of the data at a ground station. Because of the different repeat frequencies of the early and current Landsats, the path–row coordinates of scenes differ from Landsats 1–3 to Landsats 4 and 5.

Landsat-5 carries two line-scan instruments, the *Multispectral Scanner (MSS)* covering green, red, and two VNIR wavebands with an 80 m resolution, and the *Thematic Mapper (TM)* covering blue, green, red, VNIR, and two SWIR bands with 30 m resolution and a thermal infrared band with 120 m resolution. The MSS has been deployed on all Landsats and was designed specifically for global vegetation monitoring, but can address the ferric iron absorption feature in the VNIR, so that together with information on soil and rock colour and albedo it has been widely used for geological purposes. The TM aims at monitoring variation in vegetation cover and type, the iron and clay mineral content of rocks and soils, vegetation, and soil moisture content and surface temperature. It has only been deployed aboard Landsats 4 and 5. The first three Landsats carried the MSS, Landsat-3 also carrying a 30 m resolution *Return Beam Vidicon* camera (*RBV*) and a 240 m resolution thermal sensor. The near-continuous operation of the Landsat series since 1972 means that for many parts of the globe cloud-free images are available for many dates, making it the prime source of remotely sensed information for geographers, agronomists, geologists, hydrologists, and planners. The first three Landsats had on-board recorders, so that an archive of the entire globe was possible using direct relay and retransmission to US-based ground stations, as well as to foreign stations in line of sight. Landsats 4 and 5 do not have this capability, the plan being to relay data through three geostationary *Telemetry and Data Relay Satellites (TDRS)* shared with the US military. Only two of these are in orbit, and for a considerable part of the Earth's surface, data reception is limited to line-of-sight stations equipped to receive both MSS and TM data. This means that large areas in east Asia are not covered by TM data at the time of writing. By far the greatest proportion of images used in this book are products of the Landsat system. Good examples of Landsat MSS and TM images are shown by Figs 3.19 and 3.28 respectively.

Operational land remote sensing is not the exclusive province of the USA; the Soviet Union, France, Japan, and India all have systems in orbit. Most important of these is the French *Système Probatoire de l'Observation de la Terre (SPOT)*, launched in February 1986. This employs pushbroom arrays of CCDs giving 10 m resolution coverage in the panchromatic range and 20 m resolution in green, red, and VNIR bands, for 60 × 60 km scenes. The simplicity of pushbroom systems means that they can be mechanically pointed to give oblique views—up to 27° from nadir in the case of SPOT. This cuts down the revisit time to as little as four days, and also enables stereoscopic image pairs to be produced from which measurements of surface elevation can be derived with accuracies better than 30 m. Figures 2.31, 2.33, and 3.55 are examples of SPOT images.

1987 saw the launch of two high-resolution satellites by Japan and India. The Japanese *Marine Observation Satellite (MOS-1)* carries three remote-sensing systems. The *Multi-spectral Electronic Self-scanning Radiometer (MESSR)* has 50 m pixels and records similar bands to the Landsat MSS using a pushbroom detector array. The second, the *Visible and Thermal Infrared Radiometer (VTIR)*, is broadly analogous to the NOAA AVHRR, with one visible and three thermal channels, one aimed at measuring upper atmosphere water-vapour content. The third is a passive microwave instrument, the *Microwave Scanning Radiometer (MSR)*. The Indian *IRS-1* satellite also uses CCDs tuned to the same wavebands as SPOT in pushbroom mode, and is designed to capture 73 and 36.5 m resolution images simultaneously. Both these systems are currently sending data only to local ground stations.

Details of the image characteristics provided by some important operational satellites are given in Table 2.1.

Experimental remote sensing covers a very wide range of methods and applications, which are restricted to a relatively small proportion of the Earth's surface. Three methods are summarized here: radar, thermal imagery, and imaging spectrometry.

The development of synthetic-aperture radar (SAR) meant that all the advantages of active-microwave remote sensing could be exploited from orbit. The first public demonstration was from the *Seasat* platform launched into an 800 km polar orbit in June

Table 2.1 Characteristics of some orbital imaging systems

Platform	System	Wavebands	Resolution	Image width
Meteosat (ESA)		550–700 nm 10.5–12.6 μm	900 m 6.9 km	60% of hemisphere
TIROS/NOAA (USA)	AVHRR	550–680 nm 725–1100 nm 3.55–3.93 μm 10.5–11.5 μm 11.5–12.5 μm	1.1 km (LAC) to 5.0 km (GAC)	2400 km
Nimbus-7 (USA)	CZCS	430–450 nm 510–530 nm 540–560 nm 660–680 nm 700–800 nm 10.5–12.5 μm	800 m	1800 km
Landsat-1 to -5 (USA)	MSS	500–600 nm 600–700 nm 700–800 nm 800–1100 nm 10.4–12.6 μm (L-3 only)	80 m 240 m	185 km
Landsat-3	RBV	505–750 nm	40 m	99 km
Landsat-4, 5	TM	450–520 nm 520–600 nm 630–690 nm 760–900 nm 1.55–1.75 μm 2.08–2.35 μm 10.4–12.5 μm	30 m 120 m	185 km
SPOT (France)	P	510–730 nm	10 m	60 km
	XS	500–590 nm 610–680 nm 790–890 nm	20 m	60 km
MOS-1 (Japan)	MESSR	510–590 nm 610–690 nm 720–800 nm 800–1100 nm	50 m	100 km
	VTIR	500–700 nm 6.0–7.0 μm 10.5–11.5 μm 11.5–12.5 μm	900 m 900 m 2.7 km 2.7 km	1500 km
IRS-1 (India)	LISS-1	450–520 nm 520–590 nm 620–680 nm 770–860 nm	73 m	148 km
	LISS-2	As LISS-1	36.5 m	74 km
HCMM (USA)		500–1100 nm 10.5–12.5 μm	600 m 600 m	716 km 716 km
Seasat (USA)	SAR radar	23.5 cm	25 m	100 km
Shuttle	SIR-A SIR-B	23.5 cm 23.5 cm multi-angle	40 m 25 m	55 km variable

1978. For 100 days it gathered data from an L-band (23.5 cm) SAR, a radar altimeter, and a scatterometer. The SAR had a theoretical resolution of 25 m for a 100 km swathe width. Its look angle was 20° from vertical, to emphasize features of the sea surface for which it was designed. Its lack of on-board storage meant that only images around the North Atlantic in line of sight with five ground stations were recorded. Over land, its steep look angle meant that rugged terrain was expressed in a curious fashion due to the phenomenon of *lay-over*, mountains appearing to lean towards the platform. Examples of Seasat images are shown in Figs 3.14, 3.21, 4.37, 4.38, and 4.55. The radar altimeter, whose data were recorded onboard, was designed to measure accurately the mean elevation of the sea surface directly beneath each orbit. The scatterometer enabled surface wind speed and direction to be calculated for fifteen 50 km cells in a 750 km swathe along each orbit. Orbital SAR images have also been acquired during two NASA Space Shuttle missions in 1982 and 1984—the *Shuttle Imaging Radar (SIR-A and -B) Experiments*. Examples of SIR images are shown by Figs 3.22, 3.35, 3.38, 3.45, 3.46, 3.65, 3.69–3.71, 4.6, and 4.54. The early Shuttle missions also carried the simplest and cheapest of all remote-sensing systems—photographic cameras—natural-colour, panchromatic, and false-colour photographs from various hand-held cameras operated by the crew and from the *Metric* and *Large-Format Cameras* providing some of the most easily used data for a number of remote parts of the globe. The last two produced overlapping images with stereoscopic potential. Examples are shown by Figs 3.67 and 4.1–4.3.

Since the thermal properties of rocks and soils are highly discriminating, remote sensing in the region of the Earth's peak of energy emission (8–14 μm) is of great interest to geologists. The *Heat Capacity Mapping Mission (HCMM)* was launched in April 1978 to serve such research. It was a sun-synchronous polar orbiter, passing over each point on a 16-day repeat cycle at between 01.30 and 02.30 and 13.30 and 14.30 local solar time—the times of lowest and highest surface temperature in the daily heating/cooling cycle. Data from a broad 10.5–12.5 μm thermal channel and a 0.55–1.1 μm reflected channel were captured with a resolution of 600 m over a 716 km swathe. Like Seasat, only ground stations in the USA, Australia, and Europe were capable of recording data. Problems with cloud cover greatly restricted the coverage by day and night images, so necessary for investigation of the thermal-inertia properties of the surface. Examples of HCMM images are shown by Figs 3.63 and 4.20.

A more advanced sensor in the thermal region is the *Thermal Infrared Multispectral Scanner (TIMS)*, which produces data in six narrow thermal channels to highlight differences in the emissivity curves of different rocks and minerals. Currently data are only available from aircraft, but the system is scheduled for deployment from the Space Shuttle. Examples of TIMS images are shown by Figs 3.77 and 4.27.

The most ambitious system being considered for orbital platforms uses linear arrays of detectors on to which reflected radiation split into very narrow (10–20 nm) bands is directed so that detailed spectra are recorded. The linear arrays are mounted in a two-dimensional block so that using the pushbroom principle, an image of the surface is created for each of the narrow bands. This enables the spectrum for each pixel to be reconstructed and digitally compared with reference spectra from ground observations, to produce maps of different surface classes directly. The most advanced of these imaging spectrometers, the *Airborne Visible and Infrared Imaging Spectrometer (AVIRIS)*, produces images 550 pixels wide for 220 spectral bands from an aircraft. A version of this instrument is being developed for experiments on the Space Shuttle—the *High Resolution Imaging Spectrometer (HIRIS)*.

2.5 Displaying information

In the case of analogue images, such as photographs, an image is produced directly by the process involved. Opportunities for manipulating the appearance of the image are fairly limited. Digital image data exist as a string of *digital numbers (DN)* each relating to a pixel in the regular array that makes up the image. Producing an image means manipulating the array through a computer and displaying each component pixel on a monitor as a brightness signal to the electron gun of the video display system. Being in a regular array and digital form means that all the attributes of the image can be changed using a computer and a battery of statistical and geometrical software.

Most images contain *defects* due either to the sensing system or to natural processes. These include noise due to malfunction of detectors and electronic relays and geometric distortions due to the optics and the motion of the platform relative to the Earth's surface. Because *random noise* disrupts visual interpretation it must be removed (Fig. 2.27), but the methods are not important here. Distortions are generally removed by registering the digital image to a map projection appropriate to the scale of the image. This is achieved by identifying the map coordinates of easily recognized features on the image and generating a function that will allow the computer to warp the image to fit these *ground-control points*, in the manner of stretching a rubber sheet. Similarly it is possible to register one image to another by identifying common points on each.

The great advantage of a remotely sensed image is that it contains far more information than a map. It is important to express this information in as visually stimulating a form as

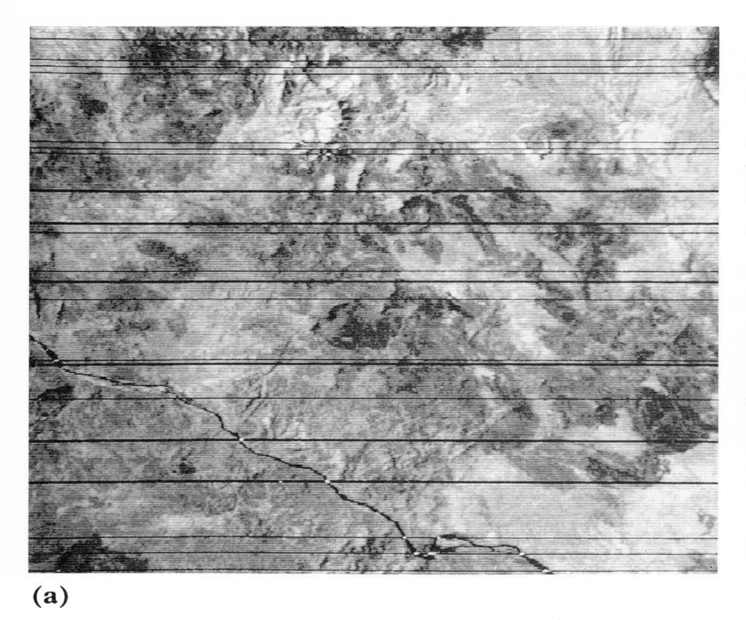

(a)

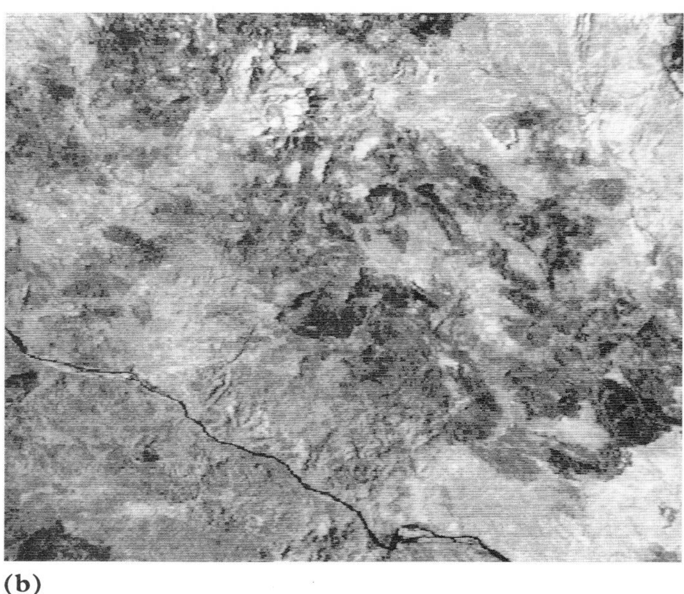

(b)

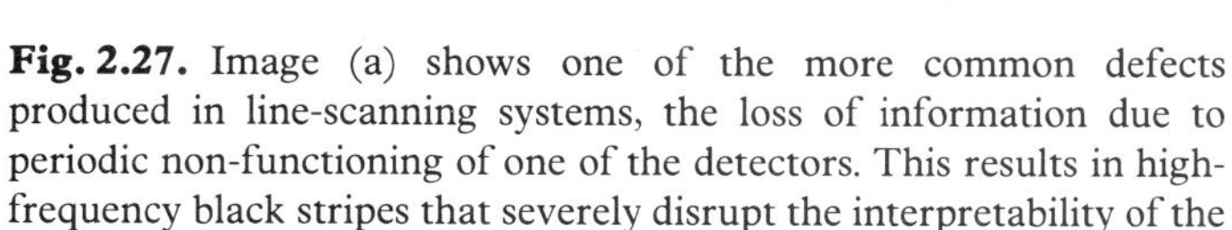

Fig. 2.27. Image (a) shows one of the more common defects produced in line-scanning systems, the loss of information due to periodic non-functioning of one of the detectors. This results in high-frequency black stripes that severely disrupt the interpretability of the image. Image (b) has had the dropped lines replaced by the average of adjacent lines. Although the missing information has not been restored, the method improves the quality of the image considerably. Width is 25 km.

possible for manual interpretation. Here *digital image enhancement* methods are outlined only briefly. The details of these methods and more complex techniques are covered by a number of advanced textbooks.

Data can be displayed in two basic forms, either one band at a time in black and white, or in combinations of three bands, one assigned to red, one to green, and the third to blue, these three additive primary colours assembling the full range of hues, as explained earlier in this chapter. Only combining bands corresponding to visible red, green, and blue in this way gives an approximation of the true colour of the surface. It is more usual to combine wavebands from outside the visible spectrum, sometimes with one or two visible bands, sometimes the colour being exclusively a result of totally invisible variations in surface reflectance or emittance. Such images are known as *false-colour composites* (FCC), the most commonly used expressing VNIR reflectance as red, visible red as green, and green as the blue component of the FCC. The colours in these images are at first confusing to the observer, because familiar objects are rendered in the 'wrong' colour. In the last mentioned case grass appears bright red, and red soil is yellow, for instance. However, this is a most powerful means of visualizing the effects of the spectral properties of different material outside the normal range of human vision. Where different FCCs appear in later sections, the colours that they display will be explained in the text.

Since vision is directly dependent on contrast for perceiving spatial detail and colour, the first step in image enhancement is to improve the contrast. There are two main reasons for this. First, the sensors in a remote-sensing system are usually set to respond to all anticipated levels of energy emanating from a scene. If they were not, something might be missed. The result is that the DN for the bulk of the scene are compressed to occupy the lower parts of the range. Second, there may be a common background energy level, such as atmospheric haze. As explained earlier, with decreasing wavelength an increasing proportion of the incoming solar radiation is scattered by the atmosphere. This scattering is uniform resulting in, for example, the uniform blue colour of a clear sky. This scattered radiation is added to the actual variations in reflection from the surface—as the sky is blue as we look up, so a blue 'haze' is added uniformly to a view of the surface from above; the higher the viewing position the more 'haze' obscures the image of the surface.

What these two factors do is best explained by looking first at the data statistics in the form of a *frequency histogram* of DN for a particular waveband. This is a plot of the number of pixels to which each of the 256 possible DN in a digital image have been assigned. It consists of bars for each DN whose height is proportional to the number of pixels with that DN and therefore to the frequency of occurrence of that DN in the image. The likelihood of finding one of these steps or DN in an image is therefore expressed by the height of the relevant histogram bar. The overall shape of the histogram tells us something about the contrast in the image (Fig. 2.28(a)). An image with a narrow histogram has a small range of DN and therefore low contrast (Fig. 2.28(b)), and a broad histogram indicates better contrast. The individual peaks indicate if the image has a continuous range of DN in the case of a single bell-shaped peak, or if it consists of several distinct categories of surface type in the case of a histogram with several peaks and troughs. Because most images of the land surface contain deep shadows, which should receive no energy in the absence of an atmosphere, a histogram which starts at a DN of zero is unlikely to represent an image with a constant background. One whose minimum value is greater than zero is probably affected by atmospheric scattering—the shadows are lit by random scattered radiation from the sky. In the latter case the image will have a distinctly bland or hazy appearance.

Improving image contrast means stretching out the histogram to fill the entire range of DN that can be displayed, hence the term *contrast stretching*. At its simplest this involves setting the lowest value in the image to zero, black on the monitor, and the highest to the maximum allowed by the display, when the monitor is said to be saturated. All other DN are shifted in a linear fashion to occupy the intervening values of brightness. This does not increase the number of discrete brightness levels, but merely makes them more different (Fig. 2.28(c)). Improved contrast improves our ability to see the individual steps and aids detection of fine spatial detail (Fig. 2.28(d)).

Where there is a background effect, for example due to atmospheric scattering, stretching still leaves a bland appearance (Fig. 2.28(d)). One way of removing this is to locate parts of the image that would normally be black in the absence of an atmosphere, such as deep shadows. If the DN associated with deep shadows are themselves set to zero, the haze effect is largely removed, so sharpening the image (Fig. 2.28(e)).

The way an image processing computer achieves a contrast stretch is very simple. A graph of input DN against the desired output is constructed according to the strategy of the operator, and each raw DN is replaced by the desired output DN (Fig. 2.28(f)). The most common of these *look-up tables* follows the 'recipe' above—setting shadow DN to zero, the maximum DN to saturation, and all those between equally distributed in the range 0–255—and takes the form of a straight line. More sophisticated stretches allow different linear stretches over different ranges of raw DN, to stretch, say, the brighter parts more than the darker, or apply various curves, such as a logarithmic distribution, to the input–output look-up table. All three bands used in constructing a colour image are stretched to suit the needs of the interpreter. Figure 2.29 shows the resulting false-colour image that contains the subarea in Fig. 2.28.

Most of the wavebands of remotely sensed data that are displayed together as colour images come from roughly the same parts of the electromagnetic spectrum. Although there are spectral features, such as the absorptions due to iron and clay minerals, and the features related to the composition and structure of green leaves, that help discriminate different surface materials when we examine spectra in detail (Section 2.2.1), by far the largest effect

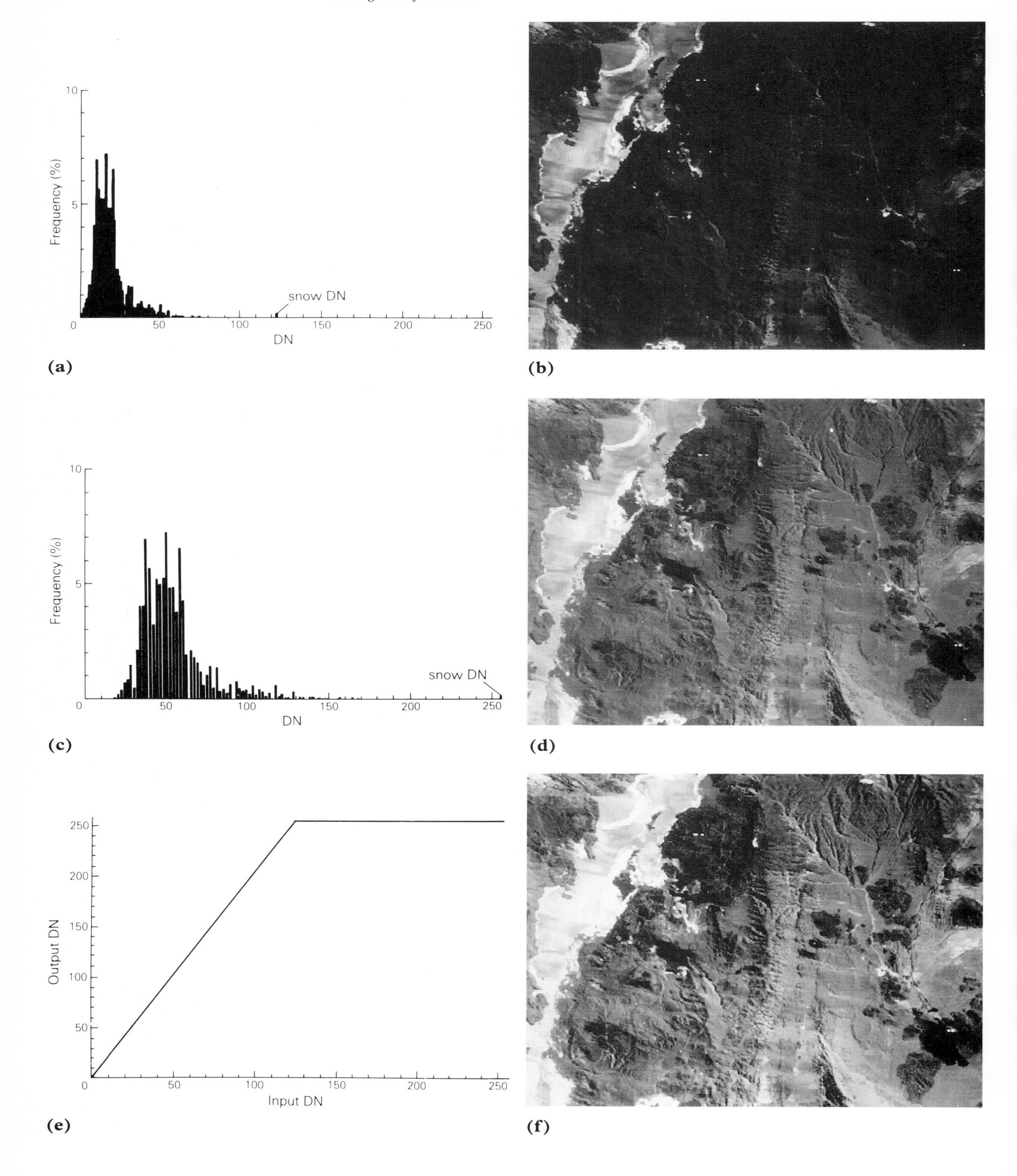

10
5
0
Frequency (%)
snow DN
50
100
150
200
250
DN
(a)
(b)
10
5
0
Frequency (%)
snow DN
50
100
150
200
250
DN
(c)
(d)
250
200
150
100
50
0
Output DN
50
100
150
200
250
Input DN
(e)
(f)

Fig. 2.29. This optimally stretched false-colour Landsat MSS image of part of the Andes includes the area shown in Fig. 2.28. It reveals many details in both light and dark terrains. Of particular importance are the compositional variations in the pale salt flats, the discrimination of lavas of different compositions and the highlighting of important topographic features, such as the giant Cerro Galan volcano to right of centre. Width is 130 km. (Courtesy of D. A. Rothery, Open University.)

on spectra is the *albedo* of different surfaces—the ability of a surface to reflect energy across a large range of wavelengths. So we have dark and light soils, and dark and light plants. In an image of an area of bare rock and soil, DN for different bands in the reflected part of the spectrum are all usually higher for a light soil than for a dark one. In other words, the different bands are *highly correlated.* This is best understood by examining a plot of DN in one band against those in another (Fig. 2.30).

The vast majority of pixels plot close to a diagonal line forming an elliptical cloud. In a three-dimensional plot of DN for three bands, the distribution is like a cigar. Data falling on a diagonal line in red–green–blue space all show as various shades of grey in a colour image. Those falling in the cigar-shaped area close to it come out as muted pastel shades that are difficult to distinguish. Only those pixels that lie far from the diagonal show in bright, easily distinguished colours, and they are rare. No matter how the data controlling red, green, and blue in the image are stretched, the cigar-shaped distribution and muted colours remain (Fig. 2.31). However, if the data are examined in another way, there is an escape. Most of the albedo variation in the scene is along the diagonal

Fig. 2.28. A histogram of the raw Landsat MSS band 4 data (a) for a volcanic terrain on the Chile–Argentina border are compressed towards low DN. Display of these data produces a dark, low-contrast image (b) in which few details can be seen. A linear contrast stretch, which sets the minimum DN (in this case 0) to 0 and the maximum (125) to 255, spreads out the histogram (c) to cover the full range available. Although such a stretch introduces discernible contrast in the resulting image (d), part of the range is occupied by DN corresponding to atmospheric scattering and to uninteresting pixels representing clouds and snow. The stretch is achieved by a simple straight-line input to output graph (e), or look-up table. A look-up table that takes account of unwanted information produced image (f), the best for routine interpretation. Width is 25 km.

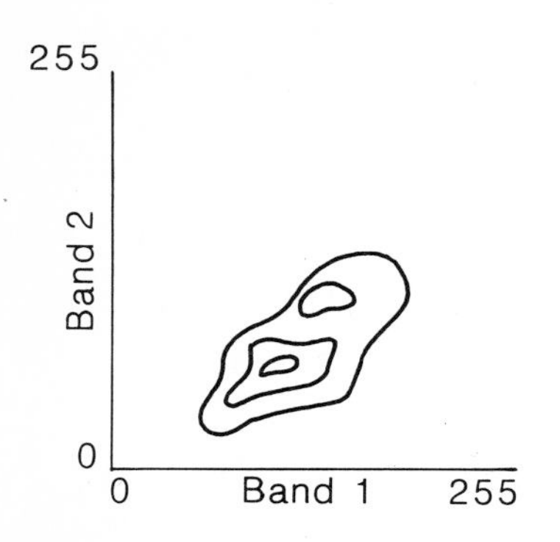

Fig. 2.30. A contoured bivariate plot of the DN from two wavebands in the visible to NIR region for all the pixels in a scene shows that as DN in one band increase there is a closely related increase in the DN for the other. Stretching the range for each band does not change the high degree of correlation in the data.

distribution (Fig. 2.30). If this diagonal is chosen as one axis of the distribution, a second axis at right angles to this contains the other variations, that are not due to the high correlation between the two bands (Fig. 2.32(a)). These variations can now be stretched so the eventual distribution is almost circular and the correlation is largely removed (Fig. 2.32(b) and (c)). This is the basis of a technique called *principal component analysis*, details of which are contained in other texts, that is used to enhance the subtle details due to minor spectral features associated with different types of surface. The newly derived axes can be used themselves in place of wavebands to produce false-colour images, or, by a rotation of the decorrelated data back to the original axes representing bands (Fig. 2.32(d)), the advantages of principal component analysis can be used in producing far more colourful images than the originals (Fig. 2.33).

Because the eye's ability to detect fine spatial detail increases with contrast, stretching by itself helps us interpret boundaries between different types of surface and the topographic features highlighted by shadows. Sometimes, however, it is necessary to enhance some of these features and suppress others, depending on their size on the image. For instance, fine high-contrast detail may clutter some important broader feature, or the small-scale features themselves, perhaps roads or small streams, may be the object of our attention. Selective manipulation of the spatial attributes of a scene is known as *spatial-frequency filtering*, and is best visualized by looking at how DN varies with distance across a scene (Fig. 2.34). The complex curve can be broken down into a series of simple sine waves with different wavelengths and amplitudes.

Filtering is basically the removal of some of these simple sine waves to leave the ones of particular interest. The simplest filter smooths the spatial variation of DN in all directions to produce an image of the broader variations (Fig. 2.35). This is subtracted from the original thereby leaving an image of all the higher frequency variations, which are generally called *edges*, because they are often boundaries between different, relatively uniform areas of brightness (Fig. 2.36(a)). To preserve some of the lower frequency information but enhance the high frequencies, this image of edges is added back to the

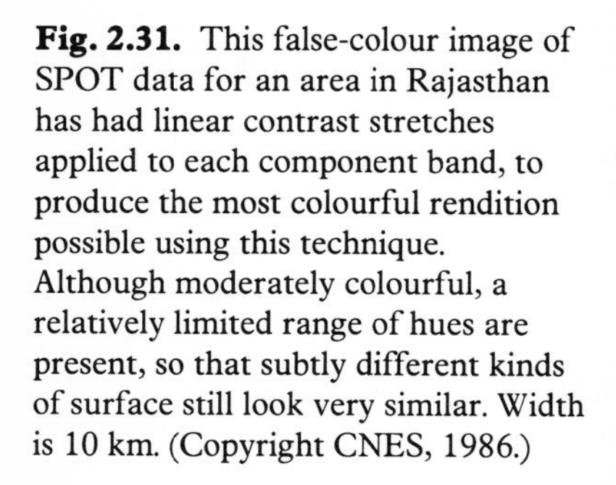

Fig. 2.31. This false-colour image of SPOT data for an area in Rajasthan has had linear contrast stretches applied to each component band, to produce the most colourful rendition possible using this technique. Although moderately colourful, a relatively limited range of hues are present, so that subtly different kinds of surface still look very similar. Width is 10 km. (Copyright CNES, 1986.)

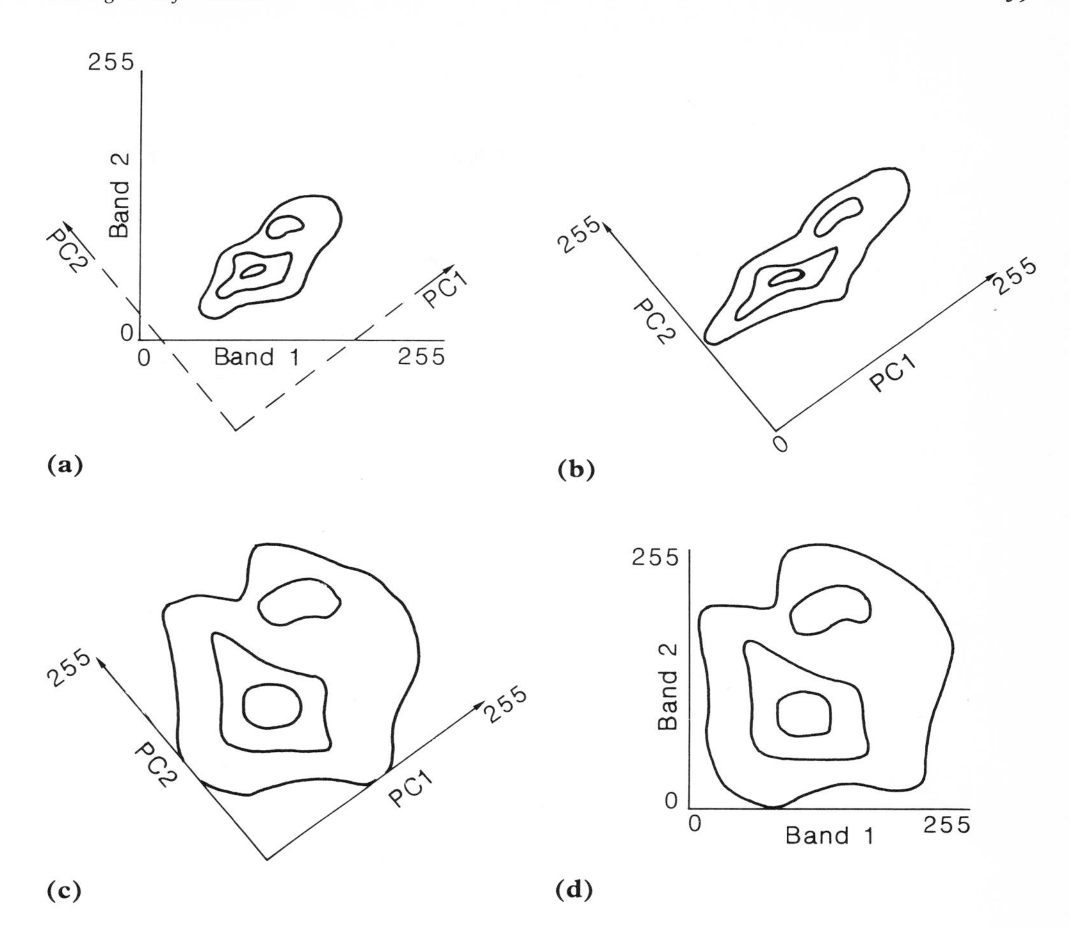

Fig. 2.32. The two-band data from Fig. 2.30 can be defined just as precisely by two other axes. In (a) these are represented by dashed lines. One, the first principal component (PC1) defines the direction of a line that fits the spread of data best. The other (PC2) is at right angles to the first, and accounts for all other variation away from the line of best fit. Linear stretches along PC1 and PC2 allow the cloud of data to be transformed to occupy nearly all the space available in two-dimensions (b) and (c). This decorrelates the data, and they can be restored in this form to the original space defined by the two-band axes by a simple rotation. Data expanded in this way is said to have been subject to a decorrelation stretch.

Fig. 2.33. This image is of the same data as those in Fig. 2.31, but they have each been subject to a decorrelation stretch, resulting in a much expanded range of colours that are more amenable to detailed interpretation. Note that the same basic hues are still present, but they are brighter and more distinct. Width is 10 km. (Copyright CNES, 1986.)

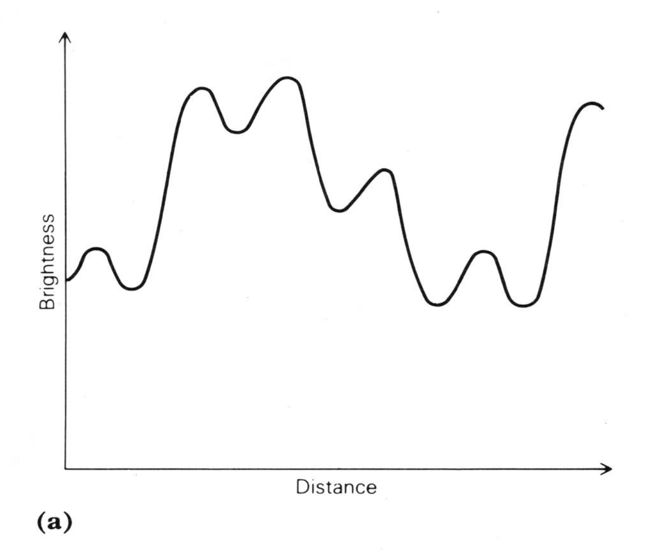

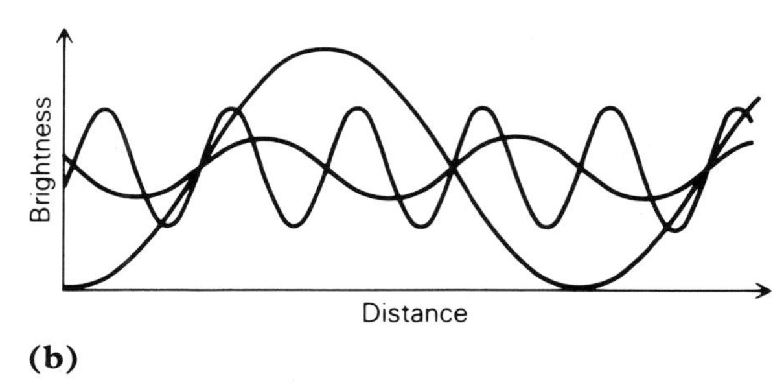

Fig. 2.34. Complex variations in DN along a line on an image (a) can be broken down to a range of simple sine waves (b) with different amplitudes and frequencies. Their cumulative effect is the same as the real spatial frequency distribution of the image.

original to produce an edge-enhanced image (Fig. 2.36(b)). More complex filters pass or suppress a specified range of spatial frequencies, so that, for instance, only roads of a certain width remain in the processed image, or medium-frequency field patterns are removed from an image leaving broad variations in soil cover and fine details of stream courses.

Sometimes the spatial detail of an image is so varied that particular orientations of some features are suppressed, such as geological fault lines in a scene dominated by random field patterns. Another form of spatial filter selectively enhances features in a particular direction by generating the gradient of DN in the direction at right angles to the one of interest. In effect this treats the spatial variation of DN as a topography that is illuminated from one side to produce shadows (Fig. 2.37).

Colour-enhanced images can be used to subdivide an area into surfaces of different types by inspection, as well as to express the overall features of the terrain. This can be done by predicting the colours that should be produced by surface materials with different spectral features in the bands employed to control the red, green, and blue components of the image. However, this limits the visual analysis to information from only three bands. By inspection of the spectra of different materials (Section 2.3) it can be seen that vegetation is characterized by high green, low red, and very high VNIR, and wet vegetation has lower 1.55–1.75 μm reflectance than dry. Red soils have higher red and VNIR than green, while grey soils have a higher green reflectance. Those soils and rocks containing clay minerals should have lower 2.0–2.35 μm reflectance relative to that in the 1.55–1.75 μm range than clay-free materials. To adequately distinguish between these simple surface types requires information from five bands. While some of the differences could be displayed by two false-colour images of band data, by using the *ratios* between bands that are contrasted for different materials, all five bands can be expressed in a single image. Using the band numbers assigned to Landsat TM data, the materials listed above could be distinguished by a colour image using the ratios 3/2, 4/3, and 7/5 to drive red, green, and blue. Such ratios can easily be generated from digital data using an image-processing computer. An example of a *false-colour ratio image* incorporating information from five wavebands is shown in Fig. 2.38.

Subdividing an image into its different component parts by visual inspection, no matter how it is enhanced and processed, is open to bias and confusion. One interpreter's subdivision will always be different from that of another. One means of overcoming this is

(a)

(b)

Fig. 2.35. Image (a) is a linearly stretched image of Landsat MSS band-7 data for an area of complex geology in Western Australia. The dark areas are underlain by metamorphosed sedimentary and volcanic rocks, the light areas are mainly remobilized masses of granitic gneiss. Image (b) substitutes the mean value for the surrounding eight pixels in the raw image for each pixel in the new version. The result of this filter is a considerable smoothing effect. As the dimensions of a low-pass filter increase, the smoothing effect becomes more obvious. Width is 50 km.

to use a computer's ability to make decisions according to a given set of rules, and to consider all the information in a data set rather than the limited range that can be displayed in a red–green–blue image. By knowing from field work that small areas in the scene correspond to the different types of surface that make up the whole area covered, the operator can instruct the computer to analyse the data from all the available bands corresponding to these *training areas*. This gives a multispectral *signature* for each recognized type of surface. The computer can then find all other pixels in the scene whose

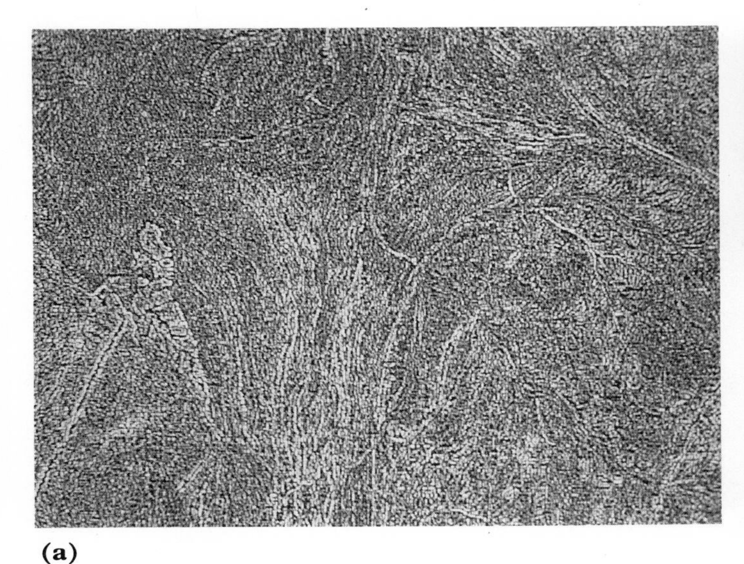

(a)

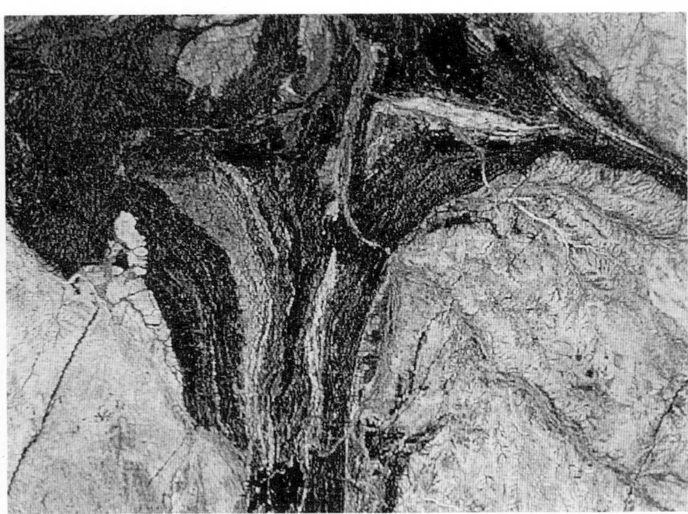

(b)

Fig. 2.36. Image (a) is the result of subtracting Figure 3.35(b) from 3.35(a) to display high-frequency features, or edges. Because all the original details of shadows and albedo variations have been removed, this image of edges is difficult to interpret. By adding these edges back to the original the edge-enhanced version shown in (b) is produced, which is more easily interpreted by a geologist. Width is 50 km.

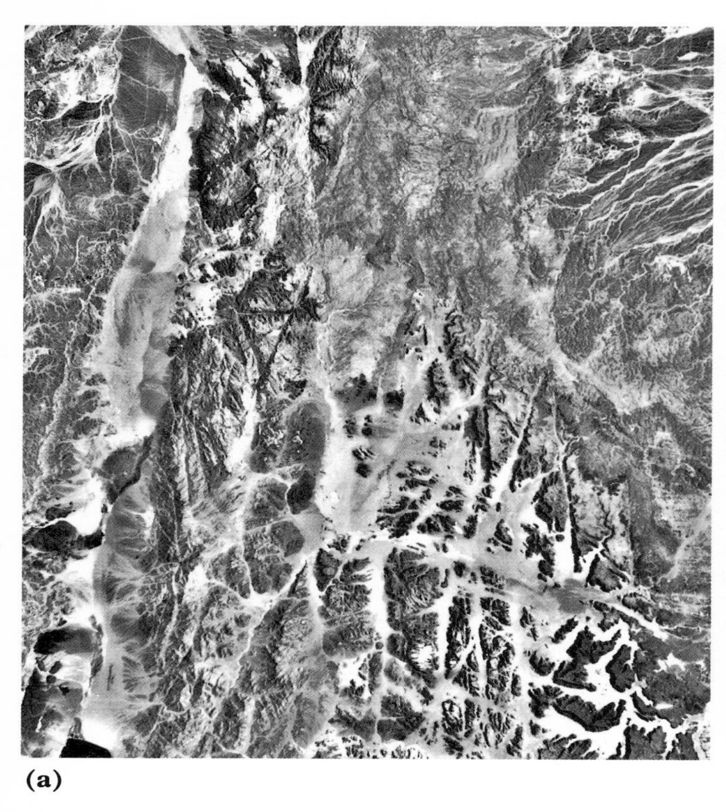

(a)

(b)

Fig. 2.37. A contrast-stretched image of Landsat MSS band 7 data (a) for part of Jordan reveals several faults which do not appear on local geological maps. In (b) the image shown in (a) has been subjected to an eastward directional first derivative filter. Although this enhancement of linear features is at the expense of tonal variations in the original image, which relate to different kinds of surface material, the illusion of highlights and shadows greatly improves the detectability of many linear features that are not so clear in (a). Width is 180 km. (Courtesy of Pat Chavez, US Geological Survey.)

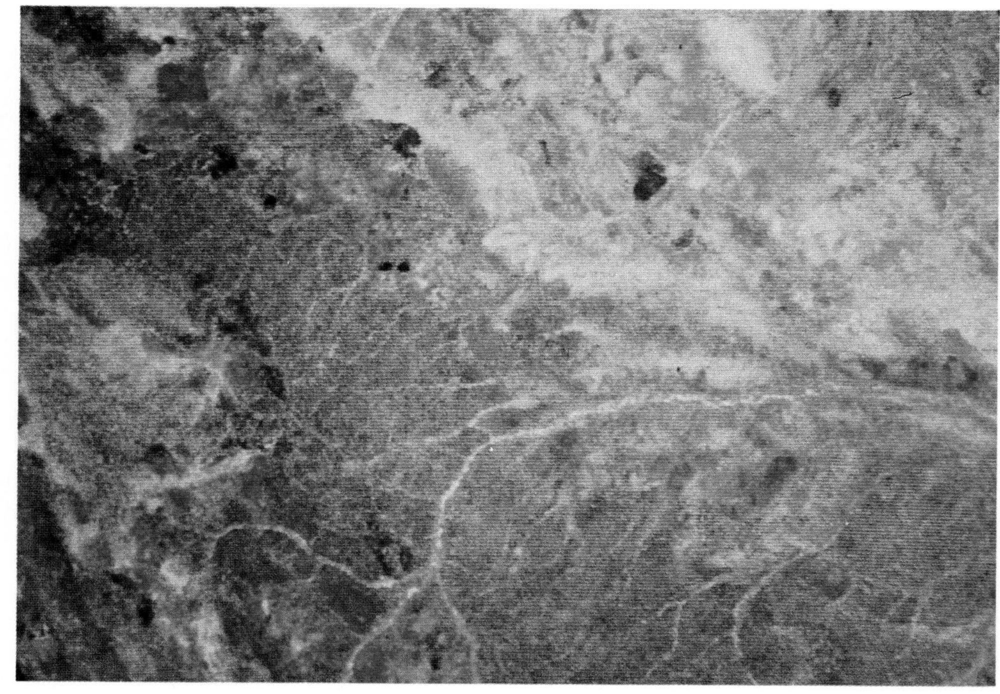

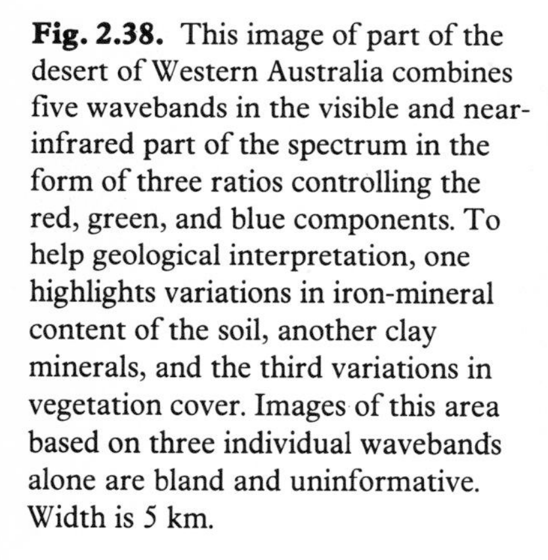

Fig. 2.38. This image of part of the desert of Western Australia combines five wavebands in the visible and near-infrared part of the spectrum in the form of three ratios controlling the red, green, and blue components. To help geological interpretation, one highlights variations in iron-mineral content of the soil, another clay minerals, and the third variations in vegetation cover. Images of this area based on three individual wavebands alone are bland and uninformative. Width is 5 km.

DN in all the bands fall within these signatures, assigning them to one of the various classes of surface that the operator has recognized. Such *multispectral classification* can be made more sophisticated by building in instructions to assign pixels to classes that they most resemble, without them actually possessing the exact characteristics of the signatures derived from the limited training areas. Another embellishment is to use the *context* of the pixel. That is best explained by the simple example of areas of bare sand—those adjacent to areas classified as water are probably beach sand, whereas those remote from water cannot be beaches. More details on classification techniques are covered by some of the texts recommended for further reading.

2.5.1 Other kinds of data

The variation of electromagnetic energy emanating from the Earth's surface is easily monitored directly in the form of an image using various electronic and optical/mechanical devices. Many other properties change equally as much, but cannot be measured in a continuous fashion. Among these are the density of rocks beneath the surface, expressed by variations in the Earth's gravitational field, their magnetic properties, topographic elevation, the chemistry of soil, the variation in vegetation density and type, and even artificial and socio-economic properties such as the value of land and the incidence of death from different ailments; in fact anything that varies in space and is measurable. Those attributes that cannot be represented directly by imaging methods are primarily recorded at points or along traverses on a map. Often they are represented conventionally in map form as contour lines joining points likely to have the same value in simple equal steps. Another type of geographic attribute does not vary but is either present or absent, like a road, a crop type, a kind of rock or soil, or the designated future use of land.

Both kinds of non-image *geographic information* can be transformed into digital images, enabling them to be combined together and with remotely sensed images in a common format. This has enormous potential in investigating the relationships between all kinds of variable in the most convenient and graphic way. Variable data located at points or along lines are first encoded as three values, two for location and the third representing the variable. A three-dimensional surface is then fitted to the data so that interpolated values can be assigned to cells in a regular grid, identical to the pixels in a remotely sensed image. Rescaling the variable to the 0–255 range of integers used by image-processing systems allows the data to be displayed as an image (Fig. 2.39). Moreover, as a *raster-format image*, it can be enhanced and manipulated in exactly the same way as a real image. This practice has great advantages over conventional contour maps, which actually omit the bulk of the variation through the simplification inherent in contouring, and in any case are difficult to understand by visual inspection.

Geographic information of the present or absent variety, the most common content of conventional maps, may be digitized in two different ways. The first represents points, lines, and areas as string of coordinates and vectors with associated 'labels' identifying different categories of points, lines, or areas. Such *vector-format maps* can be displayed as images faithfully representing their geometry and in which colours can be arbitrarily assigned to labels. The second method uses a scanning system, very like that employed in remote sensing, which measures the light reflected from or transmitted through a coloured map as red, green, and blue components to produce a three-band digital image of regularly arrayed pixels. This is identical to the raster format of a remotely sensed image. A facsimile of the map can be displayed as a red, green, and blue image through the image-processing computer. More usefully the map in this form can be combined with other sorts of data to help revision and modification, or to aid interpretation of remotely sensed images.

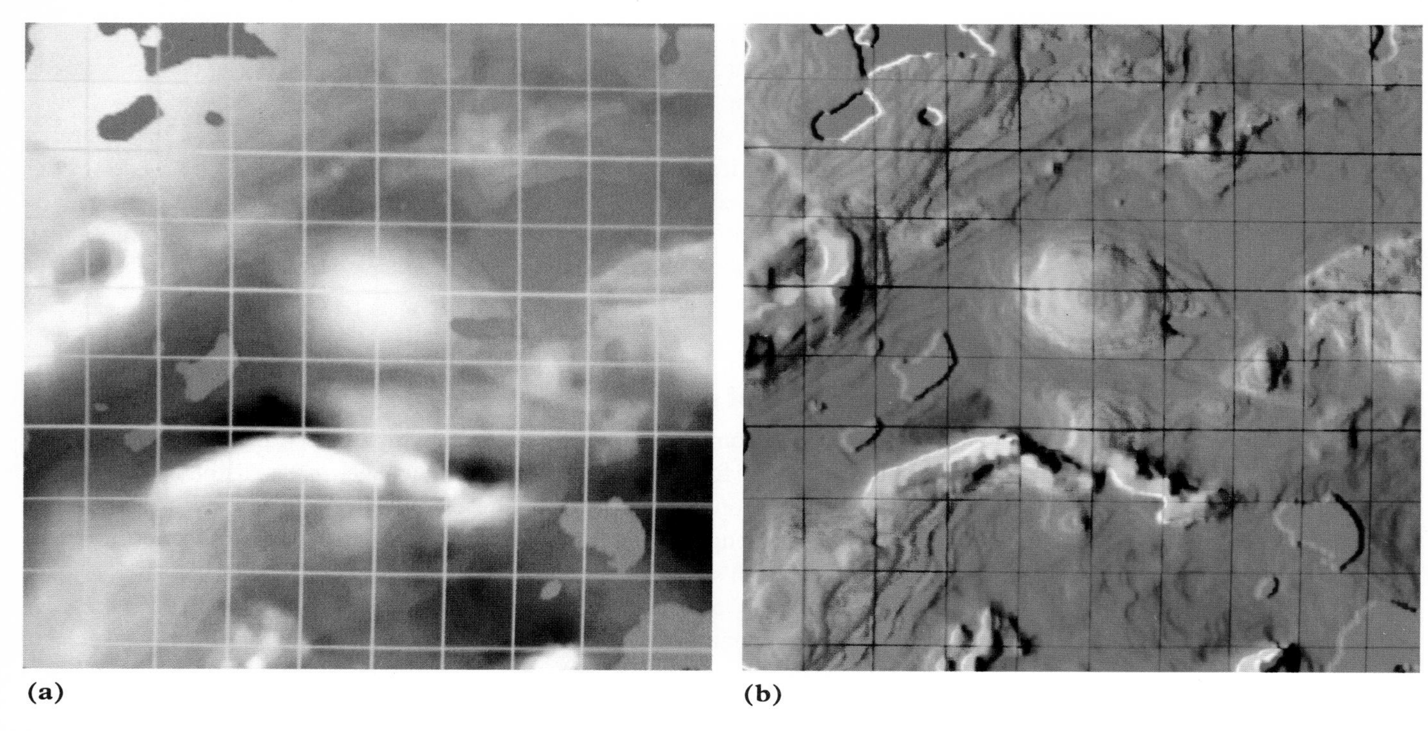

Fig. 2.39. Image (a) displays the variation in the Earth's magnetic field for part of northern England, in the form of a grid of pixels whose DN have been calculated from measurements at a series of points along lines across the area. Image (b) expresses the same recalculated information as (a) in a more easily interpreted way by simulating the effect of illuminating the data, as if it represented a topography, from the left, and assigning a unique colour to each of the 256 grey levels in (a). Width is 100 km.

A *geographic information system* assembles a great range of different data in a registered cartographic form. Because of this it becomes the potential source of entirely new kinds of information derived from the comparison and combination of the assembled data. As an example, the computer can calculate the distance of any point from any other, so producing an image of remoteness from towns, services, and physical communications. This can be combined with images of the topographic gradient and soil bearing capacity so highlighting areas unsuitable for industrial development. Further refinement can include land-ownership boundaries so that a final site for development can be decided on. The advantages are clearly those of convenience, speed, and accuracy in such a case. In the case of combining different kinds of geological and geophysical information in this fashion, unsuspected correlations between the different data sets can pinpoint entirely new targets for metal and hydrocarbon exploration.

Many of the methods discussed in this section are illustrated in Chapter 4, *Using the information*, in the context of practical applications of remotely sensed data. Chapter 3 is more concerned with showing how many of the physical attributes of the world as a whole can be seen and understood from a study of readily available images.

3 Interpreting the information

This chapter is concerned with how the many facets of the global environment are expressed from a distance. Space limits the coverage to only a few examples that demonstrate how remote sensing can show not only environmental detail, but also its widest possible context. The examples illustrate weather systems and aspects of the climate and atmospheric chemistry, features in the oceans, different landforms from the major climatic zones of the continents, vegetation communities, human use of the land, and the geological make-up of the continents.

3.1 The weather and climate

Predicting the weather is a major concern for farmers and seafarers. For the rest of us it ranges from an almost foregone conclusion in the climatically stable interiors of continents, through an eccentric obsession in the case of temperate seaboards, to a matter of life or death to those areas unfortunate enough to be threatened by hurricanes, tornadoes, and torrential rainfall. Most people's first, and often only contact with remote sensing is through the increasing use of satellite images as an aid to explaining TV weather forecasts and making them more entertaining.

Usually these images are grossly simplified from the data that are available from different meteorological satellites, such as the Tiros-N/NOAA series and Meteosat. Figure 3.1 shows the full information that could be used in a weather forecast for Western Europe and North-West Africa. It is from the NOAA-9 AVHRR system, where information from the visible, very-near infrared, and thermal-infrared bands are combined to give an impression of natural colour. The different shades of blue in the sea are due to varying contents of suspended sediments and plankton. Where the surface is bare of vegetation, as in the Sahara desert, the areas of rock outcrop are dark while sands and gravels are rendered in various yellows, oranges, and reds, somewhat removed from true colour. Snow and ice in Norway, Iceland, the Alps, and the Pyrenees show up as pale blue, while clouds are whites and greys.

The most obvious feature of Fig. 3.1 is the very detailed information about cloud distribution, density and type. From this can be inferred wind directions, the position of major frontal systems, and areas of rainfall. This in turn, combined with more conventional meteorological information about atmospheric pressure variations, enables some weather prediction to be made up to 48 hours ahead. The information is vastly more detailed than that from the usual weather map. At the top of Fig. 3.1 is a parallel system of clouds in a northerly wind flow. Within it is a system of eddies that are in the lee of the island of Jan Mayen. Though not showing the classic spiral pattern of clouds, the Atlantic to the north-west of Scotland is covered by a complex low-pressure area. Scotland itself exhibits a curious pattern of parallel cloud waves, running roughly east–west. These are lee waves controlled by the north coast of Scotland and the mountains of the Scottish Highlands. The very sharp boundary between clouds and clear air at the south of the Bay of Biscay is a major frontal system.

Fig. 3.1. Cloud patterns over western Europe and North Africa on a simulated natural-colour AVHRR image. Width is 3000 km. (Courtesy of ESA.)

Fig. 3.2. Cloud patterns over the Atlantic hemisphere on a simulated natural-colour Meteosat image. (Courtesy of ESA.)

Fig. 3.3. Atmospheric winds over the Atlantic hemisphere derived from motions of water vapour over one day, observed by the Meteosat 6.7 μm sensor.

A more generalized 'snapshot' of weather systems for the same time of year, covering the whole of the hemisphere viewed by the geostationary Meteosat, is given in Fig. 3.2. This shows a series of large frontal systems in the North Atlantic, clear skies over most of the Sahara desert, and two rotational storms near South Africa, as well as dense rain clouds over the tropical rain forests of Central Africa and South America.

Winds at the Earth's surface are a major influence over the exchange of heat between the atmosphere and the oceans, as well as the driving force for waves and oceanic currents. Between them, winds and ocean currents transport about half of the heat from equatorial regions towards the poles, and are consequently crucial factors in world climate. Though it is a tangible force of great influence and sometimes severity, our knowledge of wind is very limited. This is because the primary source of information was, until recently, from ships, and their routes are restricted, by and large, to only a few major shipping routes. Any means of broadening knowledge is therefore tremendously important.

The most obvious means of measuring wind speed and direction from satellites is to follow the tracks of drifting clouds or the movements of detectable water vapour over a period of time. This is possible on a regular basis from the geostationary metsats. Figure 3.3 is a map of part of the hemisphere covered by Meteosat where the feathered arrows indicate estimates of wind speed (number of feathers) and its direction, and the colours denote the height, derived from cloud temperature measurements. Red is for winds in the upper troposphere (up to 15 km), blue for intermediate levels, and green for low altitude winds.

While it operated during 1978, Seasat deployed a radar scatterometer that calculated wind speeds and direction from the small wind-driven waves that roughen the sea surface. Figure 3.4(a) is an image derived from seven orbits over a period of 12 hours during 7 September 1978. A complete global picture would have required one to two days of data.

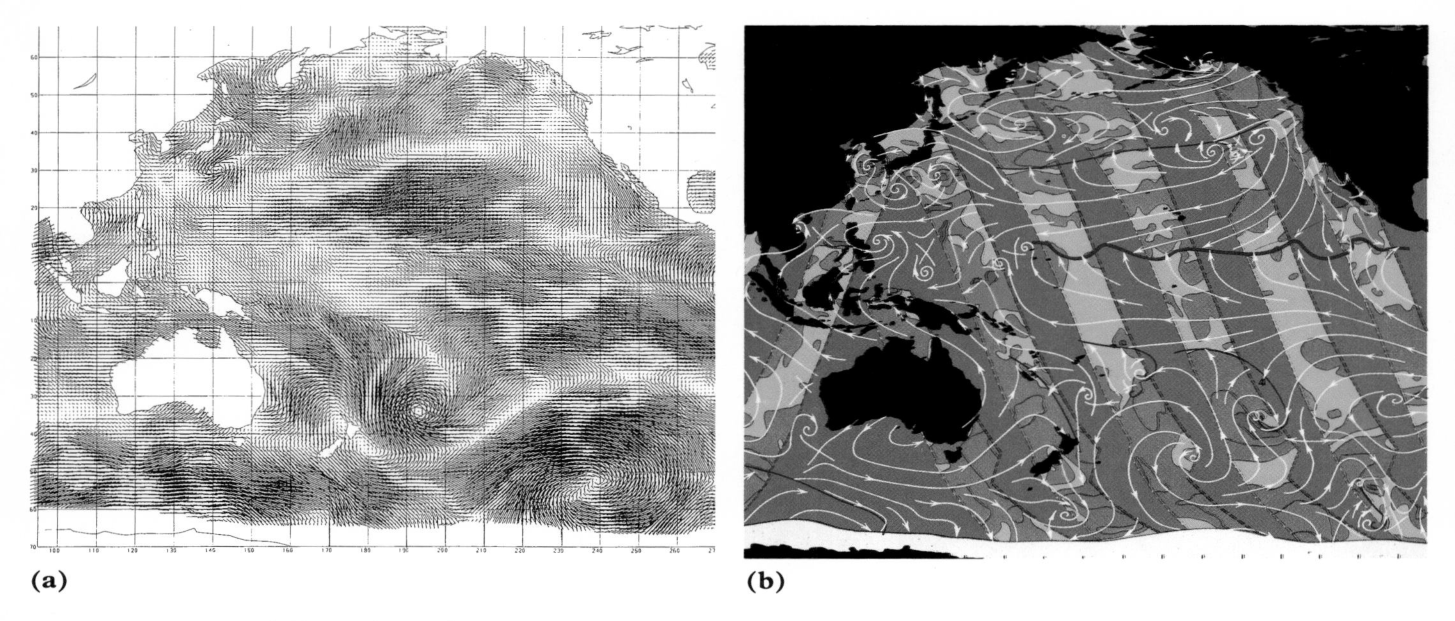

Fig. 3.4. Marine wind field over the Pacific Ocean derived from Seasat radar scatterometer data for two days. (P. Woiceshyn, JPL.)

The lines indicate wind direction, and have been extrapolated between the orbit swathes. The lengths of the lines indicate wind speed, resulting in different densities. The darker the appearance, the higher the speed. A thin, relatively pale zone, trending west from Central America is the Intertropical Convergence Zone (ITCZ) where the north-easterly and south-easterly trade winds converge. Other pale, curved lines are major frontal systems. This is the first comprehensive and accurate picture of world wind patterns at almost an instant in time.

Figure 3.4(b) is a different way of showing Seasat radar scatterometry data for the week after that shown in Fig. 3.4(a). Wind speeds are expressed in colours, so that the lowest are shown in green and yellow and the highest in pink and red. Light blue areas lack data. The white lines indicate directions of air flow, and have been extrapolated into areas lacking data. The thick dark blue line is the ITCZ, whereas the thin dark blue lines are major frontal systems.

Anticlockwise rotations in the northern hemisphere indicate low-pressure systems, clockwise indicating high-pressure circulation. The senses of rotation are reversed in the southern hemisphere. Two typhoons (1 and 2) are visible near Japan, as well as two huge cyclonic systems in the South Pacific (4). A high-pressure system is located over New Zealand. Note the association of high-pressure areas with low wind speed and lows with storms. Comparing the two images shows that wind patterns remained largely unchanged in the northern hemisphere over the week separating them, but considerable changes have affected the much stormier South Pacific. In particular the cyclonic storms have shifted to produce more complex flow systems.

Data of this kind help us better to understand the means whereby air is circulated in the lower atmosphere, by presenting a fuller picture for a much shorter time span than more conventional meteorological data. When operational, orbiting radar scatterometers will help reroute shipping to avoid rapidly developing storms and enable warnings to be broadcast to land areas towards which cyclones are moving.

As well as being the source of rain- and snowfall, clouds also exert an important control over the amounts of energy trapped by the atmosphere. Low clouds reflect solar energy

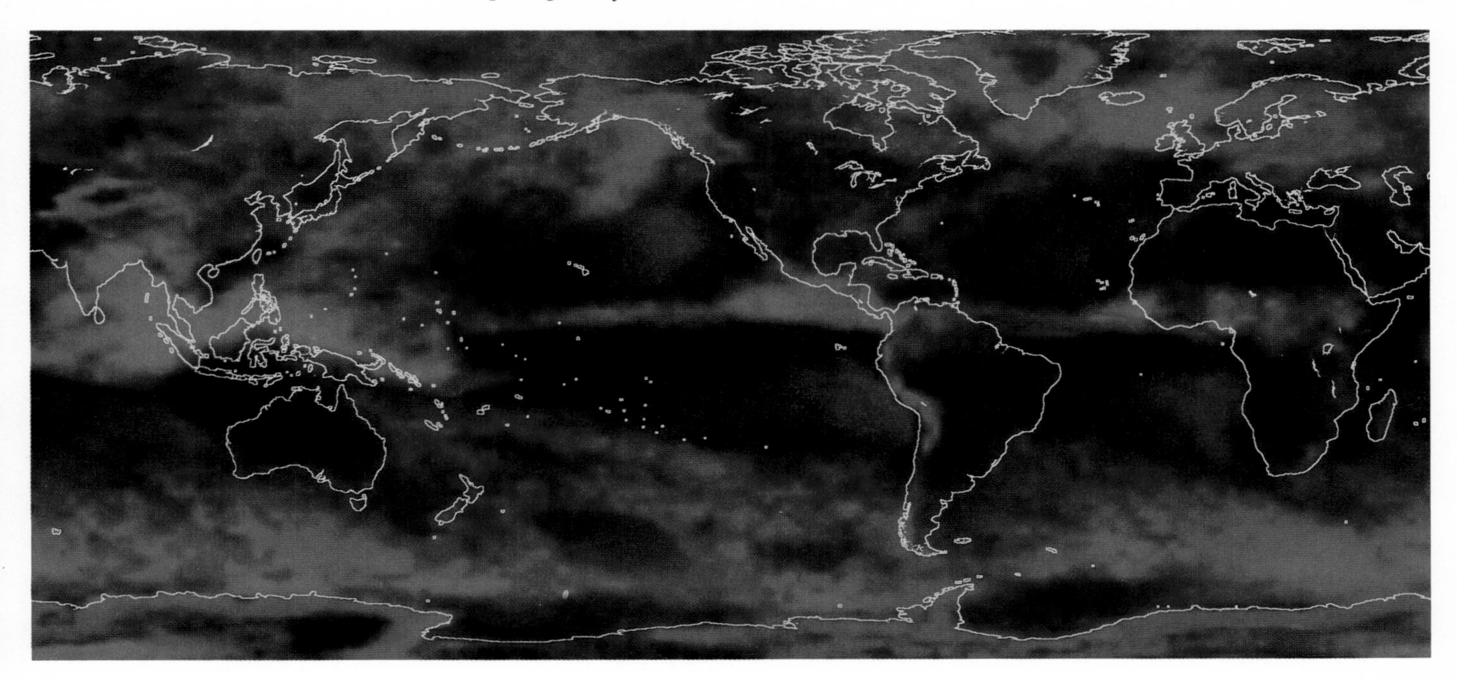

Fig. 3.5. Global mean cloud cover and altitude for July 1979, from the NOAA-6 HIRS and MSU sensors. (M. Chahine, JPL, and J. Susskind; GSFC, NASA.)

Fig. 3.6. Global mean surface temperature for January 1979, from the NOAA-6 HIRS and MSU sensors. (M. Chahine, JPL, and J. Susskind; GSFC, NASA.)

and help cool the atmosphere, whereas high clouds help trap outgoing energy and contribute to warming. Figure 3.5 expresses cloud height as red, green, and blue for high- (>8 km), medium- (4–8 km) and low-level clouds (<4 km). The brightness of these three colours depends on the degree of cloud cover. The image represents averages over one month (July 1979) derived from the High Resolution Infrared Sounder and Microwave Sounding Unit of the NOAA-6 polar-orbiting meterological satellite. For the first time the dynamics of global cloud patterns can be seen. High-level clouds are mainly associated with tropical regions near the ITCZ. The ITCZ is represented as a band of high (red) clouds with streamers extending to the north. One interesting feature is that in January, streamers of high-level cloud extend from the ITCZ to both north and south. Another is

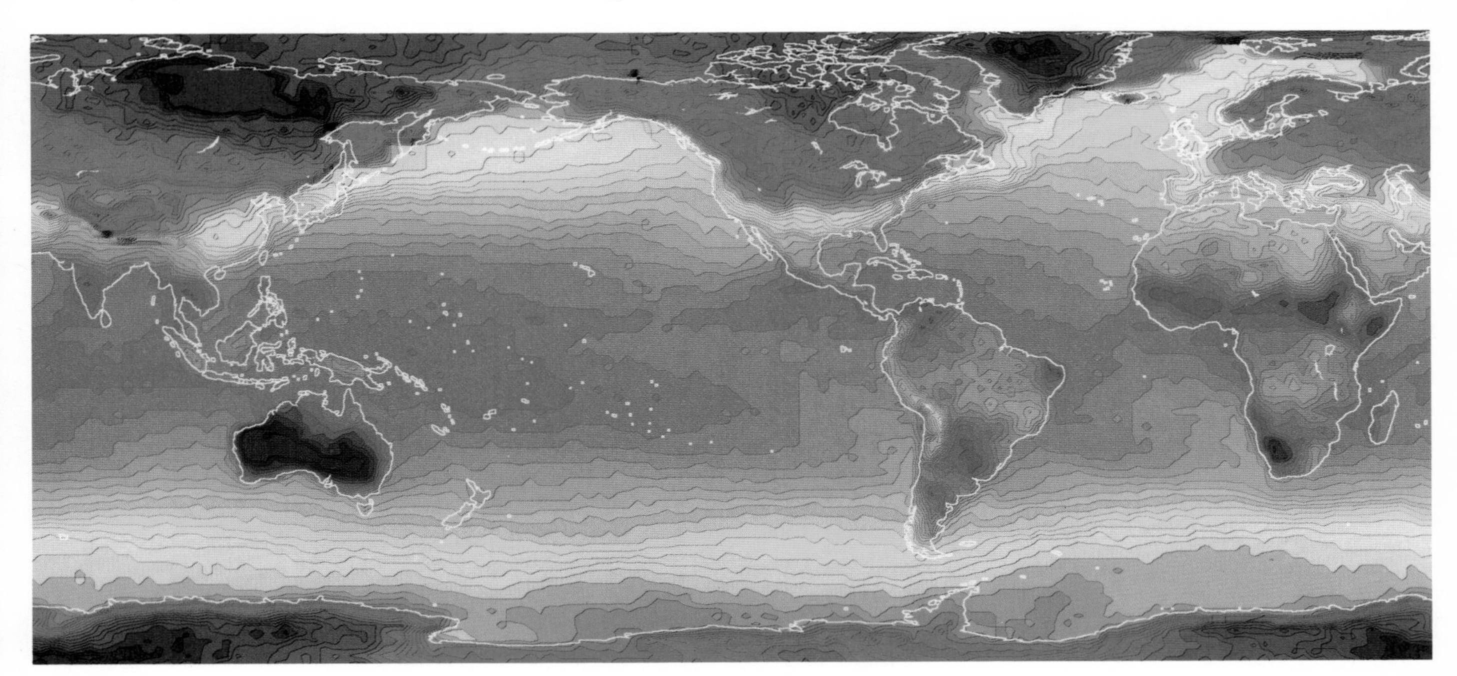

that the main cluster of high-level clouds moves from Indonesia in January into the Bay of Bengal in July, when it is associated with the Indian monsoon, shown here as a mass of high (red) clouds.

The same instruments on the NOAA-6 satellite also provided the first ever pictures of surface temperature over the whole globe, again on a monthly average basis (Fig. 3.6). Areas above freezing point are shown in yellows and reds, those below in greens and blues. The January 1979 image clearly shows the main continental cold and hot spots in Siberia and Australia respectively, together with the dramatic winter warming effect of the Gulf Stream on northern Europe. Although we often associate the southern hemisphere with higher temperatures, a July image clearly shows that the greatest area of really hot weather is in the northern hemisphere, in the Sahara, Arabia, India, Pakistan, and Afghanistan, and on the Gulf of California. On Fig. 3.6 the ocean temperatures at high southern latitudes are in a clearly defined set of zones, in contrast to the more complex patterns in the northern oceans. This is because the waters of the southern circumpolar ocean are efficiently mixed by the constant westerly winds unhindered by land masses. The image shows that in the more northerly parts of the oceans, their western sides are slightly warmer than on the east. This is due to the dominant westward ocean currents in the tropics, driven by the trade winds. Much more detailed pictures of ocean temperatures are given later in Figs 3.8 and 3.17.

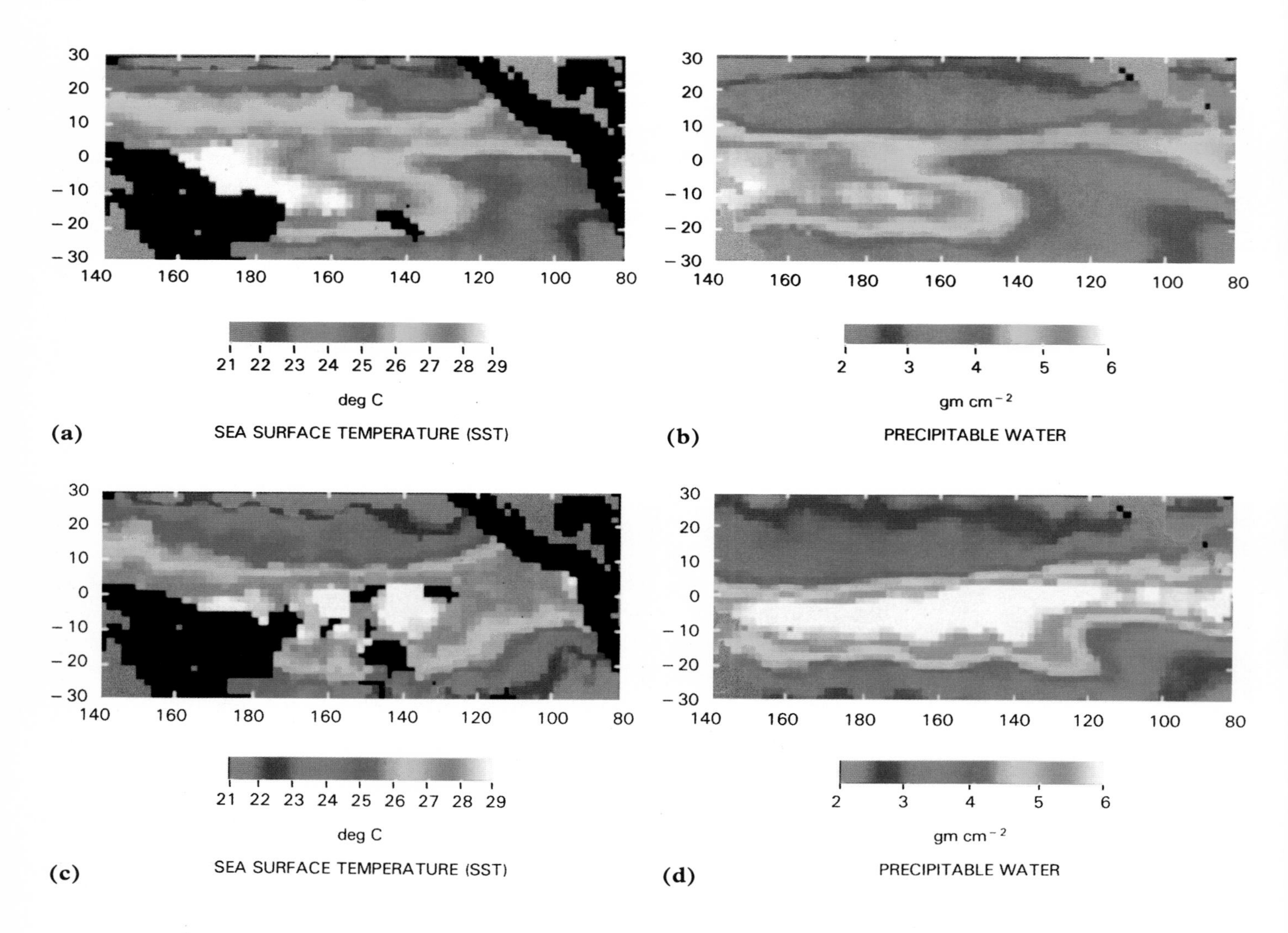

Fig. 3.7. Mean sea-surface temperature and moisture content able to fall as rain from the Nimbus-7 SMMR for the central Pacific, during January 1981 (a) and (b) and January 1983 (c) and (d), showing the El Niño phenomenon.

The ocean surface temperature is the most important climatic control in maritime areas. Remote sensing now allows it to be monitored continually, so that changes in the sea can be matched with weather patterns. Eventually this will allow detailed models to be constructed, from which better and more advanced weather forecasts will stem. An example of this is the near-global weather disruption of 1982–83, whose most publicized effect was in California, which experienced unprecedented high rainfall and wind storms. It is now known to be connected to the annual warming of the Pacific waters off Peru around Christmas, known to the local fishermen as El Niño (The Child). For most of the year the easterly trade winds push warm Equatorial surface water towards Australia, cold water flowing up the South American coast to make Peruvian waters cold and nutrient rich. This explains why they are normally the most productive fishery in the world. As the trade winds die seasonally, the bulge of water off eastern Australia—more than a metre higher than off South America—flows back, causing the warm water to float above the cold currents off Peru. This temporarily causes fish to disappear, together with the huge flocks of birds which eat them, and deposit the guano fertilizers that are so famous on offshore islands.

Seven times in the last century this reversal in surface currents has reached extreme proportions, due to an unexplained reversal in the trade winds. Figure 3.7(a) and (b) show what happened during January 1981 and January 1983. The colour steps from white through red to blue in rainbow order and to magenta on Fig. 3.7(a) identify sea-surface temperatures from 29 °C to 21 °C for both months. The same colours on Fig. 3.7(b) show the range in atmospheric water content that could fall as rain, from 6 gm to 2 gm per square cm. Black areas have no data and a crude outline of land masses is given in grey, Australia at bottom left and Central America at top right. The 1981 temperature data show the tongue of cold water protruding west from the South American coast along the Equator. In 1983 this had disappeared and warmer water can be seen to be moving eastwards. The precipitable water data shows how water vapour is highest over high temperature sea due to evaporation and convection. In January 1981 most rainfall was in the West Pacific. By 1983 high moisture content characterized the whole of the Equatorial Pacific, and was clearly responsible for the disastrous flooding in Central America that year. The shift in these two important parameters from normal in January 1981 to the extreme El Niño conditions of January 1983 is dramatically clear. This El Niño event was so anomalous that not only was the economy of Peru savagely hit by the failure of the fisheries, but the warm water spilled northwards towards California.

Figure 3.8 shows more detailed images of sea-surface temperatures off southern California for January 1982 and January 1983, from the thermal sensor on the NOAA-7 AVHRR. Blue is cool and red is warmest, while white areas are clouds. Normally, a cold current flows southwards past Point Conception (1) and northerly winds push water offshore to be replaced by deep, cool upwellings off Los Angeles (3) and San Diego (4). The 1982–83 El Niño produced abnormal onshore winds that forced the coastal sea level more than 20 cm above normal, driving sea water 2 °C warmer than normal over the cool water. Fishermen noted the disappearance of the normal commercial species, to be replaced by exotic fish usually found in the Equatorial Pacific. Although a major item of news at the time, the extremely high rainfall in California was at the margin of the effect. Elsewhere thousands of deaths resulted from floods and landslides. It is still uncertain if the El Niño has even wider effects. It is known to span a quarter of the globe, and may be connected with changes in the Indian and East African monsoons which in the succeeding years failed disastrously leading to the Ethiopian famine of 1984. Whatever, only the sort of global, accurate, and continuous information provided by satellite remote sensing can begin to produce answers and models that can warn of such vast climatic mishaps. However, the

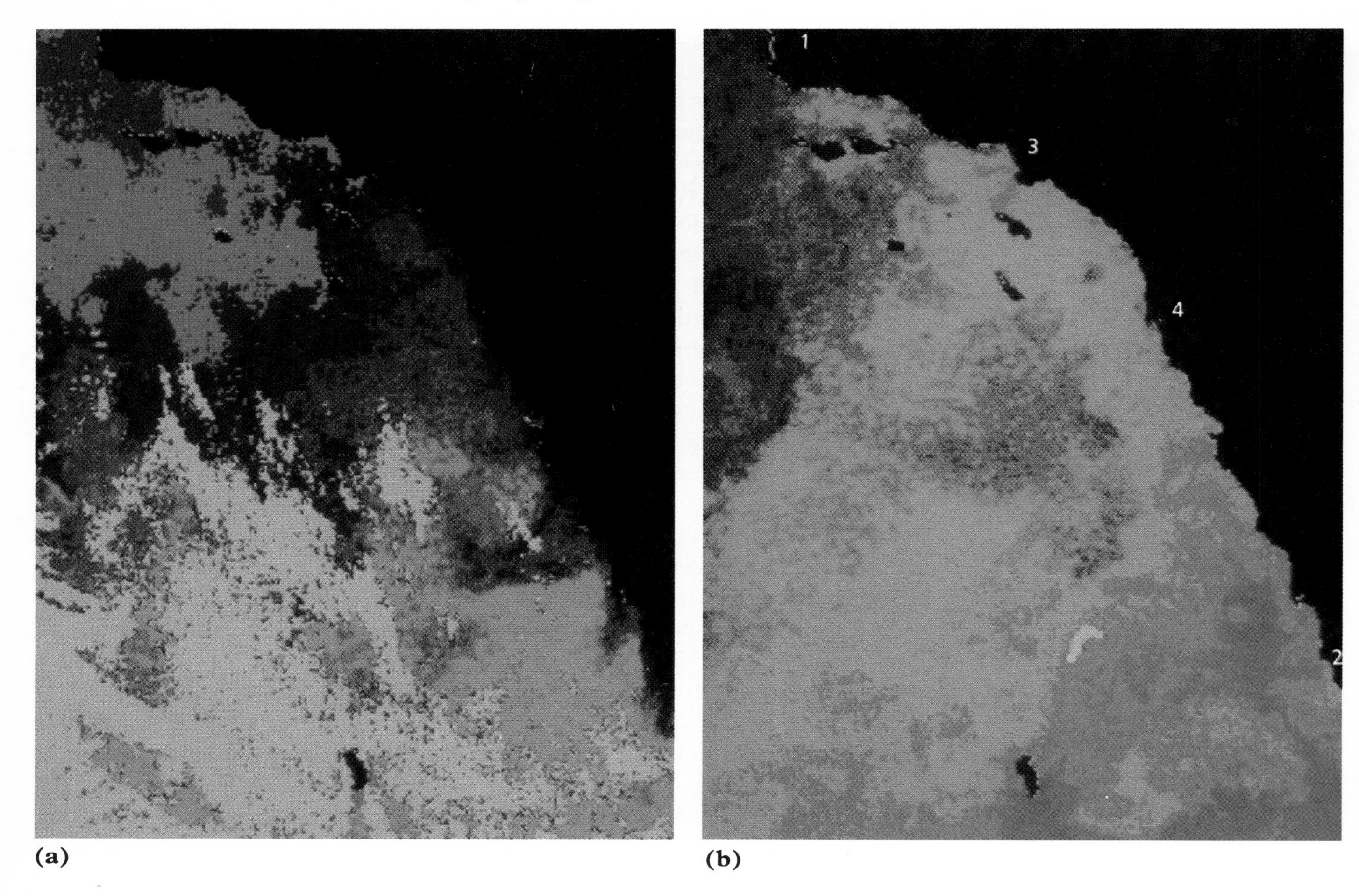

Fig. 3.8. Sea-surface temperature from the NOAA-7 AVHRR for the California coast (a) in January 1982 and (b) January 1983, showing the effects of El Niño. Width is 500 km.

climatic system involves such enormous energies that being forewarned does not mean that the affects can be escaped or modified. All that can be done is to prepare contingency plans to lessen the economic impact and earmark relief supplies and measures.

Not only the physical attributes of the atmosphere are of interest and use to the climatologist, but also its chemistry. The most obvious chemical parameter that is involved in weather is the atmosphere's moisture content, already shown in crude form in Fig. 3.7(b) for the East Pacific. To some extent it is mirrored in the distribution of clouds, but this is also a function of air temperature. Figure 3.9 shows the effect of moisture content in the atmosphere over the Atlantic hemisphere measured by Meteosat at the same time as that in Fig. 3.2. It is an image of energy emitted by the atmosphere in the 5.7–7.1 m waveband. This is within one of the ranges where incoming solar energy is efficiently absorbed by water vapour in the atmosphere and does not reach the surface (Fig. 2.11). As explained in Chapter 2, a good absorber is an equally good emitter of energy at the same wavelength. The reason an image like Fig. 3.9 can be acquired is because as altitude increases the air becomes sufficiently dry to allow transmission in the 5.0–8.0 μm band. The moist–dry interface occurs at the top of the cloud layer, around 10 kilometres up. The moister the lower air mass, the colder and higher the boundary is. In the image the lighter the signal, the higher and colder the interface is and the moister the lower air. The Earth's surface on Fig. 3.9 is masked by the swirls due to water vapour. As a means of location, Fig. 3.2 should be examined, which was produced by Meteosat on the same day. The bright (high and cold) areas on the Equator are over the tropical rain forests of South America (left) and Africa (right of centre). This was acquired on a day when the atmosphere was reasonably clear of

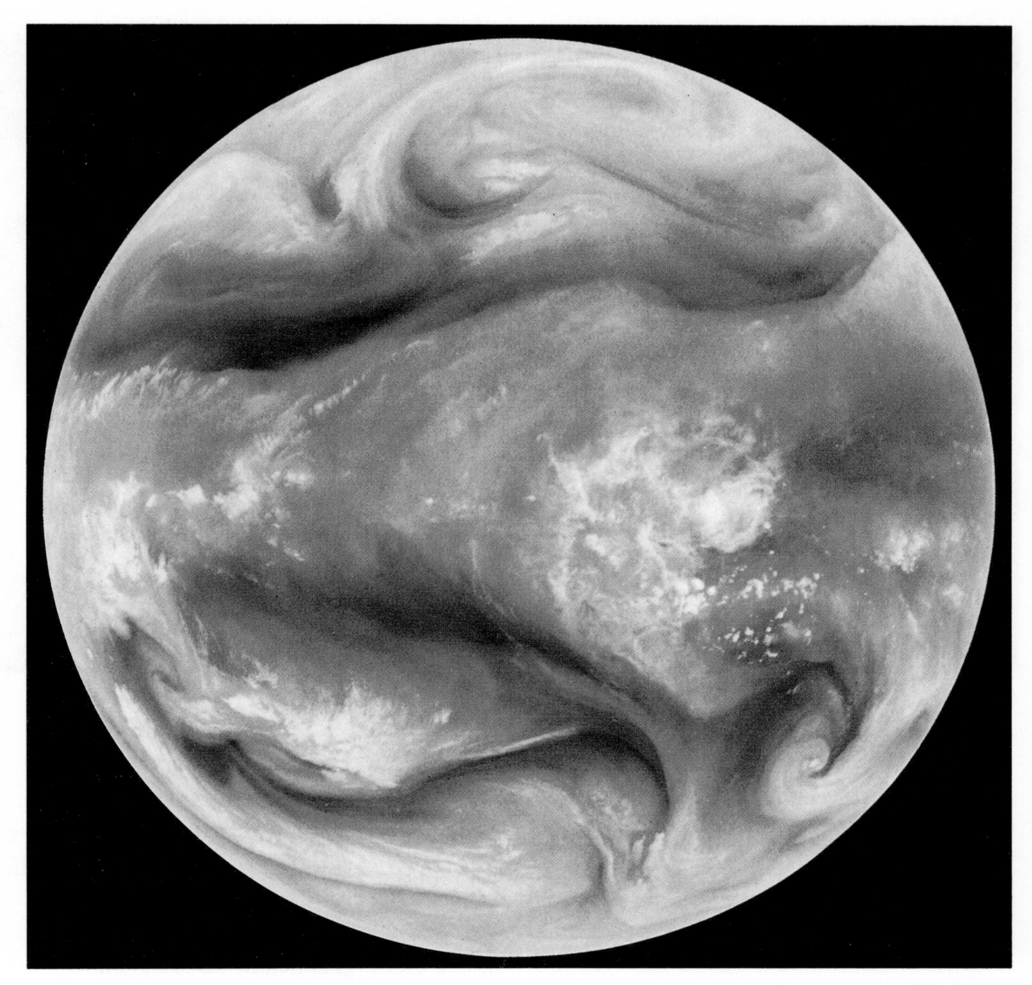

Fig. 3.9. Atmospheric moisture content over the Atlantic hemisphere from Meteosat. Compare with Fig. 3.2. (Courtesy of ESA.)

cloud, and shows that cloud cover is only one manifestation of moisture content. The swirls in the image, representing variations in the moisture contained in different air masses, can form the basis for estimating wind speeds and directions, as shown in Fig. 3.3.

The lower atmosphere, or trophosphere, is the main link between the oceans, land and the biosphere, and the rest of the atmosphere. Most of the gases, other than nitrogen, continually change because of processes at or beneath the surface. The most significant of these gases are those containing carbon, nitrogen, sulphur, and the halogens chlorine and fluorine combined with other elements. The trace gases in the atmosphere have to be measured at the parts per billion level or even lower in order to indicate details of their distribution. Moreover, it is important to measure how their concentration varies at different levels in the atmosphere. Figure 3.10 shows images derived from the measurement of nitric acid over the northern hemisphere by an infrared spectrometer (LIMS) carried by Nimbus-7. Low values below four parts per billion are in blue, and the range from green through yellow and red to white indicates concentrations up to 12 parts per billion. Image (a) is for the stratosphere where pressure is around 10 millibars, while image (b) is for deeper stratospheric levels with a pressure some five times higher. It should be noted that the nitric acid is not a product of pollution, but is formed by complex photochemical reactions between nitrogen and oxygen at high altitudes. The high concentrations near to the North Pole indicate the greater flux of charged particles and ultraviolet

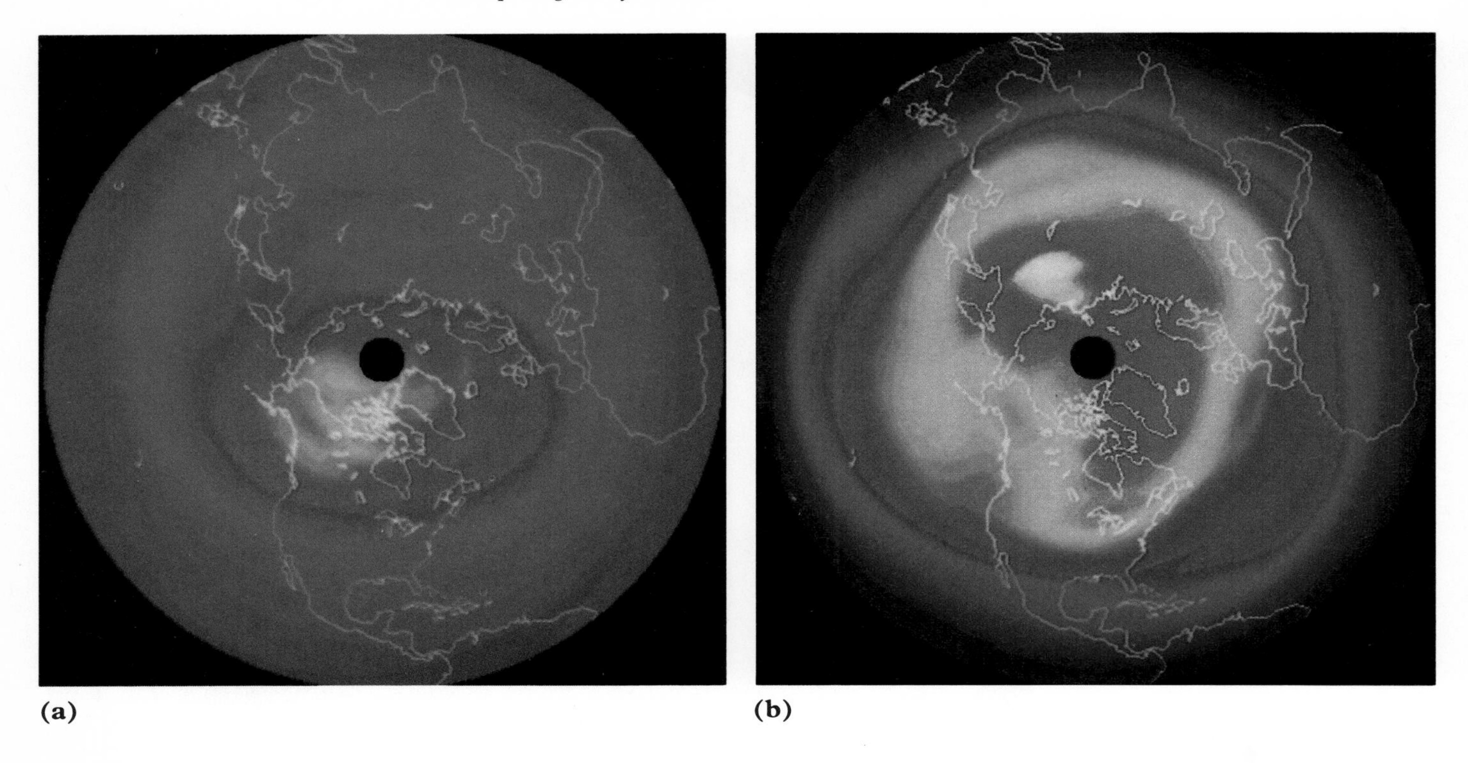

(a) **(b)**

Fig. 3.10. Nitric acid concentrations in the upper (a) and lower stratosphere (b) of the northern hemisphere, from the Nimbus-7 LIMS sensor.

radiation there. Carbon dioxide, produced by human activities, noteably the burning of fossil fuels, is currently impossible to monitor from space, but indications from ground-based measurements are that it is increasing, with the possibilities of global atmospheric heating due to the well-publicized 'greenhouse effect'.

One of the crucial gases in the upper atmosphere, as regards the health of animal life, is ozone—oxygen atoms grouped in threes as molecules, rather than the usual two. The best known property of ozone is its ability to prevent ultraviolet radiation reaching the surface, so protecting animal life from skin cancer and other rapid cell mutations. Not so well known is the fact that it may also behave as an agent to trap thermal energy in the atmosphere, in much the same way as carbon dioxide. Figure 3.11 is the evidence, from the Total Ozone Mapping Spectrometer carried by Nimbus-7, for the alarming decrease in ozone in the stratosphere over the Antarctic continent. A similar decrease in ozone has been noted over the Arctic Ocean over recent years. The main process generating ozone is the effect of ultraviolet radiation on normal oxygen. Ozone is destroyed by reactive gases, such as nitrogen–oxygen and chlorine–oxygen compounds. Compounds such as nitric acid are naturally produced in the upper atmosphere over the poles in particular (Fig. 3.10), so ozone can be expected to have low concentrations there and long-term measurements from the ground have confirmed this. However, the surprise in Fig. 3.11 is that there has been a sudden change from a stable level before 1979 to the very low values over the 1980–86 period. As yet the full reason is not known, nor are there are predictions as to what will happen in future. As well as some increase in N–O compounds, there may be another culprit involved. Seemingly innocuous carbon–chlorine compounds, used as the propellants in aerosol sprays, may leak into the upper atmosphere where they react to form chlorine–oxygen gases that destroy ozone. Since one molecule of such a compound may destroy thousands of ozone molecules, very small amounts of pollutants can have a devastating effect. Once more, more information and a greater diversity of data on atmospheric gases is needed.

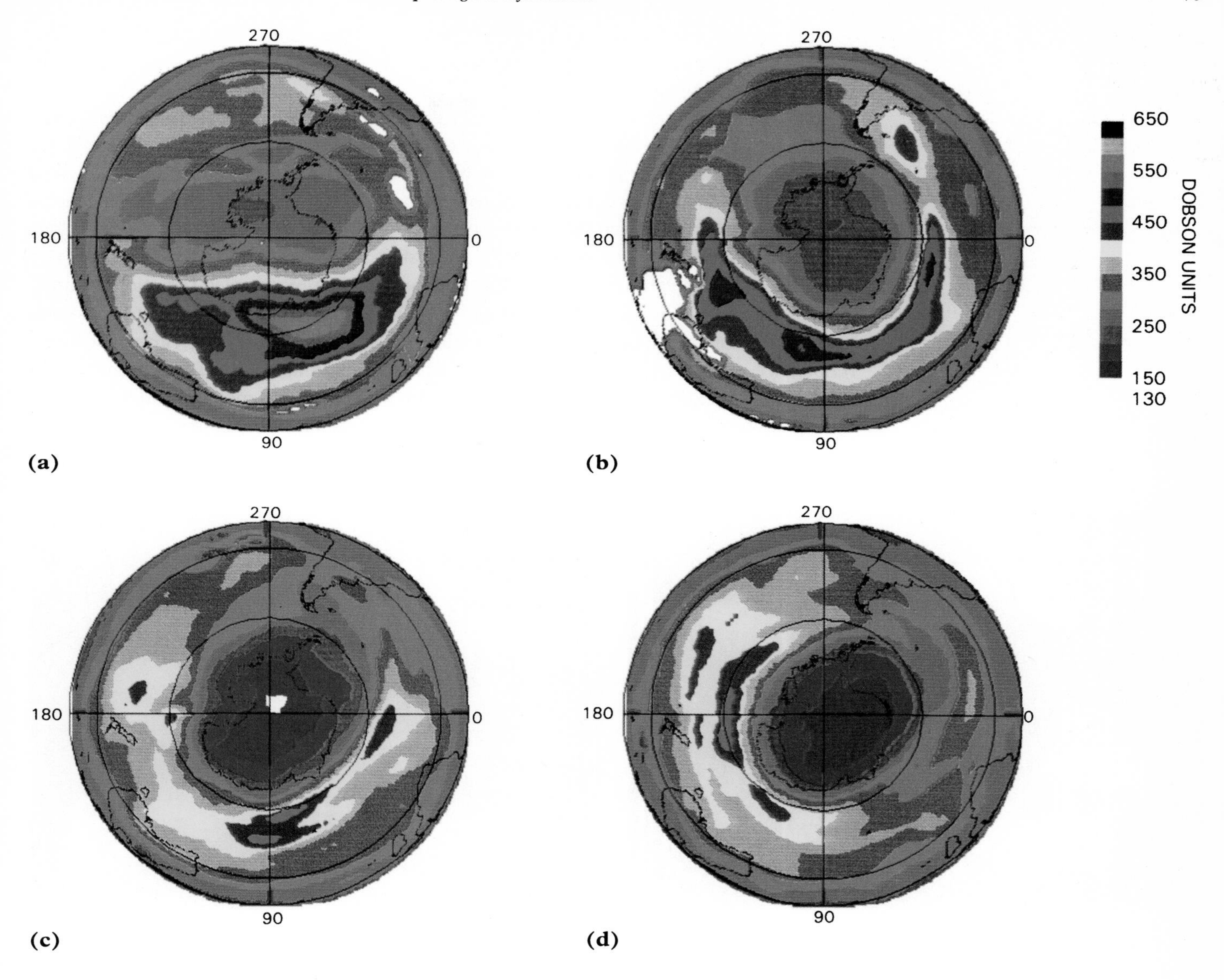

Fig. 3.11. Total ozone content of the atmosphere over Antarctica, measured by the Nimbus-7 TOMS sensor, for (a) 1979, (b) 1981, (c) 1984, and (d) 1986.

3.2 The sea

Although oceanographers are concerned with all aspects of the physics, chemistry, and biology of the oceans, and the many complex interactions that go on between water bodies and the atmosphere, historically their prime concern has always been charting the depth of the sea. Water depth has usually been measured by some form of sounding from ships, either using plumb lines or more recently by sonar. In both cases the measurements are from isolated points or along the tracks of ships, and depend to a large extent on the volume of marine traffic in any particular area. Sonar is related to echo sounding and relies on the time taken for pulses of high energy sound to travel to the sea bed and be reflected back to a microphone. Increasing sophistication of electronic recording and signal processing has recently enabled sonar to be used in an analogous fashion to radar, providing detailed images of the sea bed to either side of the ship's track. As well as showing details of sea-bed geology, sonar imaging detects and even allows the identification of wrecks. Suitably tuned and connected to an instant playback device, sonar can also detect shoals of fish. Nevertheless, conventional bathymetry and sonar imaging will take a long time to produce a complete map of sea-bed morphology, and remote sensing provides several timely options.

Because in clear water blue and to a lesser extent green light penetrate to depths around 50 m, images in these wavelengths do show some detail of the shallow sea bed in near-shore areas. Because these areas are the most heavily used by shipping and for fisheries, and because tidal and other current actions constantly shift sediment, it is important to maintain a continual watch on bathymetric conditions. Figure 3.12 shows a false colour Landsat MSS image of part of the Red Sea coast of Saudi Arabia. Offshore sandbars and coral reefs show as various shades of blue due to the reflection of green radiation from the bottom and complete absorbtion of red and near-infrared energy by water. Figure 3.13 was produced from that image by assigning different colours to different ranges of green reflectance over the sea to produce an easily understood and interpreted map of local bathymetry, at extremely low cost compared with a conventional marine sounding survey. Grey is above sea level, red is the shallowest water, and the progression towards dark blue is from shallow to deep water.

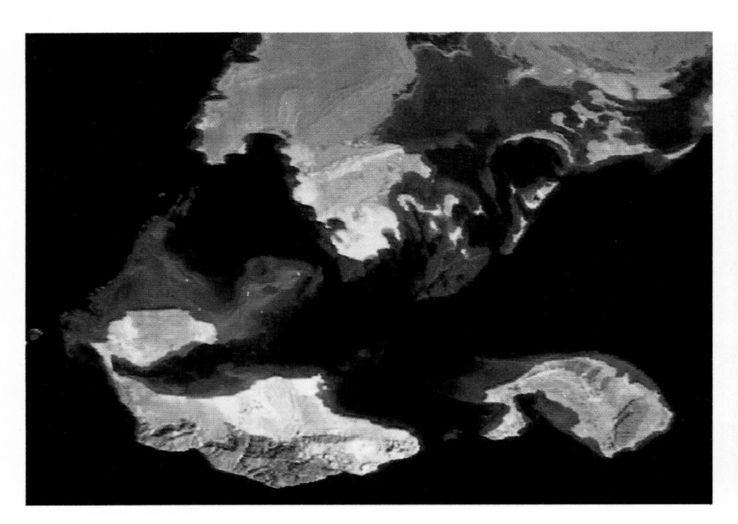

Fig. 3.12. Landsat-2 MSS false-colour image of part of the Red Sea coast of Saudi Arabia, showing different depths in clear water.

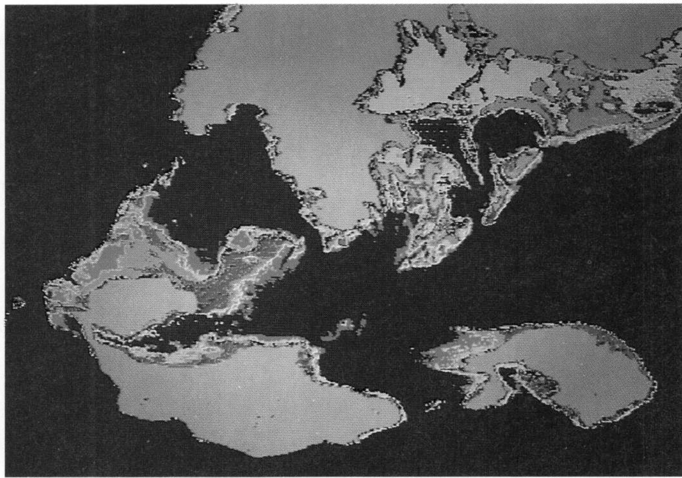

Fig. 3.13. Landsat-2 MSS band 4 data for the area covered by Fig. 3.12, colour sliced to show bathymetry. Width is 30 km.

When water is turbid due to suspended sediment or plankton, the use of reflected radiation does not allow depth or sea-bed morphology to be estimated directly by remote sensing. Using radar, however, produces some astonishing indirect results. Figure 3.14 is a Seasat radar image of part of the English Channel off the coast of northern France. In this notoriously dirty stretch of the link between the North Sea and the Atlantic Ocean the image shows very clearly the patterns of sandbanks and standing sand waves controlled by the tidal flow through the Channel. Depths are rarely greater than 30 m, and in some areas the largest ships constantly run the risk of grounding at low tide. We have already explained that radar energy is completely reflected by water, and this image must represent surface phenomena only, so how can it possibly show details of the sea bed? In the case of the Channel image, the sea-bed morphology affects tidal currents, leading to turbulence. This results in short waves at the surface. It is these variations in roughness at the sea surface that are detected and shown on the radar image, so in a very round-about way the Seasat image is reflecting bathymetry fairly precisely.

Fig. 3.14. Seasat radar image of the Straits of Dover north of Dunkirk showing sea-surface patterns produced by sandbars. Width is 20 km. North is at top right.

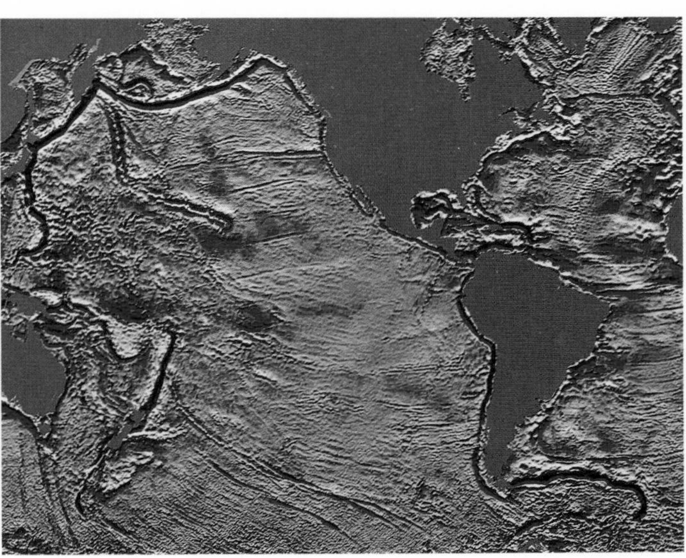

Fig. 3.15. Bathymetry of parts of the Pacific and Atlantic oceans produced from Seasat radar altimetry data of the sea-surface elevation. (Courtesy of W. Haxby, Lamont-Doherty Geological Observatory, New York.)

An even more cunning exploitation of radar and some simple principles of geophysics is illustrated by Fig. 3.15. This is clearly a very detailed image of part of the world's ocean bathymetry, showing many features such as ocean ridge systems, trenches, and submarine island chains. It stems from the Seasat radar altimeter which aimed at accurately measuring the mean height of the sea surface. As well as being controlled by the gravitational influence of the Sun and Moon in tides, the sea surface also responds to the Earth's own gravitational field. The gravitational field at any point depends on the mass of material below, down to the Earth's core. This in turn depends on the density of the materials in the Earth below. Sea water is of course much less dense than the rocks below the sea bed. So, the deeper the sea, the lower the gravitational field; and the sea surface stands lower than it does over shallow areas where the higher gravitational field attracts more water. By the very accurate measurement of sea-surface elevation using a radar altimeter, Seasat was also measuring the depth of water beneath all its orbits, thus enabling an image to be built up.

The microscopic plants that drift with ocean currents are at the base of the ocean's complex food chain. They are dependent on both sunlight for photosynthesis and a supply of nutrients dissolved in sea-water. The amount of this phytoplankton in the upper layers of sea water is therefore a direct measure of the biological productivity of an area and its potential for fisheries. These minute plants contain various pigments involved in photosynthesis, dominated by chlorophyll. The more phytoplankton, the greener the sea appears. Using those parts of the visible spectrum in which the pigments have a peak of reflectance it is possible to estimate the concentration of plant matter in the top few metres of the ocean. Such an estimation for the eastern seaboard of the USA is shown in Fig. 3.16. This is derived from the four narrow visible bands monitored by the Nimbus-7 Coastal Zone Colour Scanner (CSCZ). In the image, regions of high concentration (above $1\ mg^{-3}$) are coded in dark brown, intermediate levels in reds, yellows, and greens, while the lowest levels ($<0.01\ mg^{-3}$) are shown in blues. The light browns are land areas and the white is cloud cover. A map such as this, effectively a snapshot of one day, would be impossible

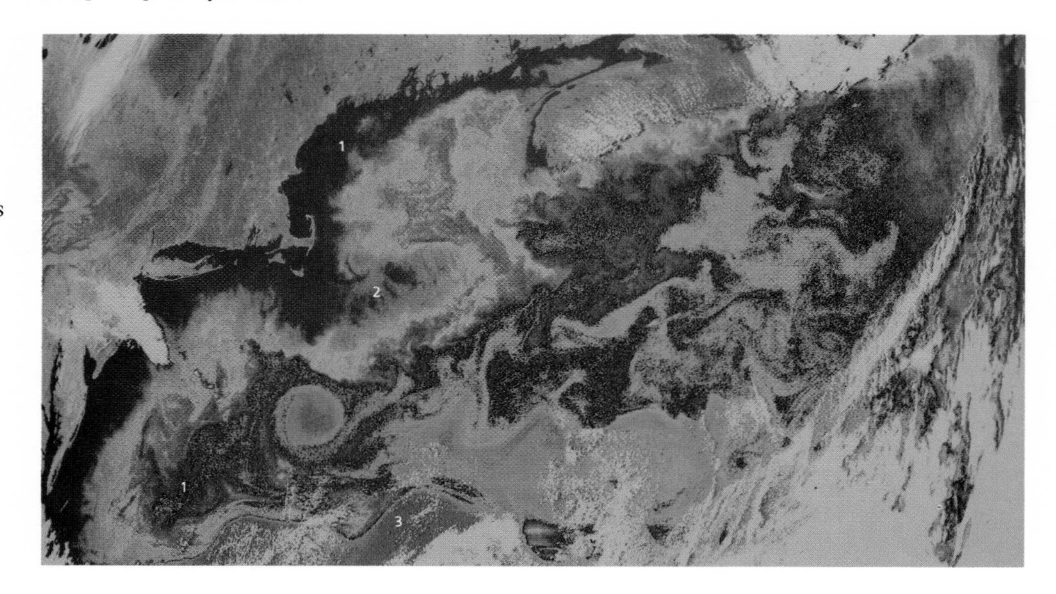

Fig. 3.16. Colour-coded image of Nimbus-7 CZCS data showing the distribution of pigments due to phytoplankton in the North Atlantic off the east coast of the USA. Width is about 1600 km. North is at top right.

using shipborne measurements, and vessels are used to gather calibration data and perform research on the results from the satellite. Where water is free of sediment, well offshore, plankton are the only contributors to reflectance and the estimates are within 30 per cent of direct measurements. The areas of high concentration near the coast contain some errors due to high visible reflectance from suspended sediment, but also show how the nutrients derived from the land and distributed by rivers and near-shore currents increase biological productivity. In this area the deep, cool waters which are nutrient rich are also brought to the surface by tides and current upwelling (1).

To understand fully the phytoplankton distribution requires information about the ocean temperature, since far offshore, the cooler the water the more nutrient is available. Figure 3.17 is a colour-coded image of mid-infrared emission also gathered on the same day by the CZCS, where temperature ranges from 25 °C in reds cooling through yellows and greens to the coldest water in blue (about 6 °C). This clearly shows the complex Gulf Stream (3) in the southern part of the image and a series of eddies between the warm, more saline water and cooler waters to the north whose salinity is slightly lower due to the

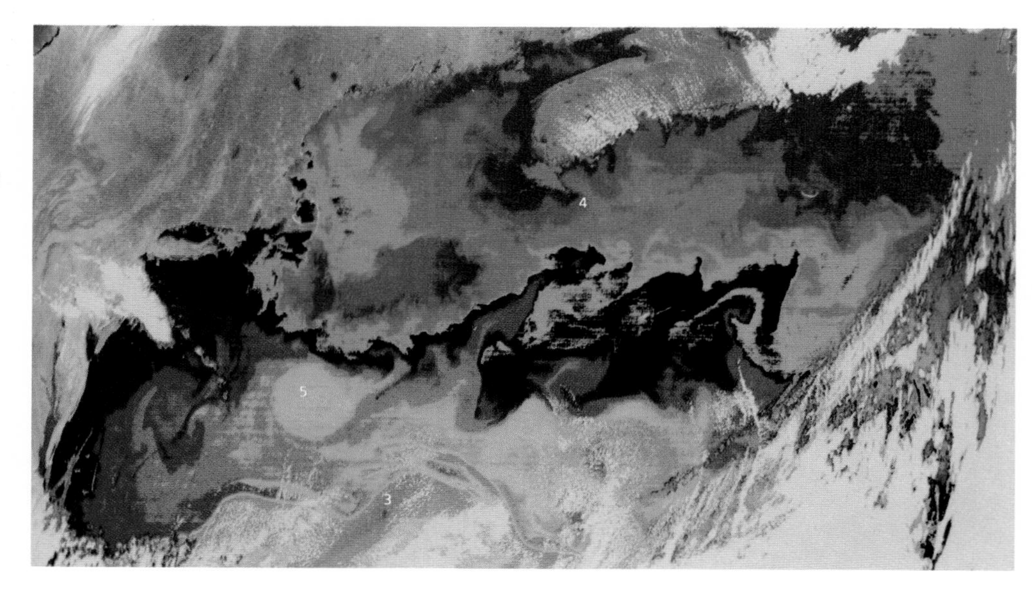

Fig. 3.17. Colour-coded image of sea-surface temperature derived from Nimbus-7 CZCS data for the same area as Fig. 3.16, obtained on the same date.

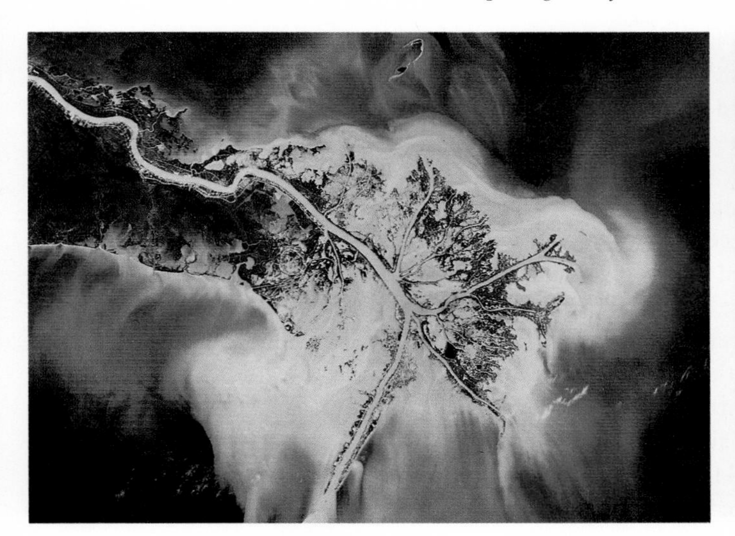

Fig. 3.18. Landsat MSS false-colour image of the Mississippi delta. Width is 90 km.

Fig. 3.19. Enhanced Landsat MSS false-colour image of the delta of the Granges and Brahmaputra, Bangladesh. (Courtesy of McDonald Dettwiler Associates.) Width is 175 km.

mixture of fresh river water. One of these eddies has developed into a circular warm core (5), which has low productivity on Fig. 3.16. As well as the high productivity along shore, the cool areas are clearly richer in plant life than the warm ones (2 and 4). As well as helping interpret images of phytoplankton distribution, because of the clear correlation between cool upwellings of nutrient-rich water and ocean productivity, thermal images such as Fig. 3.17 are useful in their own right as means of predicting planktonic blooms and their associated shoals of fish (Figs 4.18, 4.19). Thermal data from geostationary metsats enable such guidance to fishermen to be provided on a routine daily basis.

The thermal structure of the sea surface is also both an important guide to the humidity of overlying air and a reflection of very-large-scale climatic processes such as the El Niño (see above).

Around the world's coasts where rivers meet the sea, the action of saline water on the silts and clays carried into the oceans causes minute particles to clump together and fall to the sea bed to form sediments. Where the flow of fresh water is low very little suspended sediment clouds the sea water beyond a few kilometres offshore. However, where great rivers, such as the Ganges and Brahmaputra, Congo, Amazon, and Mississippi enter the sea, not only do they form massive delta systems from the coarsest sediment load but the low-density fresh water floats at the surface carrying sediment far offshore. This is distributed by currents, eventually to be deposited as constantly shifting mud-banks that pose problems to shipping, both at sea and in harbours and channels prone to silting. Figures 3.18 and 3.19 show sediment in the sea off the deltas of the Mississippi and Ganges/Brahmaputra. The Mississippi delta is shaped roughly like a bird's foot, several distributary channels prograding into the Gulf of Mexico, protected from erosion by levees of coarse sediment deposited during floods. The turbid waters offshore display swirls and sharp cut-offs due to current action and sudden mixing with saline water. The western part of the Ganges/Brahmaputra delta, at the head of the Bay of Bengal, is no longer actively supplied with sediment by the two main rivers, but has been abandoned as their courses changed. The numerous large inlets separating islands and spurs of swampland are dominated by tidal action, which redistributes the fine sediment and carries some of it offshore to extend the system of swamps. The highly active part of this huge system is still colonized by dense, tiger-infested mangrove swamps showing as dark red on Fig. 3.19.

Further inland, the delta has been drained by an intricate network of channels, and has been reclaimed for intensive agriculture, marked by the lighter reds of rice fields and pale tones of bare soil stripped of vegetation by recent floods. The channels of the Ganges/ Brahmaputra delta widen rapidly towards the open sea, a feature typical of deltas exposed to a large tidal range.

The dynamics of mixing of fresh waters discharged by large rivers with sea water was largely unknown until recently. This was because such huge volumes of water and large areas of sea were involved that conventional means of survey were too slow to match the changes involved from season to season and from year to year. Figure 3.20(a) and (b) illustrates very well how remote sensing can contribute to solving this problem. They are of Nimbus-7 CZCS data of the North Atlantic off Brazil and the Guyanas. The mouth of the Amazon is at bottom centre and the chain of islands at top left is the Windward Islands fringing the Caribbean. Like Fig. 3.16 the images show measures of the concentration of pigments in water derived from the four visible CZCS channels, although the colour coding is somewhat different (see insets). The fresh waters of the Amazon contain high nutrients together with a heavy load of organic pigments and suspended sediments, so they support high phytoplankton concentrations and show the highest pigmentation values. These natural 'dyes' allow the tracks taken by undiluted Amazon water to be followed very efficiently, as the oceanic waters support far less pigmentation and appear dark blue. Figure 3.20(a) is for the period July 1979 to January 1980, while Fig. 3.20(b) covers the January to February 1980 period. The white lines are the tracks of surface drifting buoys that allow the seasonal currents in the ocean to be compared with the water distributions. In the June–January period currents off the mouth of the Amazon swing from along the coast to an eastwards trajectory out into the deep ocean. The pigmented Amazon water is dragged out into the Atlantic as a yellow to green plume fading to light blue as mixing occurs. The currents in the January–February cycle flow more simply along the coast. As a result, Amazon water moves along the coast into the Caribbean, as shown by the wide

Fig. 3.20. Colour-coded Nimbus-7 CZCS images of the mean pigment content of water in the Atlantic Ocean off the Amazon, for (a) July 1979 to January 1980 and (b) January to February 1980. The white lines are tracks of buoys in the ocean currents. Width is 2200 km. (Courtesy of F. Muller-Karger, University of South Florida and C. McClain, GSFC, NASA.)

(a)

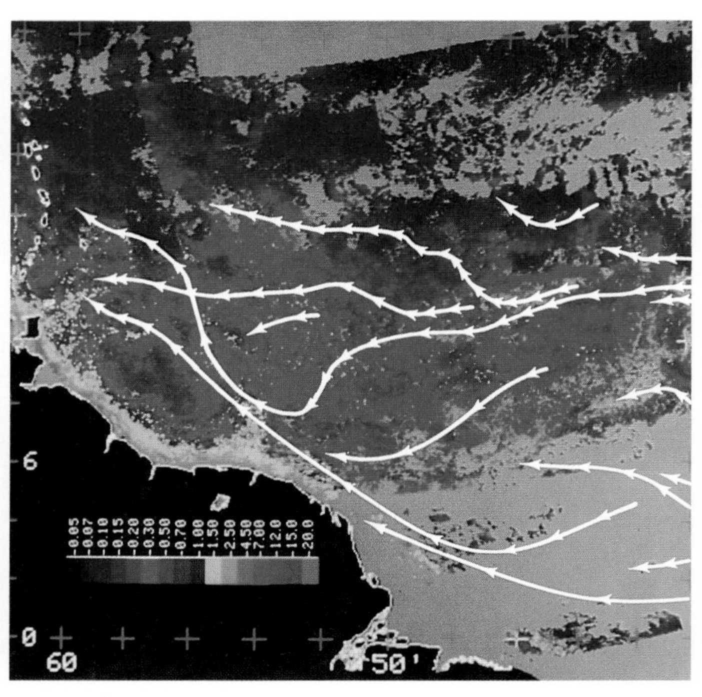

(b)

yellow–orange band along the coast in Fig. 3.20(b). There, the offshore blue to green colours indicate the region where eastward-dragged Amazon water from the previous cycle is mixed with oceanic water. In themselves, these observations are of considerable interest, but because the fresh, nutrient-rich Amazon water can support high phytoplankton contents they can potentially help target areas for deep-water fisheries throughout the year. Also, they help explain some of the features observed in the geochemistry and fossil record preserved by deep-ocean sediments in this tropical part of the North Atlantic.

Far out in the oceans, there is little commercial use for marine biological productivity, which is low anyway, and waters are clear and deep. It is there that shipping needs timely information about the state of the sea surface, which before the advent of remote sensing went largely unmonitored. The height, wavelength, and shape of waves is of vital concern to the seafarer, even for the largest ships afloat. There have been several instances in recent years of supertankers and bulk-ore carriers disappearing without warning or trace, presumably after encountering freak waves. Figures 3.21 and 3.22 shows examples of the sort of information about sea state that are now possible using orbital radar images. Figure 3.21 is a Seasat radar image of the Atlantic off the coast of Portugal. Because the deep-ocean waves are oriented roughly perpendicular to the illumination direction of the SAR system they show up very well. What is quite obvious is that the orientation of the wave crests changes far offshore. Although this is partly caused by the small islands near the centre, most is due to refraction of the waves by sea-bed topography at the continental margin, here between 30 and 100 metres deep. The wavelength decreases in places and there are several zones of wave convergence where the waves are anamalously high. When large, storm-driven oceanic swells are approaching this part of Portugal, as shown, there is a hazard to small vessels in particular.

Figure 3.22 is a SIR-A radar image of the Celebes Sea to the SW of Mindanao in the Phillipines. The variation in brightness of the image indicates considerable differences in the roughness of the sea surface. Such wave patterns that are visible are extremely irregular, and the different sea states are sharply defined. The sharp boundaries are marking abrupt local changes in wind speed and direction, typical of squalls. The bright and therefore rough areas are regions of high winds. The dark and smooth areas indicate either low wind or the effects of very heavy rainfall. Intense rain tends to smooth the water surface by damping out any wind-driven wavelets.

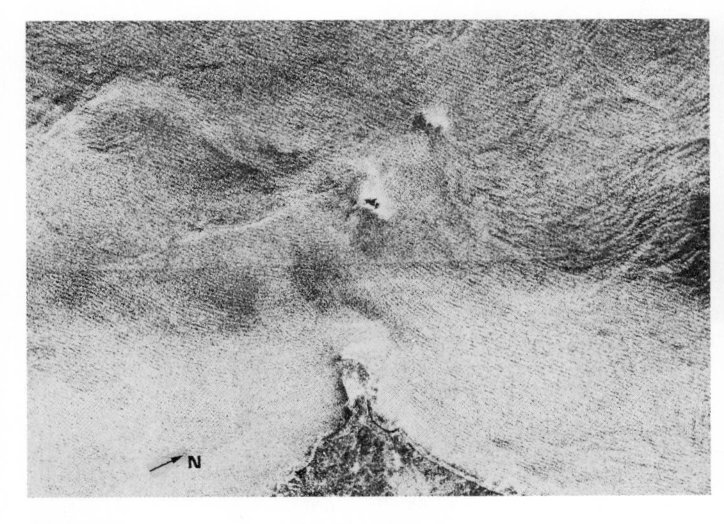

Fig. 3.21. Wave refraction by the sea floor off Portugal revealed by a Seasat radar image. Width is 50 km.

Fig. 3.22. SIR-A radar image of the Celebes Sea showing variations in roughness related to a frontal system. Width is 70 km.

Besides waves, the other great danger to navigation on the high seas is ice, both as a moving and vast hazard in the form of icebergs and as a hindrance to navigation in the case of pack-ice in Arctic and Antarctic waters. We cover pack-ice in the next Chapter (Figs 4.37, 4.38). Here the concern is with frozen oceans, both from the standpoint of navigation and because polar ice-sheets are a major controlling factor on the world's climate. They act as insulating barriers that regulate the rate at which heat can be transferred between atmosphere and oceans, and also control the polar high pressure areas that are major controls over wind circulation in both hemispheres. Although the interactions are very poorly known, it may be that a long-term change in sea-ice cover and its position may provide early warning about climatic changes.

Throughout the year, the polar oceans are exposed to wide variations in the input of solar energy, from total darkness and extreme cold in winter to continuous daylight and milder temperatures in summer. Consequently the sea-ice cover expands and contracts on a season-to-season basis. Before satellite remote sensing became possible the yearly changes

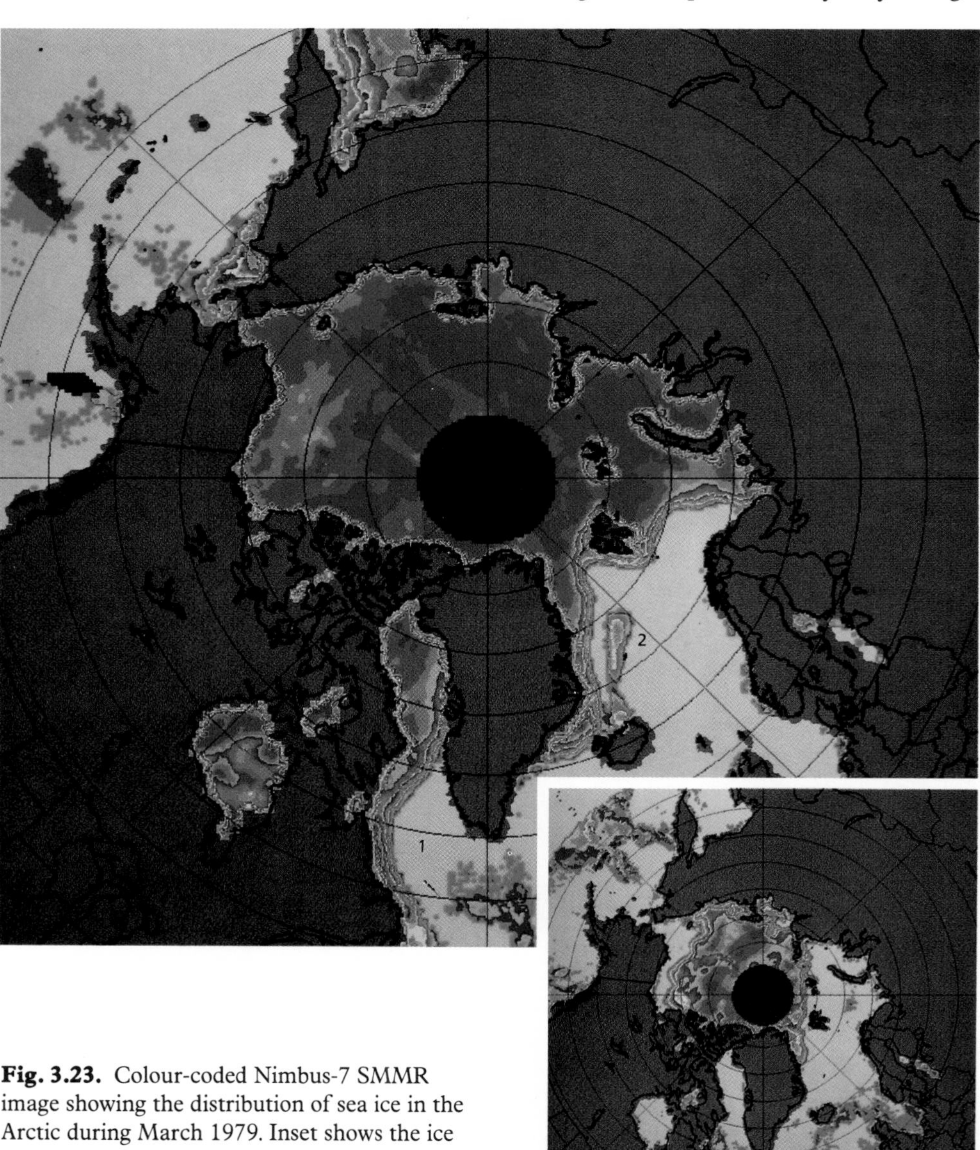

Fig. 3.23. Colour-coded Nimbus-7 SMMR image showing the distribution of sea ice in the Arctic during March 1979. Inset shows the ice conditions in September 1979. (Courtesy of D. Cavalieri, GSFC, NASA.)

in the extent of the ice were poorly known. Even with images in the reflected and thermal-infrared regions of the spectrum, from metsats and polar orbiters such as Landsat and the TIROS/NOAA series, variable cloud cover and winter darkness hindered the accurate and continuous monitoring that is necessary. In order to overcome this problem sensors were developed to detect microwave emissions that are controlled by surface temperature. These are the Scanning Multichannel Microwave Radiometers (SMMR) deployed on the NASA Nimbus satellites. Microwaves are emitted day and night and can penetrate cloud cover, so these systems allow continuous monitoring of the polar regions, albeit at a very low resolution. Sea ice and water emit different amounts of energy, due both to temperature differences and differences in their morphology. Figures 3.23 and 3.24 are colour-coded to show sea ice in a variety of colours and open water in pale blue. The images are for the Arctic and Antarctic (1979 and 1974 respectively). The inset to Fig. 3.23 shows the Arctic ice at its lowest extent the following summer. Figure 3.24 comprises images from February, April, May, June, July, and September 1974. Both sets of images are mainly self-explanatory. In Fig. 3.23 a large ice-mass at (2) has broken away from the main pack, whereas at (1) ice can be seen to be restricted to the Labrador coast because of the warming influence of a spur of the Gulf Stream.

Fig. 3.24. Colour-coded Nimbus-5 SMMR images showing the changes in sea ice around Antarctica in 1974. The sequence is (a) to (f) from February, April, May, June, July, and September. (Courtesy of D. Cavalieri, GSFC, NASA.)

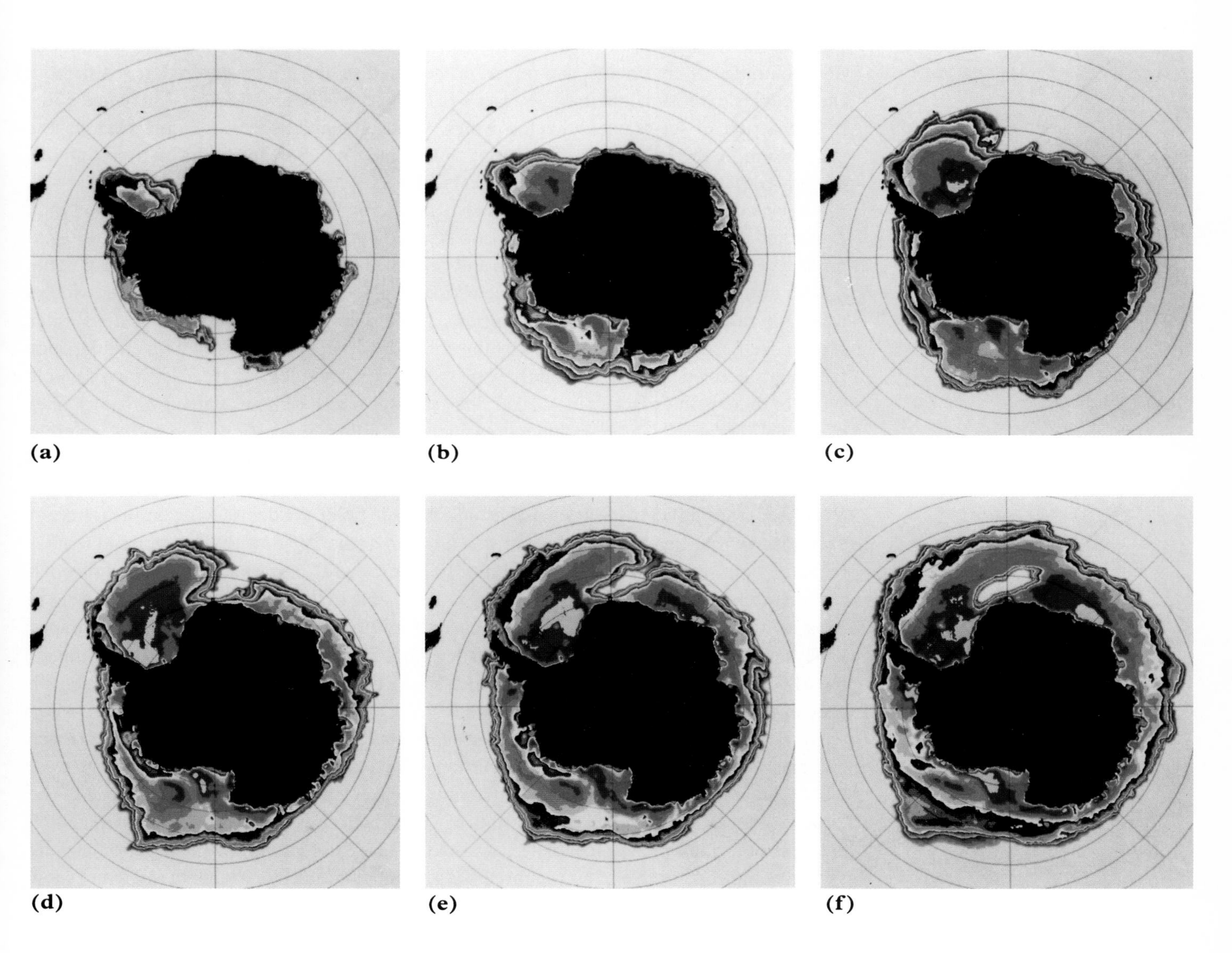

Figure 3.24 shows the progressive growth of the Antarctic pack through a full year. Beginning in March, when day and night are of equal length, the pack-ice begins to grow. By May, when much of the area is in total darkness, the rate of growth increases rapidly, until by the end of winter (September) the ice extends more than 1000 km from the shore. By the end of the brief Antarctic summer in February, more than 80 per cent of the ice has melted away so that ships can approach a few areas on the coastline. The sequence shows the envelopment between June and September of a large area of open sea in the middle of the Weddell Sea pack, known as a polynya, which is present even at the depth of winter. This feature is not always present, but when it does form it is always in roughly the same location. The reason is not known, but there must be a major loss of heat from the ocean to sustain this huge area of open water.

3.3 The land

The land surface is sculpted by many natural processes, and also, to an increasing extent, by human activities. Here we concentrate on its physical features, and there is such a bewildering variety of landforms that it is nowhere near possible to illustrate them all. Instead, the focus is on the main agencies of erosion that wear down the rock formations underpinning the continents, and on the processes of sediment deposition that redistribute the products of erosion and in some cases add to the extent of land-masses. These boil down to the action of ice-masses, flowing water, wind, and the sea. Space only permits the subdivision of the land into a few broad zones; near-polar regions and high mountains, humid temperate and tropical areas, deserts, and a few examples of coastal regions.

Regions dominated by the action of frost, snow, and ice consist essentially of two types: rocky and sometimes mountainous terrain scraped bare of soft-sediment cover by ice, and areas of sediment deposited by glaciers and reworked by the meltwater derived from them.

The most celebrated examples of the action of glacial erosion are the U-shaped valleys carved by now vanished mountain glaciers. Their shape is essentially that of the former glacier itself, the bottom ground flat by the abrasive action of ice charged with angular rock fragments derived by freezing and thawing from the valley sides. Figure 3.25 shows a whole series of these valley glaciers in Alaska. They show as bluish ribbons, striped with dark lines of medial moraines. The moraine increasingly dominates down-flow, due to the ablation and melting of ice, until the glaciers themselves become difficult to distinguish from unvegetated valley sediments. Tracing each linear moraine up-flow, it can be seen to end at a spur separating two tributary glaciers, and is made up of the debris lying on the flanks of each glacier that has been drawn out down the main line of flow. Each glacier emanates from the high snowfields in the mountains. Winter snowfall is gradually converted through several intermediary forms to crystalline ice that is able to flow slowly under its own weight. Most of the world's valley glaciers are currently in retreat, leaving the classic U-shaped valleys ahead of them. In this case, the valleys that have been abandoned by retreating ice have been invaded by lush vegetation, which shows up as bright red in contrast to the bluish grey of the rocky rivers and the snow and ice.

By far the greatest modifications of scenery by ice action took place in the northern hemisphere during the Pleistocene glaciation, which was dominated not by valley glaciers but by huge ice-sheets flowing south from major ice-caps around the Arctic Ocean. The largest remnant of these ice-caps is that of Greenland, but Fig. 3.26 shows an excellent example of a small relic in Iceland—Vatnajokull. This false-colour image was produced by special enhancement to exploit the threefold grouping of multispectral data related to snow and ice, bare rock and soil, and vegetated land. The ice-cap itself is shown in various grey

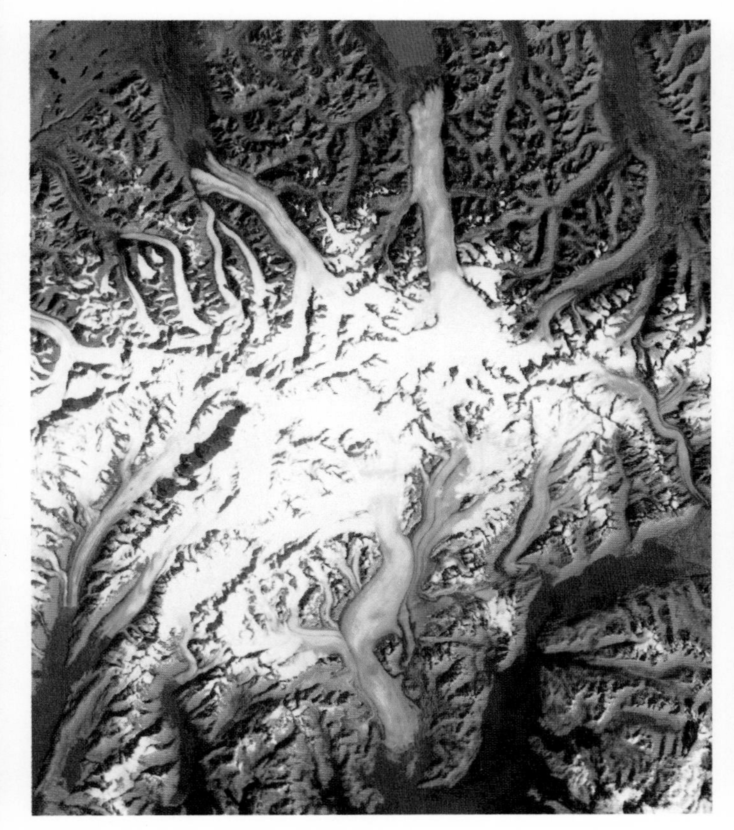

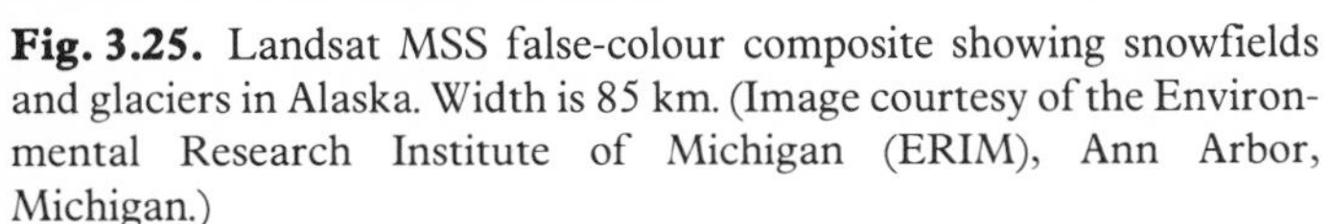

Fig. 3.25. Landsat MSS false-colour composite showing snowfields and glaciers in Alaska. Width is 85 km. (Image courtesy of the Environmental Research Institute of Michigan (ERIM), Ann Arbor, Michigan.)

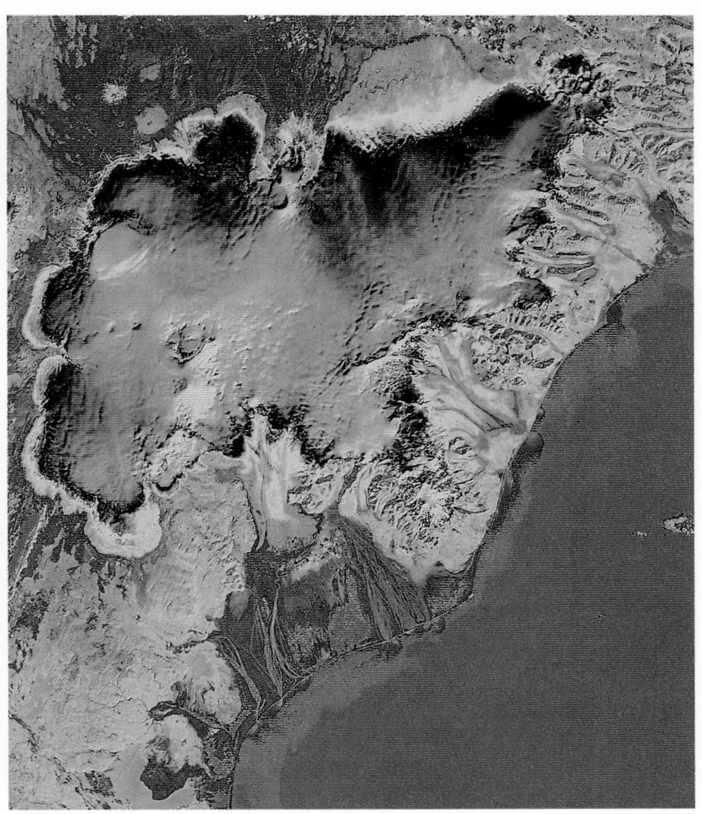

Fig. 3.26. Specially processed Landsat MSS false-colour image of the Vatnajokul ice-cap in Iceland. Width is 130 km.

tones, and has a basically smooth relief. The different shades relate to different types of snow and ice, the light grey to dry snow and darker greys to wet snow and snow in the process of being drained of water and being converted to ice. On its flanks several lobes of actively moving ice in the form of broad glaciers are coloured light blue, corresponding to blue ice that is being ablated away. The darker blue streaks ahead of the glaciers are active zones of sediment transport in meltwater, while the red areas are dark-coloured sands laid down in the fluvioglacial environment, together with areas of basalt produced by the active volcanoes in the centre of the island. Vegetated areas, that have neither been glaciated nor blanketed by volcanic rock recently, show up as yellows and brownish-grey.

Two kinds of terrain develop beneath such massive ice-sheets. Where resistant rock remains, all the intricate details of its internal structure are etched out as weaknesses and exploited by the ice flow, sometimes forming innumerable geologically controlled lakes. Figure 3.27 is of an area so sculpted in the Canadian Shield where the rocks contain extremely complicated structures formed at great depths in the crust some 1800 million years ago. The reddish colour witnesses the growth of boreal birch and pine forest since the glacial period. The products of erosion by ice-sheets eventually must be deposited, to form great tracts of boulder clay. Though sometimes relatively featureless, where these ill-sorted sediments are still experiencing intense winter cold and a short period of summer melting curious structures do develop. Figure 3.28 is an approximately natural-colour Landsat TM image of part of a tract of such subglacial and fluvioglacial sediments, along the Mackenzie River in north-west Canada. The plain of glacial sediments is now dominated by curiously

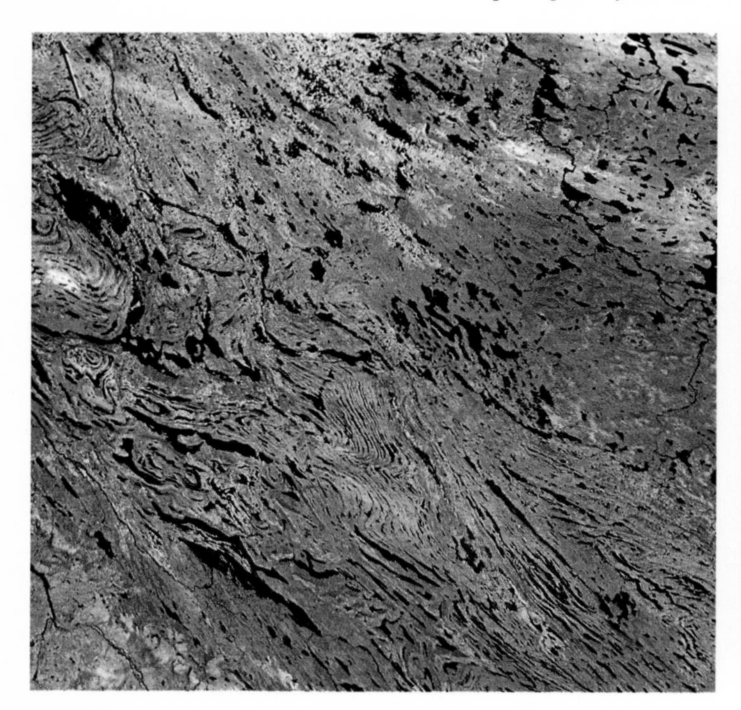

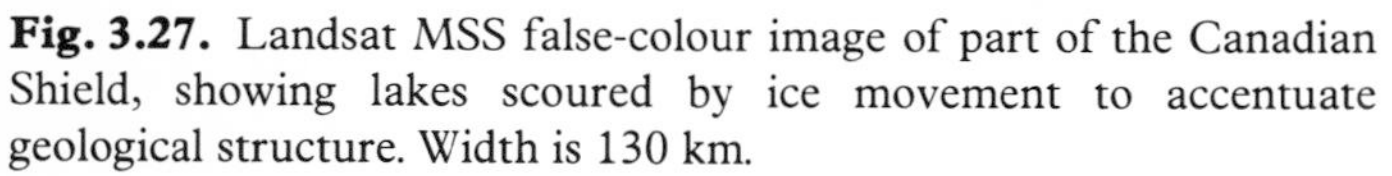

Fig. 3.27. Landsat MSS false-colour image of part of the Canadian Shield, showing lakes scoured by ice movement to accentuate geological structure. Width is 130 km.

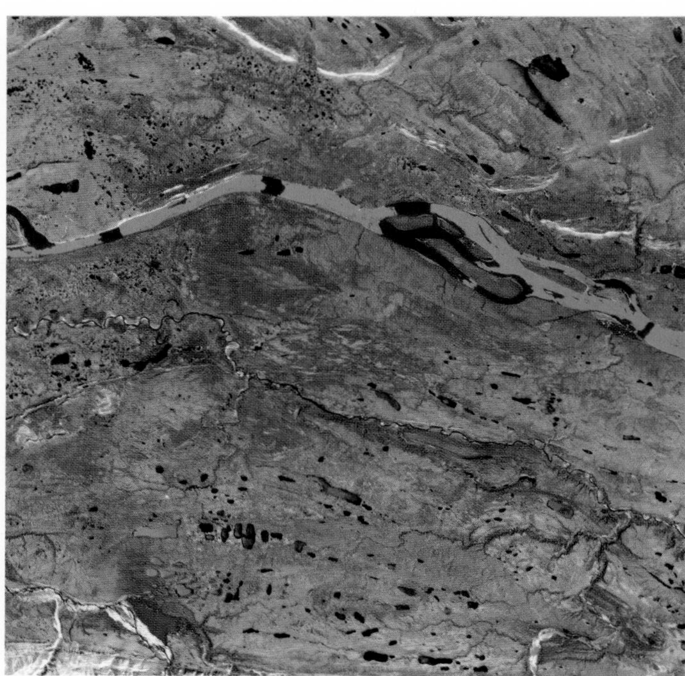

Fig. 3.28. Landsat TM image, with band 5 as red, 4 as green, and 3 as blue, of glacial lakes in the MacKenzie River valley, Canada. Width is 130 km. (Courtesy of Canada Centre for Remote Sensing.)

shaped lakes, many oriented roughly NNW–SSE. These lakes are a reflection of the effects of permanently frozen ground, where only the top few metres is ever free of ice. As the ground freezes solid at the onset of winter, the frost gradually penetrates deeper towards the permanently frozen layer. This traps water at the irregular interface, pressurizes it and so allows it to become supercooled. Lenses of ice develop in the soil profile, bulging up the surface. When they melt the surface subsides forming a small lake. Repeated action of this kind magnifies the effect, so producing a complex network of ponds. Why the ponds in this case are aligned is possibly a result of the earlier ice-flow direction. The bright blue colour of many of the lakes and partsof the river are due to ice. On the river, the ice is on the point of breaking up, the edges of the ice marked by the dark blue open water fitting together in a jigsaw pattern.

Almost by definition, the main agent contributing to landforms in humid areas, both temperate and tropical, is flowing water. The main differences between temperate and tropical scenery are due to the increased effects of vegetation and greater intensity of chemical rotting of bedrock as the tropics are approached. Dense vegetation both binds the soil and reduces the speed at which rainfall flows off the land. In densely forested areas of the humid tropics this allows steeper valley slopes to be better preserved than in more temperate areas, where the angles of slope are reduced both by erosion and the downward creep under gravity of wet soil and other surface debris.

Figure 3.29, of an area in the Appalachian Plateau of the eastern USA, graphically illustrates the erosional influence of flowing water on landscape in the humid temperate zone. The main drainage is along the Ohio River, occupying a broad valley. In the bottom right part of the image a wide valley with only an insignificant stream can be seen. This is an abandoned course of the former drainage system in the area, which was shifted due to glacial deposition during the Pleistocene Ice Age. The present course of the Ohio, is

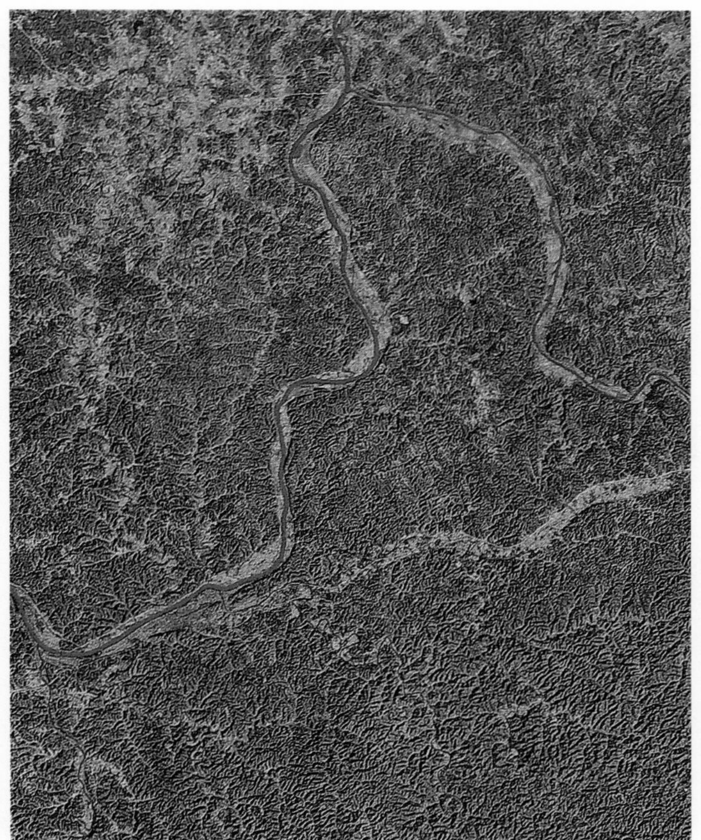

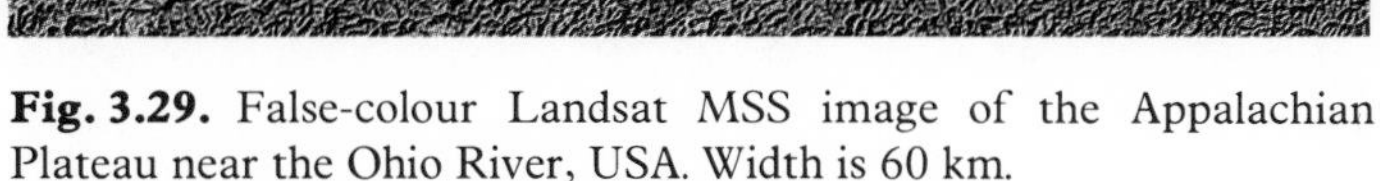

Fig. 3.29. False-colour Landsat MSS image of the Appalachian Plateau near the Ohio River, USA. Width is 60 km.

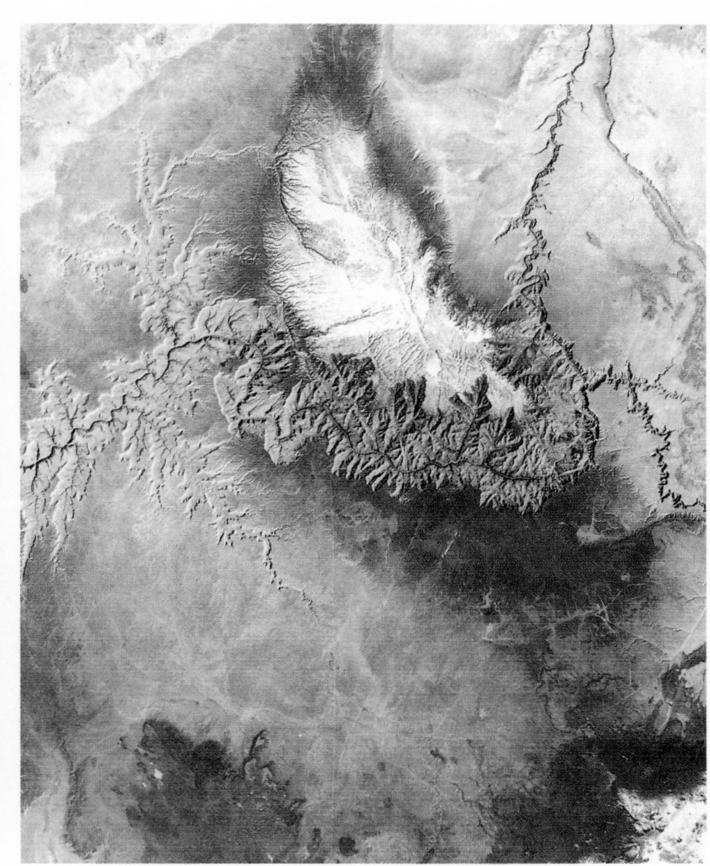

Fig. 3.30. Winter MSS scene from Landsat of the Grand Canyon, Arizona. Width is 120 km.

therefore only a few tens of thousand years old. The minor drainage in the uplands is extremely closely spaced, and has an almost random, dendritic pattern, showing that the underlying geology is quite homogeneous. The completely interlocking nature of the hilly spurs, all of which rise to about the same elevation, is an indication of what the pioneer of geomorphology, W. M. Davis, classed as a mature stage in the evolution of a landscape. In his idealized scheme, the uplift of an erosionally planed surface rejuvenates erosion. The first, or youthful, stage involves the development of isolated valleys carved into a plateau (Fig. 3.30), that expands to consume the uplifted plain by the stage of maturity. Continued erosion gradually removes all but a few relics of the former surface, so that an old-aged landscape is dominated once again by a nearly planar surface (Fig. 3.31).

The Grand Canyon of Arizona (Fig. 3.30) is probably the most spectacular example of Davis' 'youthful stage' in landscape evolution. It has formed where the Colorado River has eroded downwards at the same pace as a previously flat terrain has been warped upwards. The upwarp is marked by brownish tones indicating forest at high elevations. Part of it, the Kaibab Plateau, has a thin cover of snow. The Grand Canyon itself and several major tributary canyons form a relatively minor part of the scenery, most of which has subdued slopes. At the south of the snow-dusted Kaibab Plateau it is quite easy to visualize how the side streams of the Colorado, originating on the plateau, are gradually 'eating' their way into it. Eventually, the rugged landscape of interlocking spurs that characterizes the canyons will occupy the whole area, to produce a complex dendritic network similar to that in Fig. 3.29.

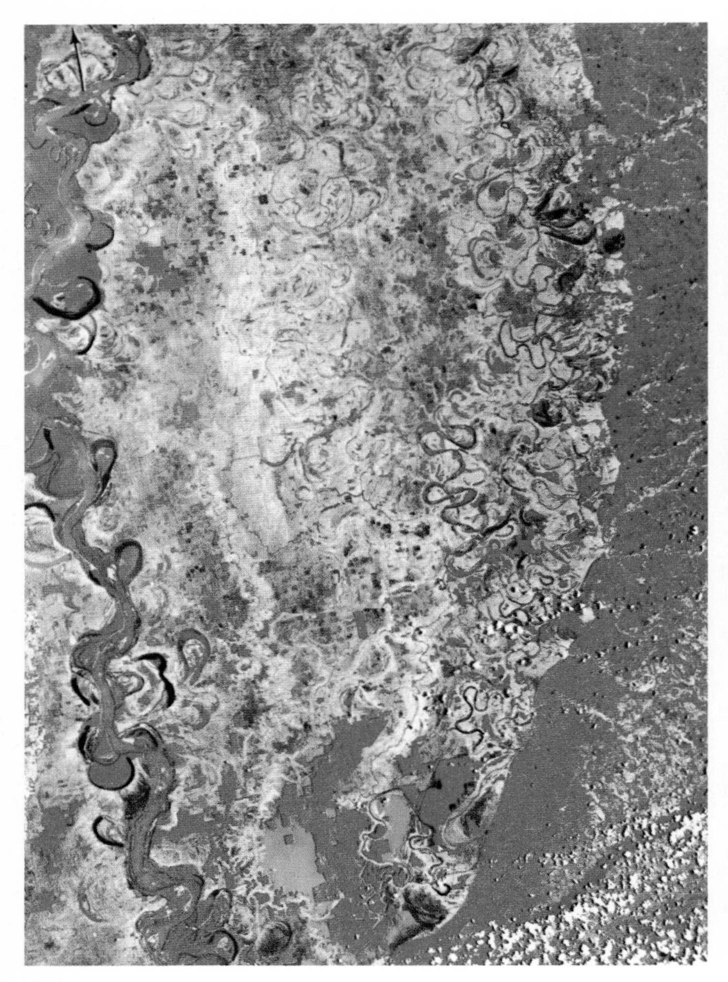

Fig. 3.31. Landsat MSS false-colour image of the lower Mississippi valley, after floods. Width is 120 km.

Fig. 3.32. Confluence of the Amazon and Rio Negro near Manaus, Brazil, on a Landsat MSS false-colour image. Width is 75 km. (Courtesy of Earth Satellite Corporation.)

Figure 3.31 illustrates the last stage in Davis' scheme, that of geomorphological senility, expressed well by part of the Mississippi valley. It was taken shortly after catastrophic floods in 1974, when the flood plain was inundated and covered by silt. The silt is picked out by light colours. The unflooded area to the east, being slightly higher, was unaffected, so that the characteristic reds of dense vegetation remain dominant. The main course of the river is on the west margin of the image. The flood plain in the centre is characterized by spectacular meandering tributaries, oxbow lakes and the complex curved ridges associated with sediments deposited by slow-moving meandering channels. Many of these features are picked out by ribbons of red vegetation growing on the elevated levees flanking the channels themselves. When the levees are breached flood water inundates the adjacent backswamps, where the finer suspended clays and silts are deposited to form the foundation for extremely fertile soils.

Figure 3.32 gives an impression of a riverine landscape in a tropical environment. It shows the confluence of the Rivers Negro and Amazon near Manaus in Brazil. The most striking feature of the image is the difference in colour between the Amazon, which is pale due to its heavy load of sediments derived from the Andes, and the black River Negro. The

Negro drains low ground and carries virtually no sediment as a result. Instead it is heavily charged with organic materials produced by the decay of vegetation. Similar black water is apparent in the swamps backing the Amazon itself. The terrain through which these two huge rivers flow is by no means flat, and there is clear evidence for an intricate dendritic drainage network, and some evidence for the control of some linear valleys by the underlying geology. What is interesting is that apart from in the main tributaries, no water can be seen. This demonstrates that the jungle here is so dense that the canopy continues right across even large streams, thereby hiding them from view. Light patches in the jungle are along major roads and where the poor soils have failed to regenerate vegetation after the forest has been cleared.

The landforms associated with arid and semi-arid areas are the products of a host of processes but are conditioned by one common factor. Where rainfall is less than 100–400 mm—depending on latitude, temperature, and hence rates of evaporation—vegetation is sparse and transient. Although the term 'desert' implies vistas of great sand-dunes, really it is derived from the fact that such areas are barren and deserted. In rugged areas of bare rock the absence of plant cover allows the rain that does fall to flow rapidly over the surface and strip it of any loose debris. Where these sediments are deposited is also unvegetated so that the debris can be continually reworked by occasional surface water, and, more important, the lack of any binding roots frees the finer sediment to be picked up and winnowed by wind. Once soil is set on the move by wind, vegetation becomes less likely to establish itself, so that inappropriate human activities, such as overcropping, the construction of ploughed fields without wind breaks, and overgrazing, can themselves lead to desertification, even where climatic conditions would not produce natural desert.

One of the places that immediately conjures up the word 'desert' is Death Valley in eastern California, part of which is shown in Fig. 3.33. The valley itself, picked out in white and orange hues, is below sea-level and has been created by the active pulling apart of the crust beneath. The mountains rimming the valley rise to more than 3000 m, yet even at these elevations rainfall is so low that vegetation is sparse. When it does rain it is in intense but brief storms. All the water falling on these bare hills rapidly descends into steep-sided valleys, carrying the loose rock broken apart by the rapid changes in day and night temperature. The valleys become nightmarish torrents up to 20 m deep charged with debris from boulders the size of automobiles to very fine silts. As they almost literally explode onto the flatter lower slopes the flood spreads laterally and slows to form a sheet. Most of the coarse debris is simply dumped as the flow spreads out to form a low-angled cone known as an alluvial fan, several of which are easily seen on Fig. 3.33. The finer material is transported to the valley floor where it settles in ephemeral lakes. As these lakes dry, the dissolved material in the flood water is left as encrustations, such as those coloured orange (rock salt) and red (A) (gypsum) on this false-colour Landsat Thematic Mapper image.

Although there are many much larger arid areas, the Namib Desert of Namibia (Fig. 3.34) demonstrates the important features of a sandy desert very well. Drainage from the rugged mountains in the hinterland along several large but ephemeral rivers transports sediment to the coast. There it is moved north-westwards by the cold Benguela current to form the large sand-spits. The dominant wind direction at the coast is from the south-west, but carries little moisture from the cold ocean, so creating arid conditions. Coastal sand together with that winnowed from the alluvial plains has been redistributed in a spectacular series of enormous dunes, here coloured yellow on a false-colour image, but which in fact are stained red by iron oxides. The linear pattern of the dominant inland dunes parallel the deflection northwards of the prevailing wind by the mountains.

One of the most arid areas on Earth lies along the Sudan–Libya border, where rainfall is less than 10mm annually. Much of the area consists of wide flat plains mantled with thin

Fig. 3.33. Death Valley, California, on a Landsat TM image, using band 5 as red, band 4 as green, and band 7 as blue. Width is 30 km. (Courtesy of ERIM.)

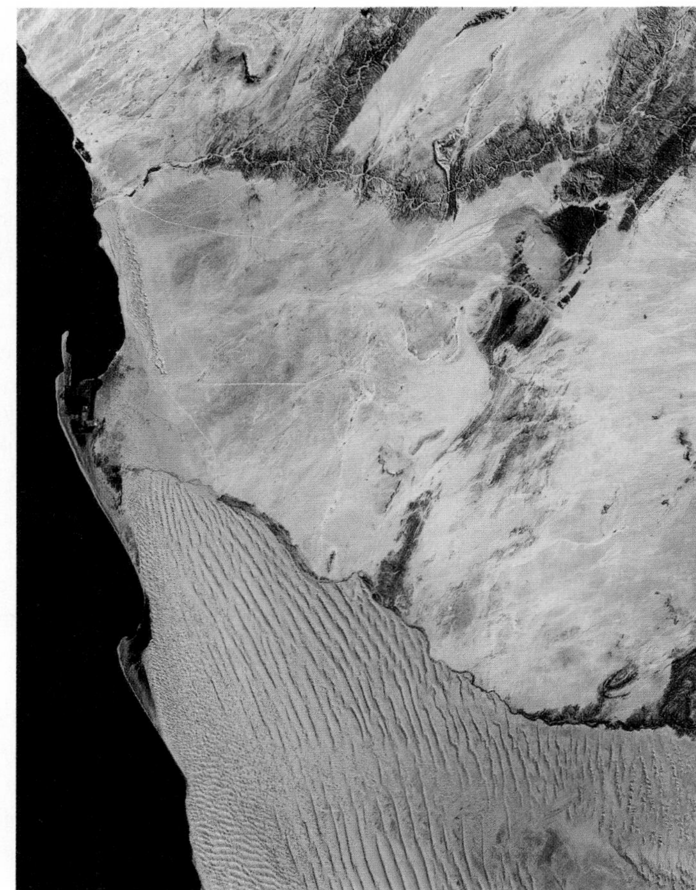

Fig. 3.34. Landsat MSS false-colour image of the Namib Desert and South Atlantic, Namibia. Width is 120 km.

sediment. On most images these areas of sand and gravel appear totally featureless, which they are. However, the very low moisture content of the sediment means that radar can penetrate up to 6 m beneath its surface. In Fig. 3.35 the result of this unique property is shown spectacularly. Although the sediment plains are dark they reveal surprising details of buried, and now inactive, drainage courses. Where tributaries enter the main courses, they form a 'V' towards the south, showing that when they were cut the flow was southwards. The very location of these hidden features caused a considerable stir when this SIR-A image was published. However, the direction of the drainage that they represent is opposed to that expected. The flow of the Nile, some 250 km further east, and all other occasionally active drainage in the Western Desert is currently northwards, suggesting that since the carving of the buried valleys, over 6000 years ago, this part of Africa has undergone subtle tilting that has reversed the direction of drainage. The detection of the buried drainage courses by radar is not clearly understood. It may be due to radar penetration and scattering from subsurface variations in granularity of the sediments. An alternative is that penetration has not occurred and the radar is picking up subtle variations in surface roughness related to the old stream courses.

The erosional features of the coast due to marine action are extremely varied, and controlled dominantly by the underlying geology, so that headlands are buttressed by resistant rocks and bays controlled by easily eroded strata. The most distinctive features of

Fig. 3.35. SIR-A radar image of part of the Western Desert in Libya, showing ancient drainage courses. Width is 70 km.

the coastline are where sediments have been added by constructional processes associated with marine action.

Silts and sands transported to the coast by river action are caught up in the currents resulting from wind, tide, and the longshore drift associated with waves oblique to the coast. Figure 3.36 is of Cape Canaveral on the flat-lying east coast of Florida. Waves are pounding the coast and their back swash is generating strong undertow for up to 5 km offshore. Sediment caught up in this is being dragged out to sea forming the scalloped pattern of pale blue turbid water in the sea. The present coast, including Cape Canaveral itself, is bounded by a beach ridge thrown up by storm action. Behind this are lagoons and swamps trapped by the modern beach ridge. These have clear shallow water, beneath which some of the bathymetric details can be seen. The traces of earlier beach ridges that have built up the coast can be seen for some 15 to 20 km inland. Indeed, the whole of the coastline shown here is formed of beach deposits. Cape Canaveral itself is dominated by the constructions associated with the Kennedy Space Flight Centre, the regularly spaced launch pads being particularly noticeable. This human activity has disrupted the beach-ridge patterns from which the Cape is constructed, but to the north-west they can be seen as a striking curved array. This pattern clearly suggests that Cape Canaveral has migrated southward along the coast over the last few thousand years. This migration is connected to coastal currents and the massive supply of sediments by rivers draining the Appalachian Mountains far to the north.

The effects of coastal currents acting on an abundant supply of sediment is spectacularly displayed by the huge sand-spits of Cape Cod in Massachusetts (Fig. 3.37). The foundation for this marine development is a series of glacial moraines, the clearest being the arcuate chain of islands in the south-west, while Cape Cod itself is also a moraine. This is confirmed by the chain of small irregular lakes on its surface, each being the product of melting

Fig. 3.36. Approximate natural-colour Landsat TM image of Cape Canaveral, Florida. Width is 40 km. (Courtesy of ERIM.)

Fig. 3.37. Approximate natural-colour Landsat TM image of Cape Cod, Massachusetts. Width is 80 km. (Courtesy of ERIM.)

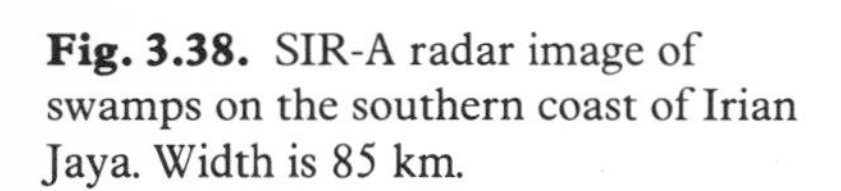

Fig. 3.38. SIR-A radar image of swamps on the southern coast of Irian Jaya. Width is 85 km.

subsurface ice in the moraine and collapse of the ground. The great hook of Lower Cape Cod formed from the reworking of the glacial sediments and their redeposition in accord with the coastal currents and wave action from the Atlantic. In the narrows between Nantucket Island and Martha's Vineyard in the south of the image can be seen shallow sand-bars and rip currents caused by the flow of the tide into Nantucket Sound.

Much quieter conditions on the tropical coast of Irian Jaya (Fig. 3.38), where large rivers drain the tectonically active mountains of the northern part of the island, are dominated by estuarine deposition. Very little coarse sediment reaches the coast here, which is not affected by strong prevailing winds or complex tidal currents. The extremely flat terrain allows the rivers to form spectacular meanders in which they gradually rework the muds, silts, and organic remains that have built up. Normally, this part of the coast of the large island, formerly known as New Guinea, is completely masked by clouds, and it is only revealed clearly here by the use of radar imagery.

Where sea water is clear and warm, marine life is able to flourish, down to considerable depths. The most spectacular evidence for its influence over constructional processes is provided by coral reefs, which grow in clear water above 18 °C. In Fig. 3.39 two of the three islands, which are part of the Society Islands archipelago in the Pacific, are surrounded by actively growing coral reefs. Each reef, seen as a white ribbon where it is above sea level and

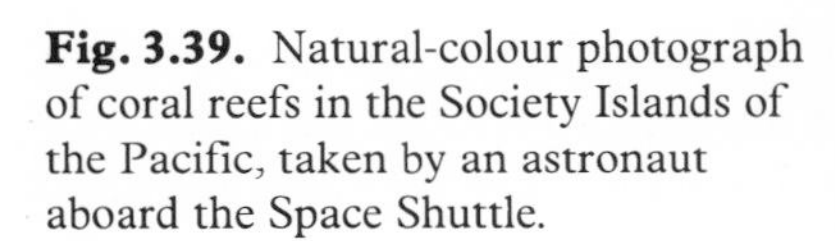

Fig. 3.39. Natural-colour photograph of coral reefs in the Society Islands of the Pacific, taken by an astronaut aboard the Space Shuttle.

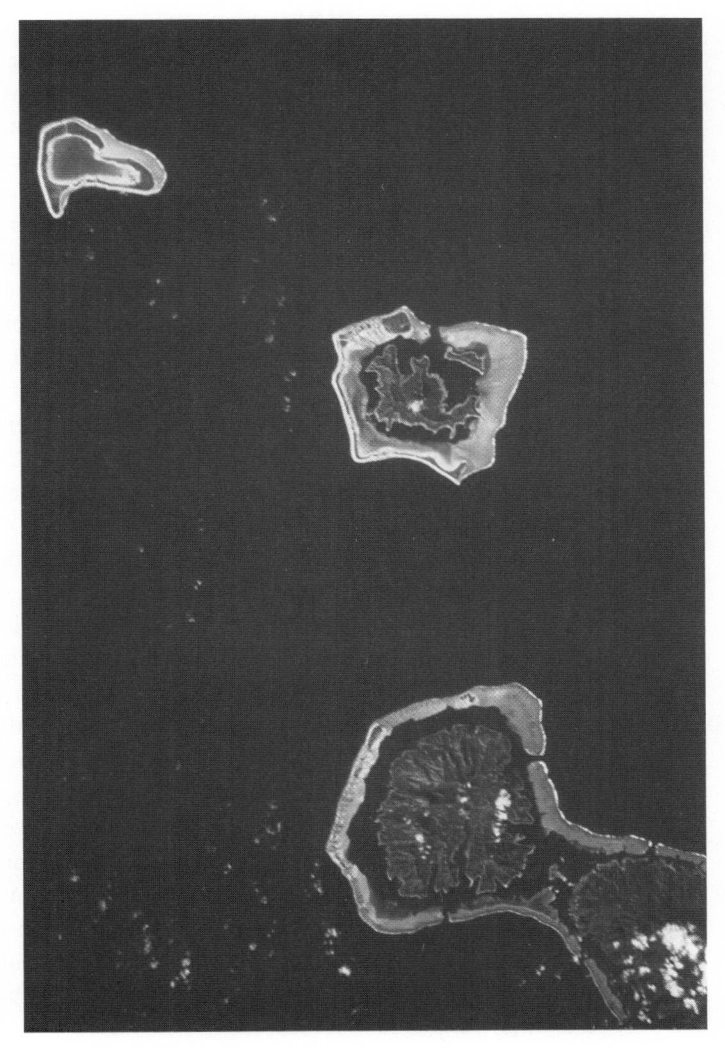

as pale blue in shallow water, is separated from the island by a narrow, concentric lagoon with a darker blue colour. The reefs are breached by tidal drainage channels, so that the islands are accessible to shipping. Such fringing reefs are evidence for an extraordinary process, first proposed by Charles Darwin. Each island is a small inactive volcano. The reefs first formed in direct contact with the volcano, but as the volcano died and cooled it began to sink slowly. Coral growth kept pace with the subsidence, so that its vertical extension eventually resulted in its becoming detached from the island itself, as shown here. The end product is the sinking of the volcano below sea level to leave a ring of coral islands around a central lagoon. Such an atoll has formed in the lower left of the image. These reef systems form a major threat to navigation, particularly as they have rarely been charted accurately. The variation in blue tones, as explained earlier in Section 3.2, is related to water depth, so images such as this can be used for very useful estimates of bathymetry.

3.4 Natural vegetation

The fundamental difference between the spectra in the visible and very-near infrared of green plants and those of all other natural materials, outlined in Section 2.2 (Figs 2.14, 2.16), means that vegetation is the most easily analysed part of the natural world using remote sensing. Indeed, the design of the sensors aboard Landsat-1 was predetermined by the need to monitor agricultural vegetation on a world scale, coupled with the relative simplicity of so doing. Most of the images with which the general reader has come into contact and which dominate this book comprise reflectance data in the very-near infrared, red, and green, expressed as red, green, and blue in a colour image. This part of the spectrum is where chlorophyll has its maximum absorption and plant cells reflect most radiant energy. The abnormally low reflectance of plants in red and their very high infrared response causes vegetation to be rendered in these false-colour composites as various shades of red, while other surfaces appear in greys, blues, yellows, and browns. Vegetation is even more strongly discriminated when the ratio between very-near infrared and red reflectance is displayed as an image. This ratio forms one of several *vegetation indices*, and is high where vegetation is most dense and low where it is absent, thereby giving a semi-quantitative measure of the density of plant cover. Moreover, the subtle differences between different species of plants in this part of the spectrum also allows differentiation between two types of vegetation.

Of all the sensors available to us, that giving the most reliable and most easily handled vegetation information for the whole planet is the AVHRR aboard the meteorological NOAA series of satellites. Figure 3.40 is a compilation of AVHRR data in the form of a standard false-colour composite for the whole northern hemisphere during July, with a 20 km resolution. The bulk of the land surface is shown in red, meaning that green vegetation is present. Although some of the image is affected by white clouds, and snow and ice around the pole, many of the pale areas are deserts. The great expanse of the Sahara and Arabian deserts is particularly well shown. Rocky deserts and high mountains come out as dark greys, and the tundras bordering the Arctic Ocean show as brown. This reflects the dominance of the cover in boreal regions by coniferous forest and low shrubs and mosses. Although quite small at this scale, the two largest areas of irrigation in arid lands, along the Nile and in its delta and along the Indus, show as red streaks in an otherwise grey to white background of barren desert. In temperate and humid tropical areas, very much the same pattern would be seen at any time of the year, but in the boreal and arid zones, vegetation is extremely restricted in its season.

Fig. 3.40. Mosaic of NOAA AVHRR false-colour images for the northern hemisphere showing summer vegetation. (Courtesy of ERIM.)

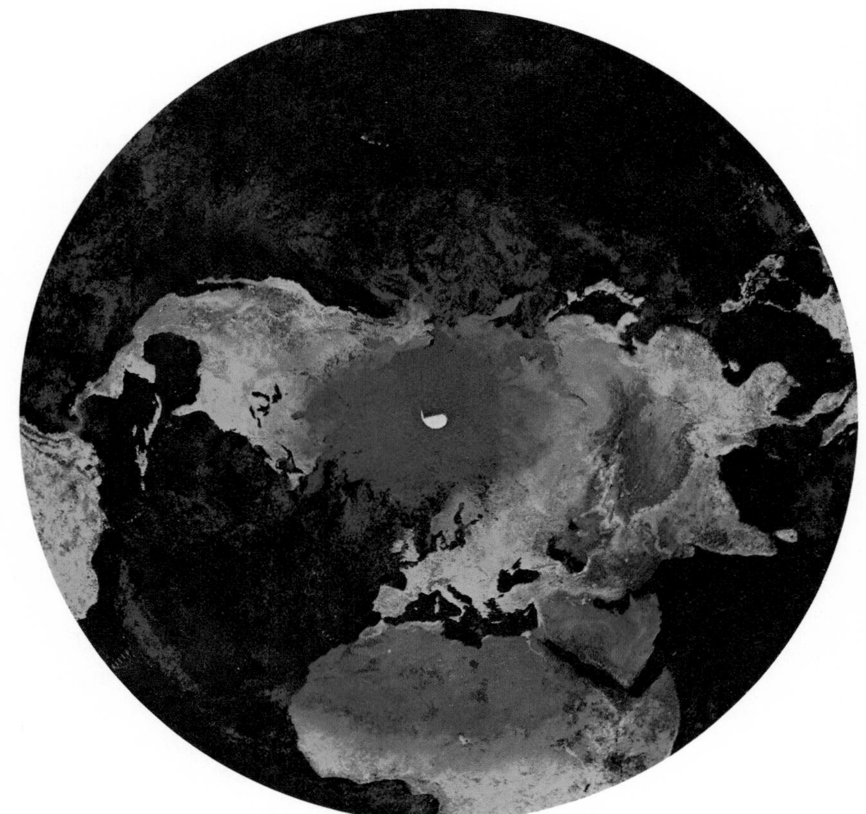

Fig. 3.41. Mosaic of NOAA AVHRR images combining vegetation indices for May and August 1983 in red and green with visible reflectance for May in blue. (Courtesy of ERIM.)

Figure 3.41 expresses this seasonal change in most graphic fashion. It too derives from the regular global acquisition of NOAA AVHRR data, this time incorporating information from May and August 1983. May is the beginning of spring at high latitudes, the end of the wet season in West and Central Africa, and the end of the dry season in East Africa. August is the height of summer, just after the main rainy season in East Africa. In the image the May vegetation index is red and the August index is green, while the blue is controlled by visible band reflectance for May. This means that areas showing as blue were not vegetated at all in the period covered, red areas were vegetated only in May, green was vegetated in August, and areas where plant cover was complete in both months show as yellows. The true deserts are very strongly picked out, as are the areas of permanent humid temperate and tropical vegetation. The magenta area surrounding the North Pole and in high mountains represents the melting of snow to reveal the meagre vegetation of the tundra. Areas expressed in reds and greens at low latitudes are those of seasonal plant cover and agriculture, often prone to drought and famine. The red swathe south of the Sahara in Africa is the so-called Sahelian belt, which together with the green area around the Horn of Africa, was devastated by famine in 1984, due partly to crop failure in 1983. Once again the great irrigation schemes of the Nile and Indus show in startling contrast to the deserts surrounding them. Images such as these are being collated on a regular basis to monitor the effects on sensitive areas of vegetation of annual climatic changes. Although they cannot be used effectively to predict and locate accurately areas likely to be hit by drought, they contain sufficienty information for strategic plans to be made for relief when famine follows in its wake. Equally as importantly, they help monitor the long-term changes in the climatic zones, which for example are causing the southern limit of the Sahara to migrate steadily southwards.

The control over vegetation by elevation and the higher rainfall that high mountains induce is well expressed by Fig. 3.42. It shows an area in the north of Tanzania, most of which is savannah grassland in the lowland terrain. At the time of year involved, January 1976, the grasses, so dense just after the wet season, are dry and lacking the typical reddish response due to cell moisture content. The sparse acacia trees that dot the landscape are bare of leaves. The lowlands appear greenish-grey, except along the stream courses where some grasses are still green and more dense trees still have their leaves. The aridity of this climatic zone during the dry season is demonstrated by the yellows, pale greys, and light blues of the salt-pans of Lake Natron at the right. There evaporation precipitates salts, mainly sodium carbonate, from the heavily mineralized waters draining this intensely volcanic area in the East African Rift. That the area is indeed volcanic is spectacularly shown by the roughly elliptical outlines of many of the mountains, the radial drainage from them and the circular craters near their summits. The mountains are picked out by the red shades that increase in intensity progressively up their flanks. This is due to the higher rainfall and lower evaporation at higher elevation, which induce denser cover by broad-leaved trees and a longer growing season for grasses. It is to some of these areas that the great herds of herbivores migrate during the dry season from the baked plains. The best known of these refuges from drought is the Ngorongoro Crater at top left, whose mottled pink and red tones reflect dense grassland. Above Ngorongoro, at the very edge of the scene, is the basin of the Olduvai River, famous for its evidence of habitation by early hominids. Not all the mountains are well vegetated. The area immediately to the south of Lake Natron is mainly blue-grey, with only a few patches of red vegetation. It is the recently active Ol Doinyo Longai volcano, whose lava fields have yet to be broken down by weathering to form soils capable of supporting dense vegetation.

Immediately after the wet season, savannah lands present a completely different picture. Green vegetation is abundant and the streams flowing with water. Figure 3.43 is from the

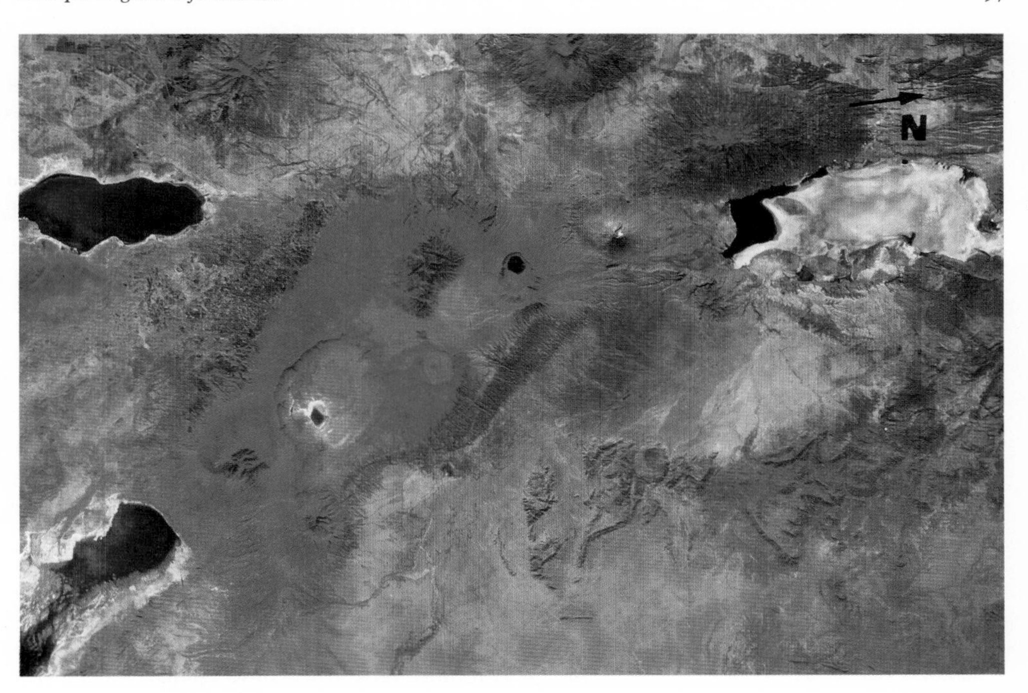

Fig. 3.42. Landsat MSS false-colour image of part of the Rift Valley of Kenya and Tanzania. Width is 180 km. North is at right. (Courtesy of Regional Remote Sensing Facility, Nairobi.)

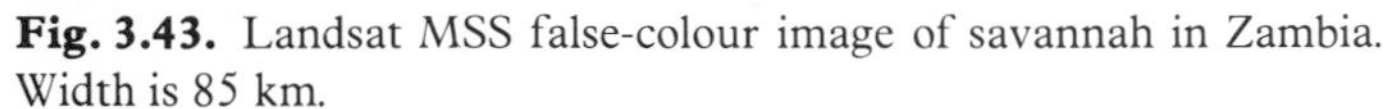

Fig. 3.43. Landsat MSS false-colour image of savannah in Zambia. Width is 85 km.

Fig. 3.44. Dry-season Landsat MSS false-colour image of the Palni Hills, south India. Width is 170 km.

north of Zambia in an area of mixed scrub forest and savannah. The pink and reddish tones clearly reveal the presence of sparse but thriving vegetation. Once again, there is a stratification of plants with elevation. The brightest reds are restricted to the rounded interfluves, which are covered with open deciduous forest. These give way to pinkish grasses nearer to the stream courses. Many of the streams are picked out by dark grey tones due to waterlogged soil in which grass has yet to develop so soon after the rains. In the central parts of the image, minor stream courses are shown in yellow, which is due to the exposure of bare unvegetated red soil as a result of erosion. The speckle of grey patches along the roads towards the top and lower centre of the scene is evidence of clearance for small-scale farming.

Pressures of population and its demands for agricultural land and fuel, and the harvesting of forest for timber have severely depleted the natural vegetation of the Indian subcontinent. One of the few remaining stands of upland tropical rain forest and refuges for wildlife in India is in the uplifted massifs of its southern states. Figure 3.44 shows part of the Palni, Anaimalai, and Cardamom Hills in Tamil Nadu and Kerala. These ranges rise precipitously from the surrounding plains to 2500 m and have annual rainfall as high as 3 m during the south-westerly and north-easterly monsoons in June and October–November. This intense precipitation has carved deep gorges and given the terrain an extremely jagged form. The remaining areas of natural jungle show in this dry-season image as intense reds in the most rugged and least densely populated areas. They stand in sharp contrast to the relatively bare plains, where yellow surfaces are due to red, iron-rich lateritic soils, especially in the Kumbum Valley in the centre. Darker grey areas are underlain by fertile, highly organic black cotton soils. The brownish area of hills to the west of the main mountains are plantations of rubber, cardamom, and coffee. Irregular areas in the central and northern parts of the main massif have been logged in the first case. In the second, rolling grasslands at the highest elevations have been planted with dense stands of Eucalyptus, an entirely alien set of species, whose vertically hanging leaves reflect a much lower proportion of solar radiation than those of the indigenous trees, dominated by broad-leaved species, mainly figs. Other red areas in the scene, in the plains, are irrigated areas of rice.

One of the problems confronting the remote senser interested in the great tropical rain forests, such as those of Amazonia, West Africa, and Indonesia, is that they are frequently covered in thick cloud. Radar imagery is often the only available source of information for large areas. In this case useful information is not provided by the spectral reflectance of the radiation used, in terms of the molecular structure and composition of the surface, but by how its roughness determines the amount of radar energy that is scattered back to the radar antenna. Rough surfaces return a high proportion of the energy, while those that behave as smooth to the particular radar wavelength used reflect most of the energy away to the side. Figure 3.45 shows part of the coastal forests of Equador, and the effects of agricultural development. Two versions are displayed, the left being the raw radar image, the right having been specially treated to express in colour the variation in roughness. The brightest areas in the left-hand version relate to the buildings in the city of Guayaquil and appear yellow in the colour version. Areas of intermediate roughness are coded in magenta, while the dark, smooth areas range from green to black. The purple and yellow areas outside the city are natural forest, the yellow indicating stands of large tress with irregular canopies while magentas relate to smaller trees with a less irregular canopy. The green areas are nearly all cleared areas where grass is managed for cattle rearing, and are clearly demarked by the colour-enhanced radar image.

Even where skies are frequently clear, radar images can provide useful information about vegetation that is unavailable from any other source. Figure 3.46 is a radar image

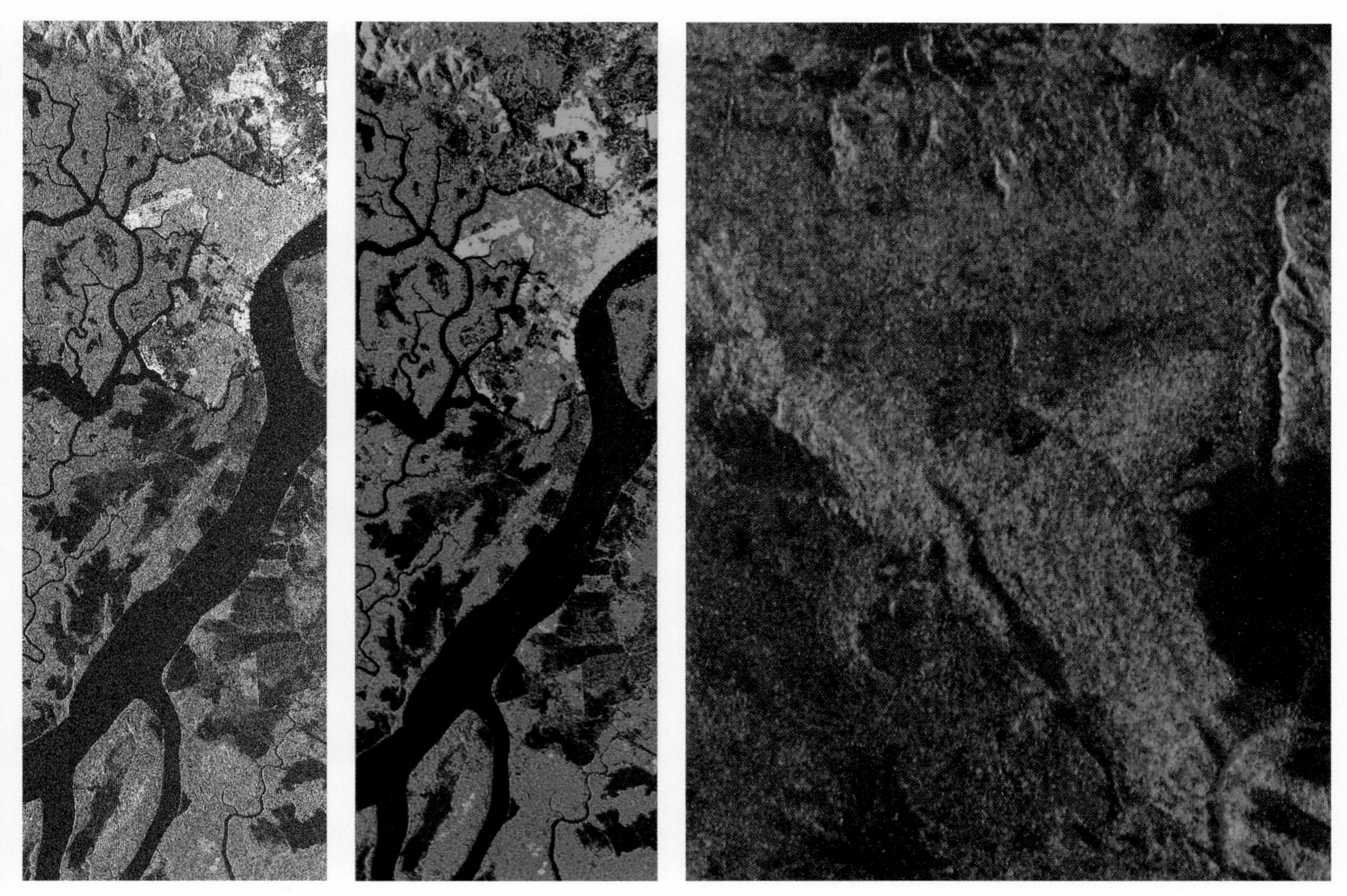

Fig. 3.45. SIR-B radar image of tropical rain forest around Guayaquil in Equador. On the left is the raw data, on the right an enhanced image expressing roughness of the vegetation canopy in colours. Width is 20 km.

Fig. 3.46. SIR-B image of wooded savannah in Argentina, in which the red, green, and blue components are controlled by radar image data produced by different illumination angles. Width is 15 km.

compiled from data produced by illuminating the surface of an area in Argentina at different angles. The steepest illumination image controls red, that at an intermediate angle is used for green, and the smallest angle of illumination is in blue. Different canopy structures among the various forest types produce different effects at different incidence angles, giving rise to the colours on the image. The reddish, pale blue-grey, and darker grey broad areas are characterized by different species of the same genus of tree. The white returns from the valleys are willows with a good water supply. The black area at the right is a mountain with snow cover. The discrimination is not based on leaves, which were absent at the time the data were captured anyway, but on the branch structure of each tree. This is often a surer means of identifying a species of tree than the spectral properties of its leaves.

The richest areas of vegetation cover and animal habitat, but also the most changeable, are wetlands. They are subject to annual or occasional flooding, and equally are prone to drying out during periods of drought. Because they form areas of swamp, yet have both high water supplies and fertile soils, they often form targets for reclamation and economic use, and are thereby threatened. Figure 3.47 shows part of the course of the Niger River, which rises in the humid tropical highlands of Guinea and flows north-eastwards into the semi-arid desert of Mali, shown here, before making a great sweep southward to drain into

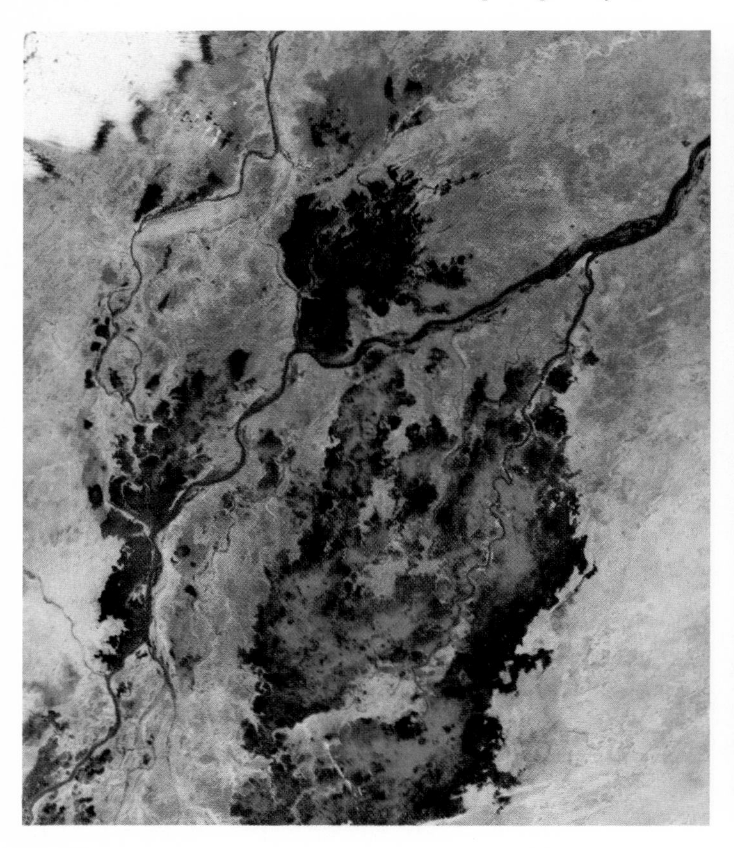

Fig. 3.47. The inland delta of the Niger in Mali, shown on a Landsat MSS false-colour image. Width is 120 km.

Fig. 3.48. Mangrove swamps of the Coburg Peninsula, northern Australia on a Landsat MSS false-colour image. Width is 120 km.

the Atlantic in Nigeria. In Mali, floods from the Guinea Highlands cannot be contained by the channel of the Niger, flowing from south-west to north-east, and spread out over a large area to form an inland delta. Formerly the area was a huge lake, which subsequently spilled to carve the present course of its lower reaches. The area lightly watered by occasional floods is tinged red, in marked contrast to the pale creams of the surrounding desert. The main area of perennial swamps is in reds and dark reds, and has an intricate network of channels. Currently the inland delta of the Niger is shrinking as desertification in this part of the Sahelian region spreads.

Coastal wetlands in the tropics are dominated by mangrove swamps, seen on Fig. 3.48 of the Coburg Peninsula in northern Australia as the dark red patches at the heads of some of the irregular inlets on this drowned coastline. Here even the mangroves, which are highly tolerant of saline water, are restricted in their extent by extremely high evaporation in the shallow tidal inlets that produce hypersaline water. The lighter areas fringing the coast are salt-encrusted tidal flats. Inland, very little drainage is visible due to the extremely low relief and low rainfall. The red-fringed pinkish areas with an irregular outline are areas of higher ground underlain by a thin sandstone, which has a sparse Eucalyptus and Acacia cover. Darker patches are areas of residual laterite soil.

Perhaps the largest remaining areas of little-disturbed natural vegetation are those of the northern continents above about 50 degrees north, in the boreal forest and tundra. They dominate the areas formerly covered by ice sheets from about 2 million until 15 000 years ago, when all previous vegetation was stripped and the land resurfaced by glacial debris. Much of this terrain has low relief, but because the ice exploited many subtle features in the bedrock beneath, it often has an intricate structure, as described in Section 3.2 (Fig. 3.27). Figure 3.49 is of the boreal forest in Northern Canada, where the overall orange tinge indicates that the land surface is comprehensively coated with vegetation of one kind or another. However, the bright reds so typical of grasslands and deciduous forest are not present. This is because the main cover is moss and lichen, with low chlorophyll and an anomalously low very-near infrared reflectance. The tree cover is dominantly coniferous, the needles of which give a low response on such basic false-colour images. The most dense tree cover is indicated by the slightly darker orange tones of open black spruce, the very pale areas being mainly sandy material with low vegetation cover. Lakes divide into two types: those with clear water which show as black; while those with a high algal content are rendered in various blue-greens. Some of the smaller dark patches are brownish in colour, and are bogs formed in what were small ponds, which support a dense cover of dwarf black spruce.

Visitors to the Highlands of Scotland are often convinced that the vegetation, which forms one of the main elements in making them unique, is entirely natural. This is not so, most of upland Scotland having been covered by open coniferous forest until only a few hundred years ago, when progressive deforestation produced the present terrain bare of trees. Regrowth has been prevented by the intensive browsing of tree seedlings by sheep and by the probably permanent alteration of soil chemistry by extensive leaching. Figure 3.50 however, reveals the great variability of the ground cover in one of Britain's only 'wild' areas. This specially processed image based on principal component analysis of the six

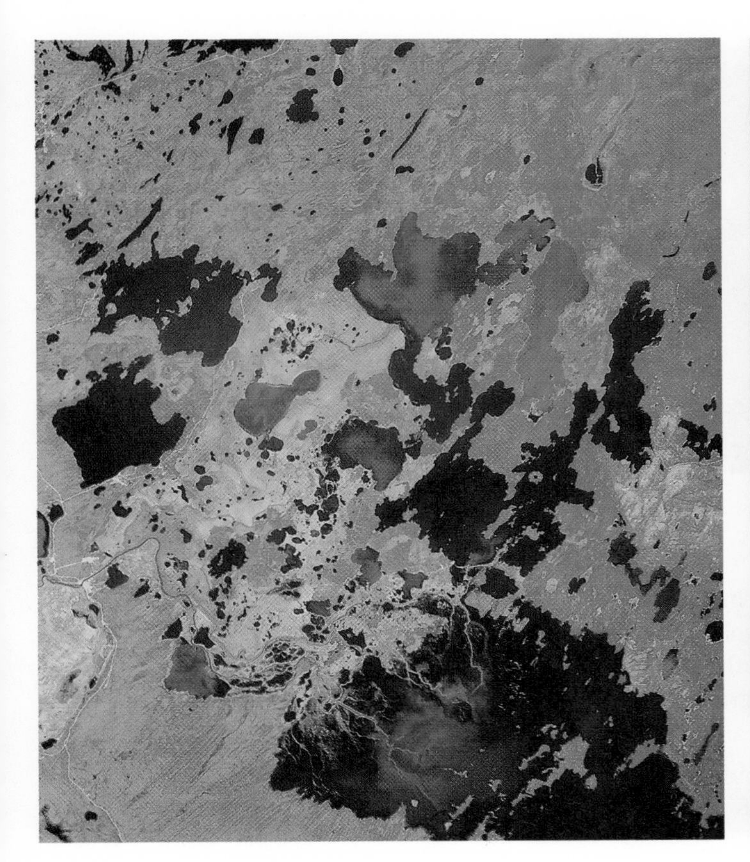

Fig. 3.49. Part of the boreal forest of northern Canada on a Landsat MSS false-colour image. Width is 110 km.

Fig. 3.50. Specially enhanced image of airborne multispectral-scanner data from northern Scotland, based on the results of principal component analysis. Width is 3 km.

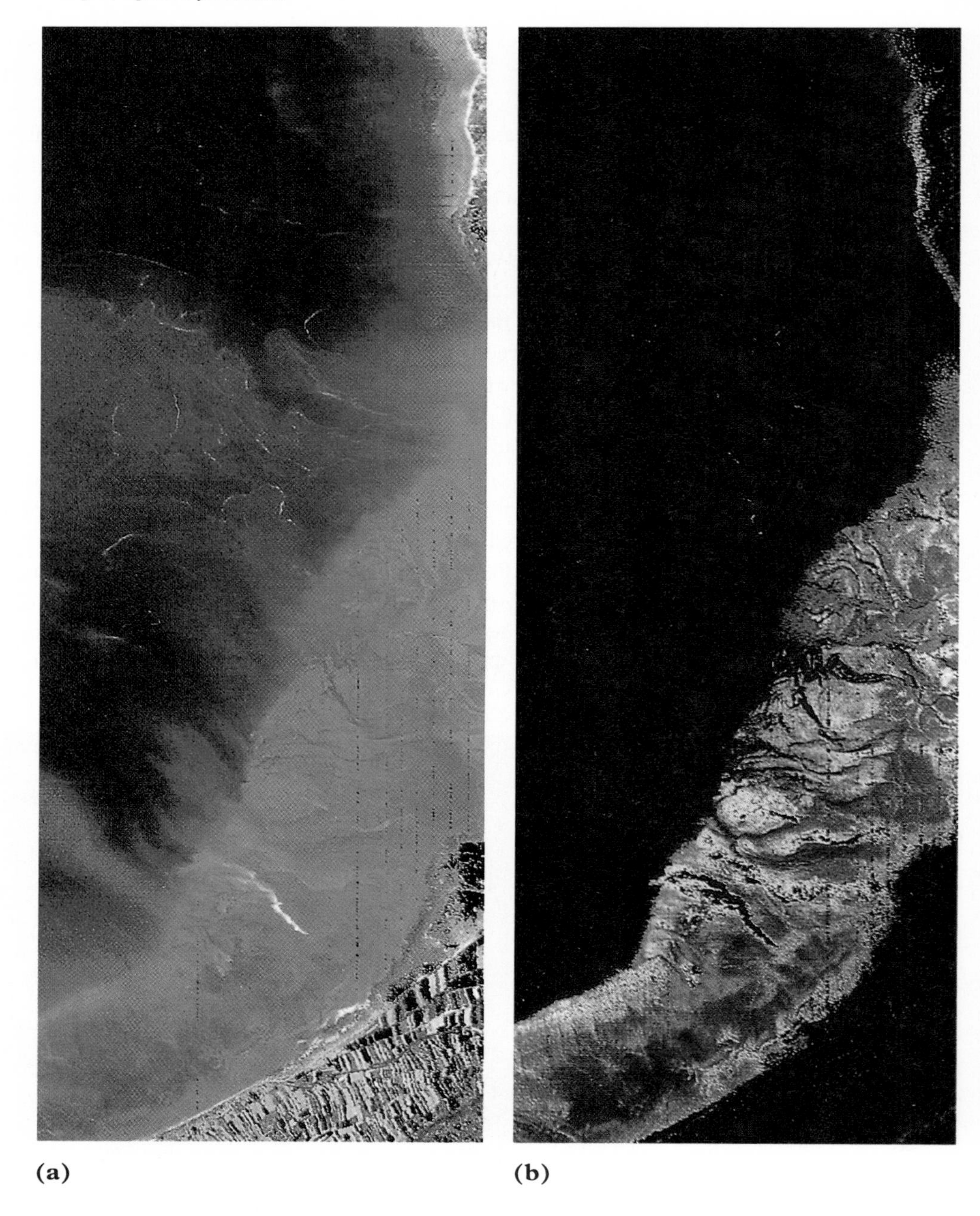

Fig. 3.51. Part of the coast of Brittany, France, shown on a simulated SPOT image. On the left is a normal false-colour image. The right-hand image has been subject to special enhancement to highlight the variations in algal cover in the intertidal zone. Width is 3 km. (Courtesy of CNES.)

reflected bands captured by the Landsat Thematic Mapper renders the various vegetation types in very different colours from those to which the reader will by now have been accustomed. Various heaths covering poorly drained ground are shown in greens, dark browns, and dark blues, grasses appear in a variety of magenta, orange, and yellow tones. Bare rock and scree appears as bright cyan.

A little appreciated fact is that a significant proportion of the plant matter living on the Earth is distributed along the coastal fringe as seaweed and other types of algae. These plants do not respond in the same way as land plants in the reflected part of the spectrum, partly because they are covered by water at high tide, and partly because of their different pigmentation compared with land plants. Figure 3.51(a) is a standard false-colour image of the coast of France in Brittany, near Mont St. Michel. The agricultural vegetation and newly ploughed fields, and the patterns involved, show as a patchwork of reds, creams, and blacks at bottom right. The irregular fringe of red along the coast is seaweed left as the tide falls, but whatever variations in algal cover are present below the tide level are obscured by the green tinge imparted by shallow water. The right-hand image has been selectively contrast-stretched to enhance the variations there and to remove the spectral effects of the shallow water. In a dramatic way it reveals the complex distribution of many different algal species, including stands of oar weed in red, growing on rocky reefs, and algal stratification on the tidal flats through which several distributary channels meander. The greens are algae developed on silts, while the blue colours are thinner, more slimy algae developed on the muds in the lower energy part of the coastal strip.

3.5 Use of the land

A myth born of the Apollo programme is that the only evidence from the Moon of human activity on the Earth is the Walled City in Beijing. I personally doubt it, although this ancient and huge enclosure is a geometric square, and so pretty obvious. From the altitudes of the satellites that monitor the surface it is difficult to miss abundant evidence of our activities. The land is used more or less intensively wherever we look, except for the polar areas, deserts, and some of the great mountain ranges. In the case of pastoralism and some kinds of forestry the natural environment is exploited in its 'raw' state. Although this induces changes—often dramatic, as described in the next Chapter—few regular patterns are imposed. So, at any one time such activities can go unnoticed, as the irregularities of vegetation and terrain are undisturbed by obvious artificial patterns. Because the eye is attuned to smooth lines, simple shapes, and sharp boundaries, roads, canals, fields, and clearings are very easily seen. In some cases, the cultural feature need not necessarily be as large as the minimum size specified by a system's resolution, provided it is very different from its natural surroundings or arranged in a clear geometrical shape.

One of the largest and oldest centres of civilization is in one of the most arid regions on Earth. The reason for its rise and continued existence is the continuous flow of the Nile river, fed by rainfall in Ethiopia and the eastern part of Central Africa. Annual flooding naturally irrigated its narrow floodplain and the classic triangular delta at is outlet into the Mediterranean, as well as giving a continual supply of fertile soils. At the apex of the Nile delta stands the ancient city of Cairo, shown on Fig. 3.52 as an irregular dark patch with a tracery of roads. The Nile shows clearly as a black ribbon. On this simulation of natural-colour produced by the Landsat-4 Thematic Mapper, the limits of land that can be irrigated are picked out sharply by the greens of thriving crops and the pale tones of the barren desert at elevations beyond the reach of water supplies from the river. A large number of irrigation channels, both ancient and more recent, are clearly visible. The smaller ones subdivide the agricultural land into a patchwork of regular fields. Two features highlight the span of social history. At the lower left the Pyramids of Gisa are just about discernible by their triangular shadows, while at the upper right can be seen the arrow shape defined by the runways of Cairo International Airport.

The Nile delta was not the only cradle of civilization, but all the others in China, Mesopotamia, and elsewhere were associated with great rivers. The Indus, on which the Harappa civilization developed more than 4000 years ago, drains the western Himalaya, but formerly most of its flow debouched into the Arabian Sea, only a narrow strip along its course being vegetated as it crossed the Thar desert, the driest region on the Indian subcontinent. At the time of the Raj, a massive programme of canal building and irrigation transformed the desert into a highly fertile plain (Fig. 3.53). Without it the modern state of

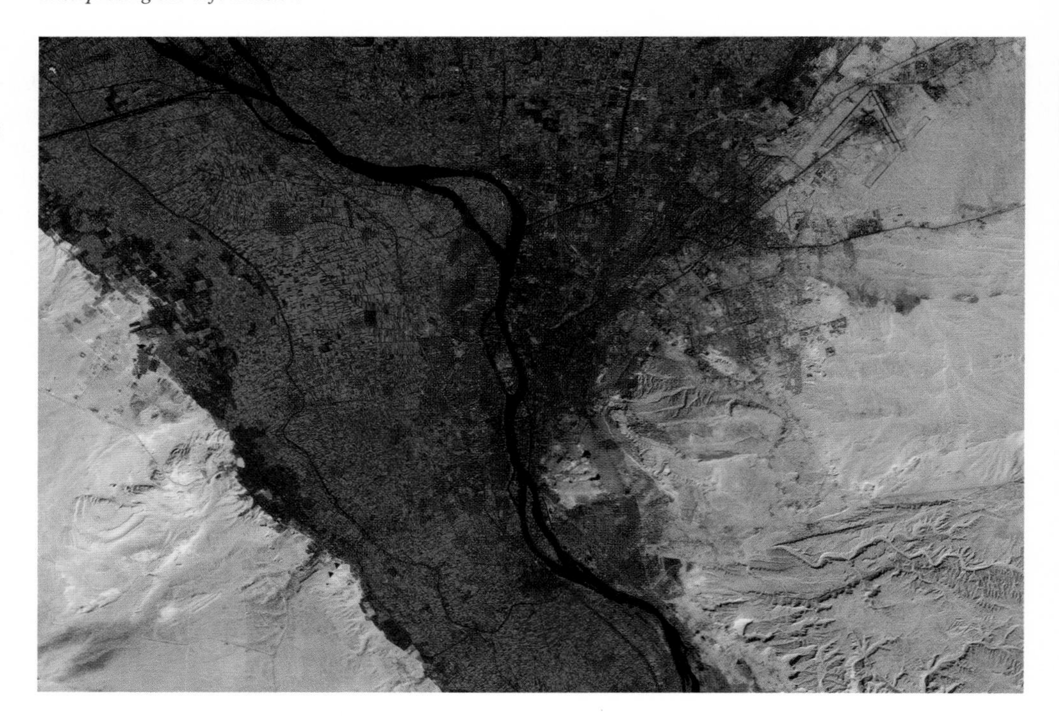

Fig. 3.52. Simulated natural-colour Landsat TM image of part of the Nile delta near Cairo. Width is 70 km. (Courtesy of ESA.)

Fig. 3.53. NOAA-6 AVHRR false-colour image of Pakistan, and parts of Iran, Afghanistan, and India, showing the irrigated Indus plains. Width is 1500 km.

Fig. 3.54. Centre-pivot irrigation scheme near Riyadh, Saudi Arabia, shown on a false-colour image from the experimental German MOMS system, flown on the June 1983 Space Shuttle. Width is 15 km. (Courtesy of DFVLR.)

Pakistan could never have existed. Although the resolution of this false-colour image is low, the broad red area of flourishing crops stands in contrast to the stark, pale-yellow sands of the desert.

Another method of irrigation in arid lands is based on tapping and pumping of ground-water by deep bore-holes. Although the flow can be canalized to irrigate conventional fields, an attractive alternative is to use the water directly from the bore-hole. This is achieved by a system called centre-pivot irrigation, where a straight, rigid pipe is carried on a series of wheels driven around a central pivot by motors. This radial arm sprinkles water over a circular area of crops. Such a system developed in the very arid central part of Saudi Arabia, near Riyadh, is shown on Fig. 3.54. Each of the curious circular fields is fed by a bore-hole penetrating deeply into the permeable rocks beneath, shown outcropping as a series of tonal strips. The main crop is lucerne, used as fodder for cattle and sheep, although the system is suitable for growing vegetables and cereals. Such irrigation is costly, both for establishing the wells and in driving the sprinkler system around. Moreover, when used in such an intensive way in an area of extreme aridity, pumping depletes the available supply of ground-water and may exhaust it. The danger is that eventually the investment in equipment is lost. Since ground-water in arid regions is often saline, continued irrigation gradually enriches the soil in salt. Unless heavy rainfall flushes this contaminant away, the soil eventually becomes infertile, except for salt-tolerant types of crop.

Seasonal climates, such as the monsoonal areas of the Indian subcontinent are often characterized by two types of farming. During the rainy season the areas between valleys—the interfluves—become saturated. This allows the planting of crops there, such as wheat, sorghum, millet, and maize. Only one rain-fed, interfluve crop is usually possible, which comes to maturity early in the dry season. In the southern part of the Indian subcontinent, two are sometimes possible, watered by the south-westerly monsoon in May and by that from the north-east in October and November. In Fig. 3.55, showing the Mewar plains in Rajasthan underlain by granitic gneisses, thriving vegetation (mainly wheat and vegetables) shows in reds, in a complex pattern defining the dendritic stream systems in this poorly drained area. The image was taken during the dry season, and the crops are watered by shallow ground-water and some surface flow in the ephemeral streams. Rajasthan, being served by only one weak monsoon, supports just one wet-season crop on the interfluves and a single crop in the valleys during the dry season.

In the wetter Mysore Plateau of South India, shown in Fig. 3.56, a dry-season image reveals much the same dendritic pattern of valley cultivation, in this case dominated by rice. The large red patches at bottom right show the benefits of irrigation canalized from the large Krishnarajasagar dam across the Hemavathi River, to the north of Mysore City. In this gently rolling scenery, straight canals are not easy to construct. Instead, the flow is along contours on the north bank of the Hemavathi drainage basin. Irrigation water is channelled into existing small valleys, greatly increasing the area of land under cultivation during the dry season. The main crops are sugar cane and rice. Although the area has a high population density, and most of the land is farmed, at this time of year all the interfluve crops have been harvested to reveal the nature of the underlying soil. The yellowish areas are underlain by gneisses, the white represents granitic intrusions, and the dark streaks are complex bodies of sediments and volcanic rocks that have been caught up in the gneisses and intensely deformed.

Fields and gardens smaller than about 50 m across are virtually impossible to map and evaluate from orbit. In fact they may not be detectable, masquerading as natural vegetation. However, in most developed countries, fields are large and easily recognized. How the land is divided governs the shapes and patterns of fields. In Britain, where most of the fields date back to medieval times, boundaries are irregular, though sharp. In Fig. 3.57

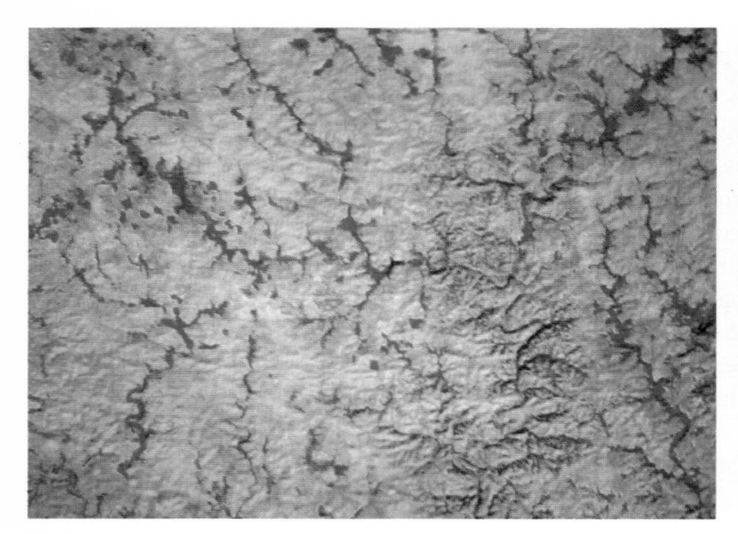

Fig. 3.55. Valley agriculture of Rajasthan, India shown on a SPOT false-colour image. Width is 10 km. (Copyright CNES 1986.)

Fig. 3.56. Irrigated area on the Mysore Plateau, South India, on a Landsat MSS false-colour image. Width is 50 km.

Fig. 3.57. Irregular fields, roads, railways, and airfields in Hampshire, England, shown on a false-colour Landsat TM image. Width is 20 km. (Courtesy of NRSC, RAE Farnborough.)

Fig. 3.58. Contrast in land use across the USA–Canada border, shown on a Landsat MSS false-colour image. Width is 85 km. (Courtesy of CCRS.)

this irregularity is easily seen, although before the enclosures of the late Middle Ages the land in each field was broken up into narrow strips. The pasture fields and some of those planted with winter cereals show up as bright reds. Crops just beginning to sprout are in oranges and yellows, whereas newly ploughed fields have a distinct blue tinge. The latter are confused with the many towns and villages in the scene. Patches of green are areas of remaining natural woodland in an area almost totally dominated by human activities. As well as the field patterns, many other signs of human activity are visible, such as the straight railway running ENE–WSW, the highway close to it, and an airfield at top right.

In North America, agricultural development is much more recent and well-organized than in Europe, the land having been apportioned in a regular system defined by latitude and longitude. Each holding is approximately a geometrically exact square. Figure 3.58 is of an area straddling the Canadian–US border separating Alberta and Montana. Outside the deeply incised rivers, whose flanks are affected by soil erosion to form intricate dendritic badlands, the square holdings and divisional roads are easily seen. Each farmed holding is divided into parallel strips with crops at different stages of growth. The differences in population density and farming practice in Alberta and Montana result in the 49th Parallel standing out sharply in the otherwise identical topography. Areas of natural vegetation along the streams are brighter red than the farmland. In the unfarmed parts of Alberta vegetation differences define a NW–SE grain to the land reflecting the direction of glacial flow during the last Ice Age. Several of the larger river valleys are flanked by irregular bright areas. These are zones of active soil erosion forming badlands. The marked linearity of one of the rivers is probably controlled by a fault in the underlying bedrock.

Rangeland, used for grazing livestock, is usually in areas of natural grassland, except in lowland parts of intensively farmed terrain, where stock are generally grazed in individual fields of managed pasture. Consequently, actively grazed rangeland is difficult to discriminate from unused areas. However, in some areas, attempts are made to improve the browse, resulting in easily recognized signs. Grasses in savannah lands grow rapidly in the rainy season, often reaching two or three metres in height. This growth is tough, and when dry almost useless as fodder. Moreover, the fallen dead grass hinders regrowth after rain. A common solution is to burn off the dead grass at the end of the dry season, so allowing new tender shoots to grow freely after rain. The scars left by this burning are quite distinctive, as shown on the right side of Fig. 3.59, an area of nomadic pastoralism in southern Sudan. Fires spread towards the direction of the prevailing wind, so forming elongated streaks with roughly straight sides and a scalloped windward edge. Recent burn scars are marked by tones (8), while earlier ones are pale or pinkish as new grass begins to grow. The major sites of grazing from mid- to late-dry season are marked by bright red tones (11), and are perrenial grassland subject to prolonged flooding from tibutaries of the Nile. The yellow area (9) is thick dry grass that has not been grazed, probably because water for stock is scarce. The main areas of active grazing (7) are marked by numerous burn scars, and are nearly devoid of vegetation. They are based on a clay-rich soil that is only wet after the rains. Those areas from which virtually all vegetation has been stripped by overgrazing (6) appear bright blue. The pink area in the bottom left is deciduous forest with green-leafed canopy that is unsuitable for stock grazing.

Most of the upland areas of the British Isles have soils that are too poor and rainfall that is too high to support any farming other than free-range sheep-rearing for wool and meat. Figure 3.60 shows part of the Pennines of northern England. Newcastle and Teesside are marked by dense grey patches along the east coast. The red hues of the lower ground around the greyish Pennine hills represent well-managed pasture. As the east coast is approached, more recently ploughed fields show up as a patchwork of greys among the reds. This indicates the gradual change in land use from the pastoral areas of

Fig. 3.59. Rangeland and welands of the southern Sudan on a Landsat MSS false-colour image. Width is 140 km. (Courtesy of NRSC, RAE, Farnborough.)

the high-rainfall land in the west to the more dominantly arable farming in the drier eastern lowlands. The hills of the Pennines stand out very sharply, as do the patterns in them controlled by the underlying geology, particularly the Pennine Fault towards top left. The greenish and pale-brown colours in the hills signify the presence of upland grass, dry at this time of year (late winter). These grasslands form the main rangelands for hill sheep, but are broken up by large dark-brown patches dominated by heathers. These heathlands grow on areas of organic soil or peat, often controlled by particularly acid and poorly drained soils. They form extremely poor pasture, and consequently attempts have been made to clear the heather and improve the soil. This has led to the present patchy development of this formerly very extensing cover.

Areas of dense natural forest form important resources for timber and a source of wood pulp for paper and man-made fibre manufacture. More open woodland is of less use for intensive industrial exploitation but forms a source of fuel, either wood or charcoal, for domestic use in many developing countries. Forestation depends partly upon soil type but more importantly on rainfall, and so is characteristic of areas where precipitation exceeds evaporation. These conditions are provided in cold boreal areas at high latitudes, in maritime terrains, and in the humid tropics.

Finland is one of the largest producers and exporters of timber products in Europe, being densely cloaked in coniferous softwood forest. Timber operations are closely managed. Figure 3.61 shows part of the forested area in southern Finland, together with the effects of timber operations and the clearance of land for agriculture. Because conifers carry a canopy of small needles with low chlorophyll and moisture content, they do not appear in the same way as most other vegetation on standard false-colour images. Dense stands of softwoods appear dull brownish-green, whereas the cleared areas of agriculture show as bright reds to red-browns. Felled areas which have been replanted to being regeneration of

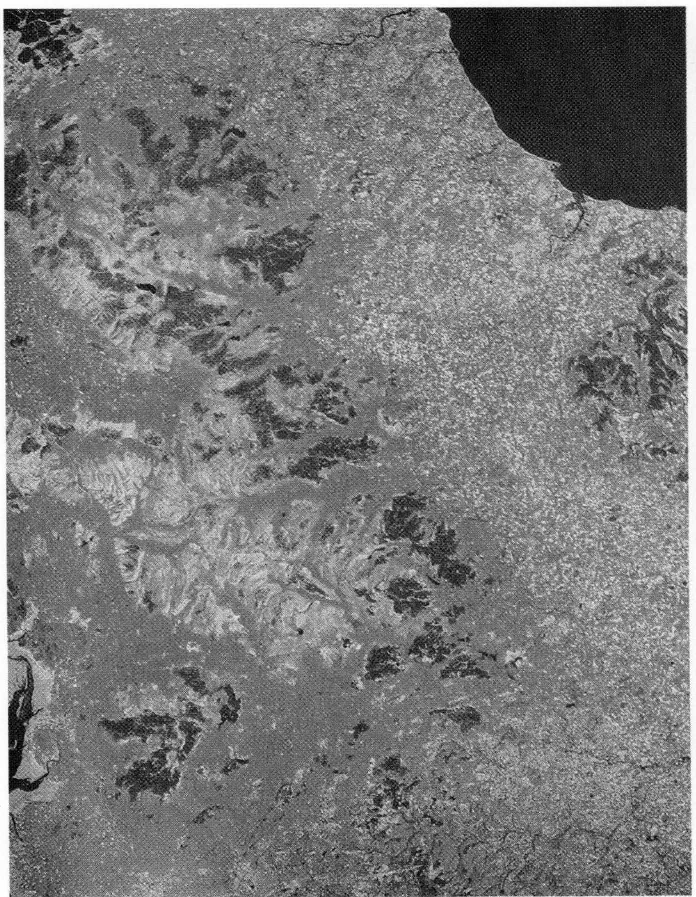

Fig. 3.60. Farmland and uplands of northern England on a mosaic of Landsat MSS false-colour images. Width is 100 km.

Fig. 3.61. Landsat TM false-colour image of managed forest in Finland. Width is 30 km. (Courtesy of Technical Research Centre of Finland and National Board of Survey, Helsinki.)

the forest are shown in yellows. The upper central part of the image contains an irregular area of pale tones that is open peat bog, possibly once a lake that has become filled by floating mosses. Other lakes appear as the irregular and angular dark areas.

Although from an aircraft most of the features of cities, towns, villages, and even individual houses are immediately visible, on the remotely sensed images available publically from satellites such details are rarely apparent. Detecting settled and urbanized areas depends much more on the patterns and textures that human activity creates, and sometimes on the changes wrought on the spectral properties of natural surfaces. The tiles, slates, and concrete of roofs and the surfaces of roads when viewed vertically from above have more in common with rocks and soils in their interaction with electromagnetic radiation. On a standard false-colour image, urban areas appear in the same bluish-greys as bare rock and soil (Figs. 3.44, 3.57, and 3.60), thereby standing out from the reds associated with vegetation. Where resolution is coarser than about 50 m, towns are often confused with areas of bare soil, although roads and railways, by their very sharp alignment, are easily defined (Fig. 3.57).

Figure 3.62 has an effective resolution of around 10 m, due to the combination of Landsat TM with airborne radar data. Particularly because of the 'gridiron' layout of this typical North American city, even the road system of the suburban areas stands out

Fig. 3.62. Landsat TM false-colour image combined with airborne radar imagery to improve spatial resolution, showing part of the Detroit–Windsor conurbation on the USA–Canada border. Width is 10 km. (Courtesy of ERIM.)

sharply, as well as features such as freeways and bridges. Most cities consist of a mixture of buildings and roads, together with open, vegetated spaces. Because of the different spectral properties of these two gross components, it is possible to assess the density of buildings. Here, the densely packed urban centres and industrial areas are in light blues, whereas the more open residential areas contain a brownish component due to gardens and parks. The parks are distinguished from agricultural fields by their more mottled appearance because of their irregular mixture of trees and grass. Patterns and shapes give clues to some of the detail; for instance, the darker area with regular bright spots at top left is part of one of the automobile factories for which Detroit is famous, and the small circular feature at the tip of the island is some kind of monument.

As well as displaying textures, patterns, and spectral features that are unmistakeably the products of human construction and communications, cities and towns have another signature. Their dominance by buildings and streets constructed from stone, concrete, asphalt, brick, and tiles means that they absorb solar energy more efficiently than their surroundings. The construction materials store some of this in the form of heat, so that conurbations become 'hot spots'. Because of the great and highly concentrated consumption of energy in modern cities, some of this artificially released energy also contributes to the thermal anomalies which they comprise. Figure 3.63 demonstrates this unnatural property in a very graphic way for the urban areas of Belgium and the southern Netherlands, using a day-time thermal image from the Heat Capacity Mapping Mission with a resolution of 600 m. Areas with increasingly high surface temperature are shown ranging from green through yellow to red and magenta. The most prominent 'hotspots' are Ghent, Antwerp, and Brussels.

The bulk of the world's population does not live in cities or towns, but in innumerable villages centred on areas of fertile agriculture. In some cases these small centres of population stand out clearly in contrast to their vegetated surroundings, but in others the

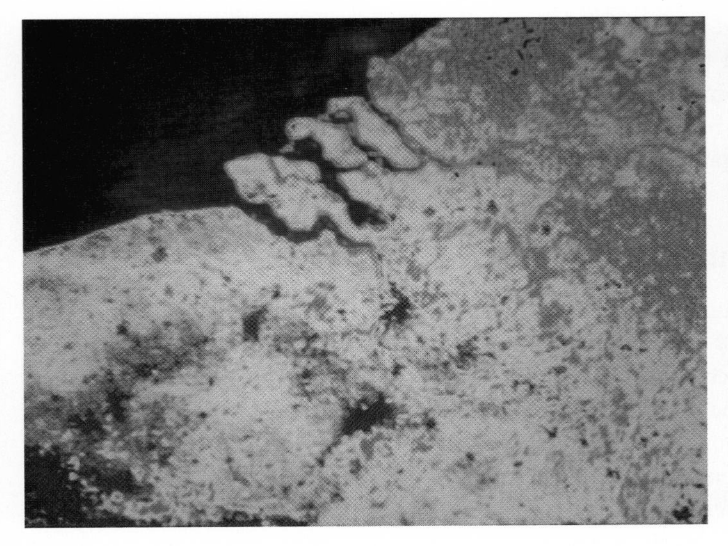

Fig. 3.63. Colour-coded night-time temperature image from the HCMM, showing conurbations in Belgium. Width is 200 km.

Fig. 3.64. Landsat MSS false-colour image of part of Andhra Pradesh, India, showing villages as small grey patches. Width is 50 km.

use of local building materials in villages surrounded by bare soil or their cover by shade trees completely hides them from view.

In Fig. 3.64, showing part of the Cuddapah Basin of southern India, the fertile rain-fed organic soils in the dry season appear almost black, crossed by a dendritic network of sand-filled and sparsely vegetated stream courses. Villages, because of the absence of suitable clays for brick construction are made from quarried stone. As a result they show very clearly as light, almost evenly spaced patches, even though the unpaved tracks that serve and radiate out from them are invisible.

Where villages are hidden by trees or blend imperceptibly with the soils surrounding them, one property that enables even these to be detected is the usually angular nature of buildings and fences. As explained in Chapter 2, corners bounded by flat surfaces at right angles act as extremely efficient reflectors to radar energy. This property is exploited in Fig. 3.65 to reveal the closely spaced network of small villages in the intensively farmed and

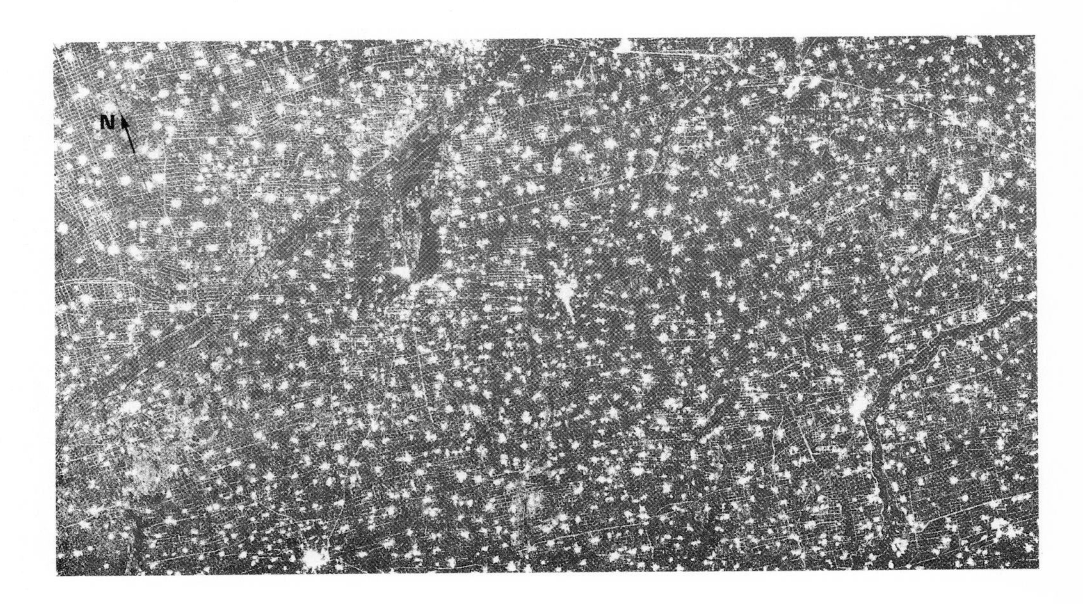

Fig. 3.65. SIR-A radar image showing villages in the densely populated plains of eastern China. Width is 85 km.

densely populated lowlands of coastal China along the Huang He, which replenishes the soil with rich silts derived from the west during annual floods. The density of villages is almost one per square kilometre. Also shown are details of the field systems associated with rice cultivation, enhanced by the effect on the radar signal of the small embankments used to control the flow of irrigation water.

3.6 Beneath the surface

Details of how the crust underlying the Earth's surface is constructed are in most cases hidden beneath a cover of vegetation and soil. Even in arid areas where the surface is composed almost exclusively of bare rock, the distinction between the different rock types is usually obscured by a rind of weathered material, whose colour does not vary a great deal from rock to rock. This is one of the reasons why geologists carry hammers—classifying a rock in the field depends on seeing a fresh surface. Despite all these difficulties, geologists form probably the largest group of consistent users of remote images simply because they give such a wide view of the surface. Seeing the connections over large distances between geological features on the ground is hindered by topography and vegetation, so any clues to continuity and variation over wide areas that an image from the air or from space reveals make the work of a geologist easier and more efficient.

Rocks of different compositions respond to weathering and erosion differently, depending on the chemical composition and physical structure of their constituent minerals, how they are assembled in the rocks, and on the climate. This means that the intricacies of the land surface are frequently controlled by the underlying rocks and the structures in which they are disposed. The longer and more intense the weathering and erosion the greater the geological control over scenery. Crystalline igneous and metamorphic rocks are the most resistant to weathering and erosion, and remain as mountains

Fig. 3.66. The fold mountains of the northern Appalachians in the eastern USA, shown on an airborne radar image. The image is 50 km across. Width is 45 km. North is at top right. (Courtesy of Intera Technologies, Calgary, Canada.)

and hills for much longer than sedimentary rocks. The latter may be cemented or unconsolidated, again producing different responses in the scenery. Rocks that are compositionally layered in a regular fashion, such as sediments, most metamorphic rocks, and some kinds of igneous rock, have the subtleties of their banding picked out by weathering, imparting a more or less clear grain to the scenery. The steeper the layering the straighter and clearer the grain of the country is. Resistant layers form escarpments while weaker rocks form valleys.

Figure 3.66 gives a spectacular example of the control over landscape by dipping rocks with different resistances to erosion. It comes from the Appalachian Mountains of Pennsylvania and Virginia, USA. The topography is picked out by the shadowing effect inherent in radar images, whose illumination is from the top of the figure. The ridges which curve and repeat themselves due to folding are underlain by resistant sandstones. Their bright signature stems from the upland forest with a rough canopy. The darker valleys are dominated by smoother pasture, and have developed in shales and limestones that have been eroded more easily.

Fractures in the crust, in the form of faults across which large masses of rock have moved relative to one another, produce zones of altered and shattered rock that are easily exploited by erosion. Where faults are steeply dipping they outcrop as sharp, almost straight lines. Because regular patterns are so rare in nature, steep faults are probably the most readily identified geological features on remotely sensed images. When they dip at shallow angles, their outcrops are more intricate and do not exert such a strong control on the appearance of the terrain from above.

Figure 3.67 shows one of the great faults that has taken up the crustal motions associated with the driving of the Indian subcontinent into Asia as a result of sea-floor spreading in the Indian Ocean. It is the Kun Lun Fault, many hundreds of kilometres from the actual site of the collision along the Indus–Tsang Po Valley. The displacement that has been taken up along the Kun Lun Fault is mainly sideways slip of blocks of crust past each other. The line of the fault is weaker than the rocks in the blocks that it separates because of intense fracturing, and so it has been picked out as a narrow valley. Moreover, movement is still

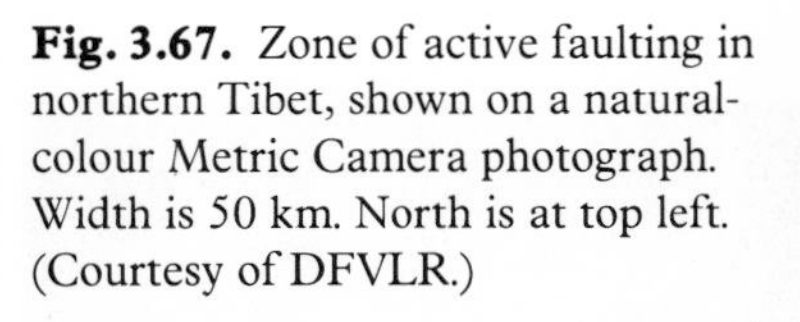

Fig. 3.67. Zone of active faulting in northern Tibet, shown on a natural-colour Metric Camera photograph. Width is 50 km. North is at top left. (Courtesy of DFVLR.)

going on along the fault, as can be seen from the numerous examples of small stream courses that appear to have been displaced. Spurs from the mountains bounded by the fault terminate suddenly at the straight line, indicating that they have been sheared by faulting as the landscape has been evolving. In a few cases, small ridges are present along the line of the fault that are directly across the trends of stream courses. These *shutter ridges* are slices of resistant material that are being transported laterally by the fault movement, partly blocking drainage. In other cases, such ridges may dam lakes.

The keys to unravelling the geological evolution of an area using remote images are those traditionally used by field geologists. A fault or an igneous intrusion cutting through a series of rocks is a clear marker of relative timing. Rocks and structures that are cut are older than the fault or intrusion. Where other rocks overlie or mask the time markers then they are younger, and any folds, faults, or intrusions affecting them in turn are younger still. Sometimes it is possible to work out the angle and direction at which a layered sequence dips, and thereby distinguish between layers that are deep and older and those which were laid down higher and later in the sequence of events. In the case of several episodes of crustal deformation, it is often possible to distinguish earlier structures from those produced at a later date from the overall shape of the outcrops that have resulted.

In Fig. 3.68, which is of part of the state of Andhra Pradesh in India, the darker area is a basin of 1500 Ma old sediments in the Cuddapah Basin, the curved features being boundaries between different sedimentary layers all dipping gently to the ENE. Red patches are areas of healthy vegetation. In the SW corner is a patch of yellowish surface crossed by a number of dark lines. This is much older rock cut by vertical sheets of igneous material, or dykes. The dark dykes stop abruptly at the curved boundary between the yellow and dark surfaces. The boundary dips gently to the ENE, so the dark rocks are above it and the yellow ones are below, and are older. The truncation of the dykes at this boundary means that they were intruded after the formation of the yellow rocks, but were eroded down to a flat surface before the deposition of the dark rocks. The boundary is an *unconformity*—a relative time marker. Just above the prominent white sediments of the river at left centre, a series of irregular blue rocks in the basin is displaced by a nearly E–W straight fault line. So this fault is younger than the sediments. On the right-hand side of the image, features in the sediments run NNW–SSE, and include several prominent folds, showing that the basin was affected by a compressional event after the sediments in it were

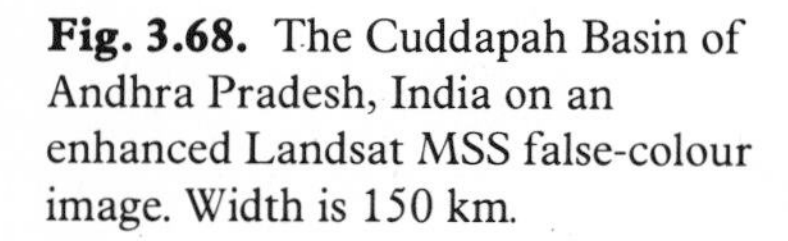

Fig. 3.68. The Cuddapah Basin of Andhra Pradesh, India on an enhanced Landsat MSS false-colour image. Width is 150 km.

laid down. Clearly, this image can help reconstruct quite a lot of the geological history of the area without the need for detailed field mapping.

The composition and internal structure of rocks often control the finer details of the surface, particularly the drainage patterns that develop upon them. Rocks that are impermeable to surface water, such as clay-rich sediments or crystalline igneous and metamorphic rocks, retain all rainfall at the surface and so have fine, intricate drainage patterns. Rocks that allow some water to seep into them are less dissected by the flow of streams and have a more open network of channels. Some rocks are extremely permeable, either because they formed with wide cracks and open cavities, as in the case of some lavas, or, in the case of limestones, have been partly dissolved away by weakly acid rain-water to form underground cavern systems. Surface water soon disappears below ground and no continuous streams are able to carve regular drainage systems. Because of these simple differences in the response of rocks to the flow of surface water it is commonly possible to recognize important geological boundaries by inspection of the varying texture of the land surface. Figure 3.69 is a SIR-A radar image of part of the densely forested uplands of Irian Jaya, generally enveloped in cloud. The gross texture of the landscape is very well displayed.

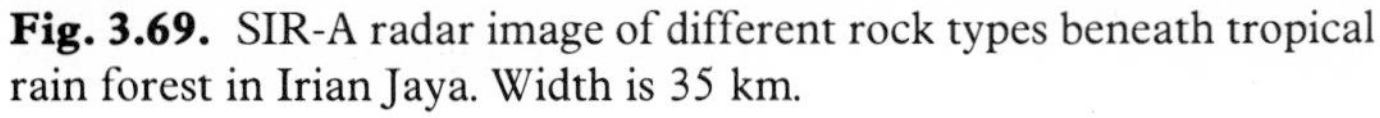

Fig. 3.69. SIR-A radar image of different rock types beneath tropical rain forest in Irian Jaya. Width is 35 km.

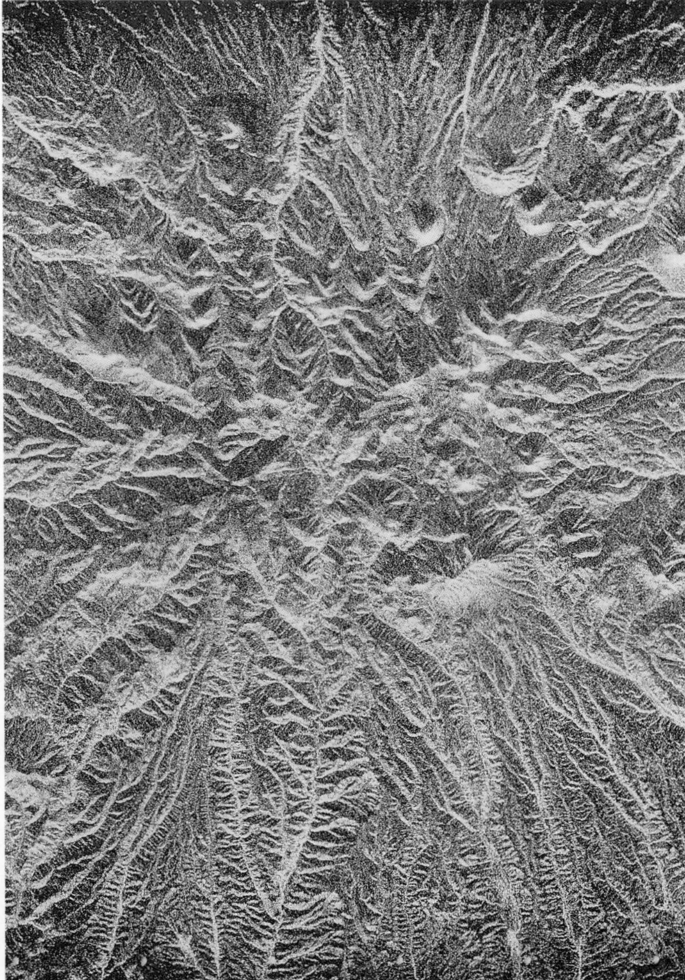

Fig. 3.70. SIR-A image of volcano in north-western Iran. Width is 35 km.

The central portion has virtually no valleys on it and is pitted, being an area of cavernous limestones and underground drainage. The impermeable rocks have a well-developed system of valleys, especially to the WNW of the limestone terrain.

Drainage patterns also give a clue to the structures in which various geological units are disposed. Uniformly dipping, layered sequences of differently resistant rock often develop a parallel or trellis-like pattern of streams (Fig. 3.66). Outcrops of homogeneous rock on the other hand show no clear arrangement, having instead a drainage network like the veins in a leaf—a so-called dendritic pattern (Fig. 3.29). Large upstanding masses of resistant rock, such as near-circular granite intrusions, but more frequently the remnants of conical volcanoes, are drained by streams that define a radial system of valleys. This is well shown in another SIR-A radar image (Fig. 3.70) for the partly dissected volcano Sahand in Iran. At the top right of the image are a series of smaller more recent volcanic cones, each with their own radial drainage in miniature.

As well as controlling the larger elements of landscape, such as drainage networks, the composition and internal structure of rocks also have an effect on the local products of erosion. Some rocks produce quite smooth weathered surfaces, while others are characteristically very rough. As explained in Chapter 2, one of the benefits of radar remote sensing is that the energy returned from the surface to the sensor is dependent on surface roughness, thereby enabling this important parameter to be uniquely monitored. A combination of this property and the excellent emphasis of large-scale surface features by radar images makes them extremely useful in mapping geological structure.

In Fig. 3.71 special processing, involving analysis of the high-, medium-, and low-frequency topographic features has been applied to a SIR-A image of part of Irian Jaya. The areas with dominant high-frequency features are tinged red, those with medium frequency are green, and low-frequency features are highlighted in blue. Together with the brightness of the image, depending on invisible surface roughness of the surface, this rendition enables carbonates to be picked out approximately in reds, shales and sandstones in green, and metamorphic rocks in blue.

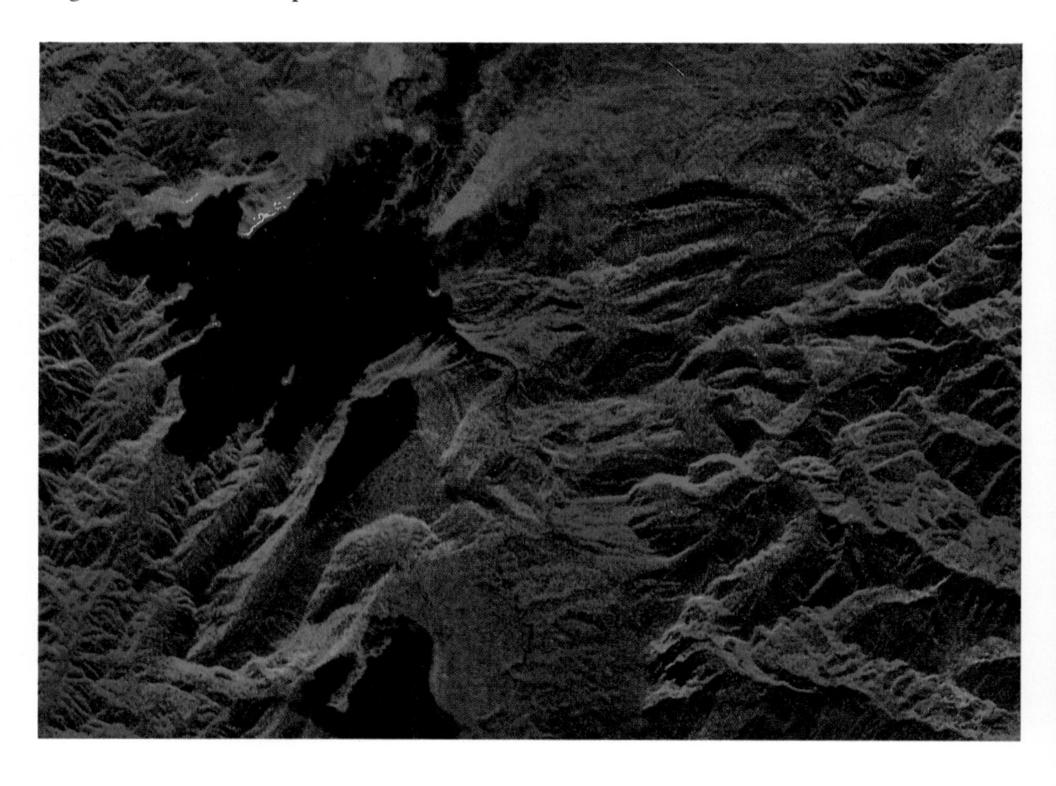

Fig. 3.71. Coloured SIR-A radar image of forested area, where the red, green, and blue components are controlled by images expressing the variation of high-, medium-, and low-frequency information in the radar image. Width is 20 km.

A much more geologically complex area is displayed as a SIR-A image in Fig. 3.72, over the ancient sediments of the Hammersley Range of western Australia. The complex structure of the area is picked out by variations in the surface roughness caused by different rocks varying response to erosion. Bright bands are units which weather to form rough, blocky debris, whereas the darker ones develop a smoother surface. The elliptical area with a mottled surface at the centre is underlain by granite, upon parts of which has developed a smooth sandy soil, giving the dark returns. The irregular light areas in it are weathered patches of an iron-oxide cemented soil called laterite. Crossing the elliptical granite from N–S is a straight, bright feature which is an igneous dyke from which a bouldery soil has been derived. The dyke is restricted to the area of the granite, and the surrounding concentric series of bright and dark units are younger sediments all dipping away from the central dome of granite. As well as the dome, the variegated units surrounding it define a synclinal structure in the lower part of the image. The uppermost unit at the core of the syncline is dark because a fine smooth soil has formed upon it, due to its relatively low resistance to erosion.

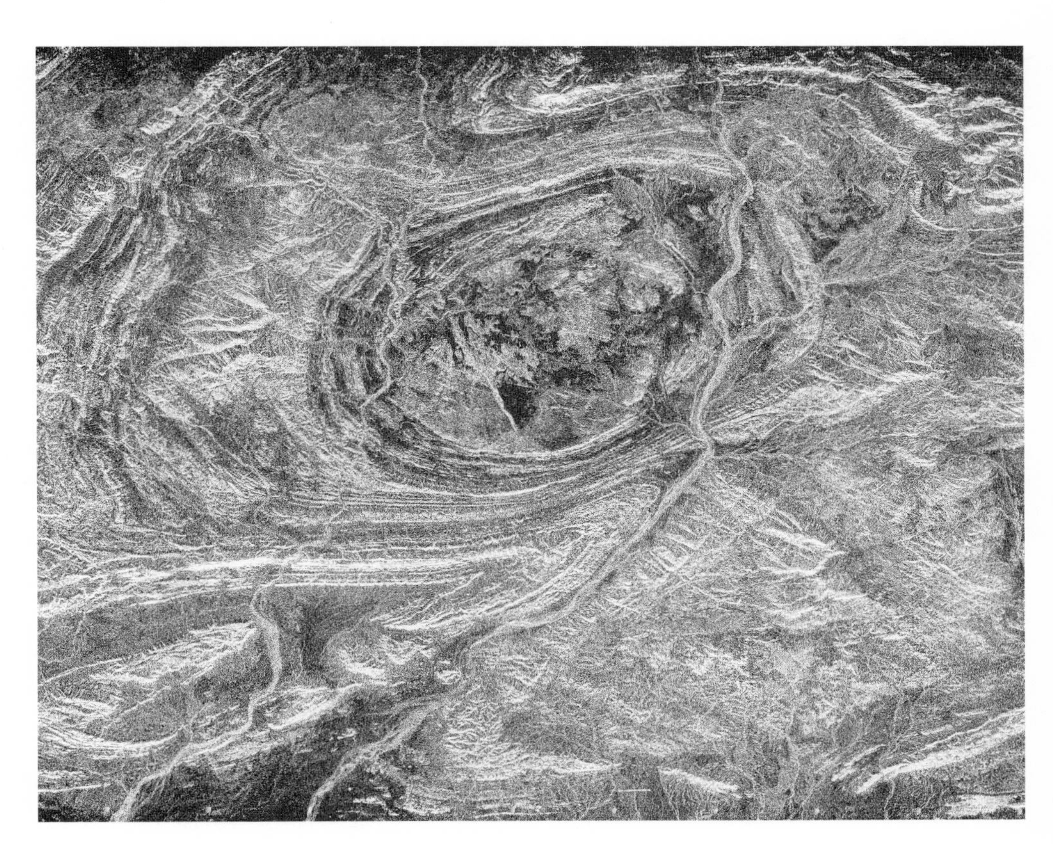

Fig. 3.72. SIR-A image of folded rocks in Western Australia. Width is 65 km.

Even when a rock surface is bare and clean, a natural-colour image rarely allows us to tell the difference between different kinds of rock. Although geologists use the colour of a rock in identification, much more emphasis is laid on the minerals that make it up, particularly the presence or absence of really quite minor components. Of course, this depends on being able to see the minerals with the naked eye. The bulk colour of a fresh rock depends on the relative proportions of usually colourless or light minerals—quartz and feldspars—and darker coloured, usually black minerals—olivines, pyroxenes, and amphiboles. In a crude way this allows igneous rocks, and some metamorphic rocks to be divided into acid (light), intermediate (greys), and basic to ultrabasic (dark grey to black).

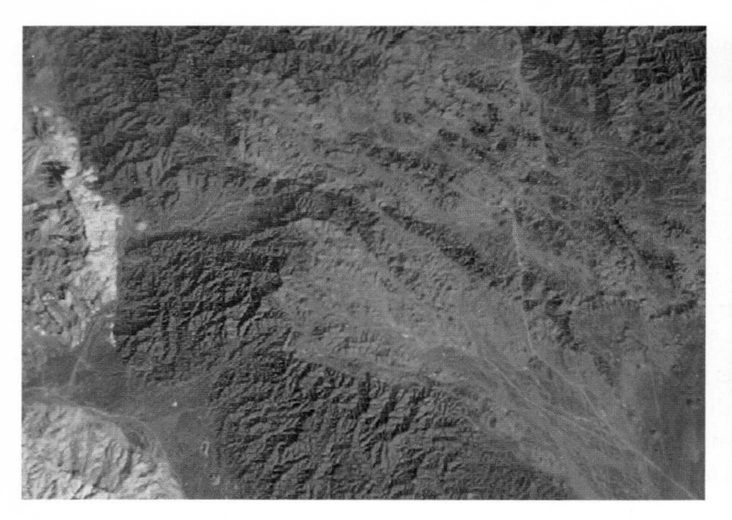

Fig. 3.73. Natural-colour Landsat TM image of part of the Oman Mountains. Width is 15 km. (Courtesy of D. A. Rothery, Open University.)

Fig. 3.74. False-colour Landsat TM image of part of the Oman Mountains, using bands 7, 5, and 4 as red, green, and blue, after decorrelation stretching. Width is 15 km. (Courtesy of D. A. Rothery, Open University.)

Coloration of sediments, composed mainly of different proportions of quartz, clays, and carbonates—all light minerals—depends very much on the presence of minor amounts of various iron minerals. Even quite small amounts of iron oxide and hydroxide (reds, oranges, and browns) or sulphides (dark grey) can confuse red or black sandstones, shales, and limestones, for example. Where outcrops are partly masked by vegetation, lichens, and weathering products, visual discrimination is even more difficult.

Remote sensing overcomes the restrictions of the visible parts of the electromagnetic spectrum, as explained in Chapter 2. The very-near infrared contains information relevant to the different types and densities of vegetation cover. The 1.55–1.75 μm and 2.1–2.35 μm wavebands in the short-wave infrared exploit the spectral reflectance features due to water and hydroxyl ions in minerals. A natural-colour Landsat Thematic Mapper image of an arid mountainous area in the Oman (Fig. 3.73) is typical of the poor discriminatory power of the visible part of the spectrum. All that can be said is that there are light and dark units, which correspond to sedimentary rocks and a basic–ultrabasic complex thrust over them. Figure 3.74 is a false-colour composite of bands 7, 5, and 4, also from the Landsat Thematic Mapper, expressed as red, green, and blue, after an extreme contrast stretch, as described in Section 2.5 (Fig. 2.32). The different spectral properties of the rocks in the near infrared express themselves so dramatically that an excellent subdivision of all the rock types is possible.

Although vegetation generally masks the true nature of the rocks and soils beneath, especially when the vegetation is agricultural and restricted to a few crop types, natural vegetation sometimes varies according to the different soil chemistry or water content associated with different rocks. Different kinds of vegetation often express themselves best when they are at the height of their annual growth cycle. In a strongly seasonal climate this is usually just after the wet-season rains, when all living plants are in full leaf. Figures 3.75 and 3.76 show the contrast between the end of the dry season and the end of the wet season for an area in the Transvaal in South Africa. Apart from a few areas of irrigated crops, the vegetation in the dry-season image is dead and only a few gross geological features are visible. The wet-season image shows considerable variation in colour due to vegetation differences between the rock units and varying soil-moisture content. Immediately apparent are the circular Pilansberg intrusion at top left, the broad blue sweep of a huge

Fig. 3.75. Dry-season Landsat MSS image of an area in South Africa. Width is 110 km. (Courtesy of ERIM.)

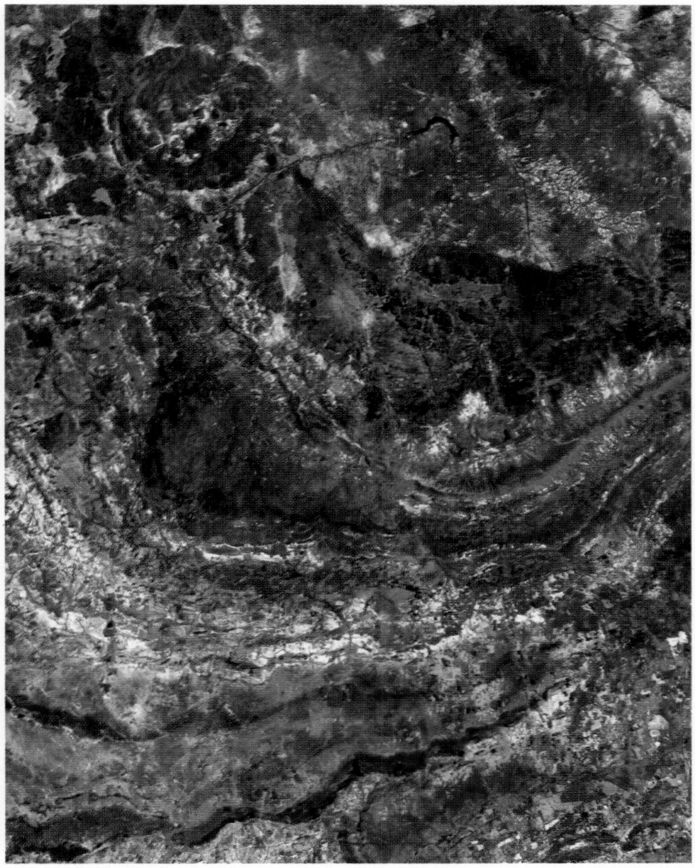

Fig. 3.76. Wet-season Landsat MSS image of an area in South Africa. Width is 110 km. (Courtesy of ERIM.)

basic igneous intrusion called the Bushveldt Complex, and many variations in the sedimentary rocks that it has intruded in the lower part of the image. In other climates, vegetation differences are expressed at other times of year. For instance, in temperate deciduous forest the different tree species show greatest differences in their leaf canopies just after bud break and immediately before leaf fall when the leaves display different colours as well as changing at different times. In monsoon climates, as in India, rock differences are best displayed in the dry season (Figs 2.33, 3.56, and 3.68).

In the thermally emitted parts of the infrared (8–14 μm), remotely sensed data allows us to use the differences in thermal properties of rocks as a means of exploiting their surface temperature at different times of day as an indication of lithological differences. This is even more useful if the thermal infrared is divided into narrow bands when the subtle deviations of different minerals from the ideal of thermal emission potentially allow even more detail of rock mineralogy to be extracted (Section 2.2.2). Figure 3.77 is a false-colour image displaying different thermal-infrared bands from a thermal-infrared multispectral scanner as red, green, and blue. Inspection of the thermal spectra of important rock-forming minerals, given in Fig. 2.18, suggests that with this combination of bands, rocks rich in quartz should emit more strongly in band 5 than the others and therefore appear red, whereas carbonates have peak emittance in bands 1 and 3 and should be tinged blue and green. In the arid area of the Wind River Basin of Wyoming, USA, shown in Fig. 3.77, the existing geological maps simply divide the rocks into sandstones and carbonates. Sure

Fig. 3.77. False-colour image, using data from narrow bands in the thermal-infrared part of the spectrum, of an area in Wyoming, USA. Width is 15 km. (Courtesy of H. Lang, JPL.)

enough, the image shows red and orange sandstones and blue-green carbonates, but much more information is available from the thermal image. The carbonate unit to the left proves to contain numerous, irregular layers and patches of sandstone. The sandstone unit occupying the right half of the image is far more varied than originally thought. The red and orange units are indeed pure quartz sandstones but the magenta and yellow units contain high proportions of carbonate in their cement.

As already mentioned, radar images provide a new dimension related to the micro-topography associated with different kinds of rock, and the soils and vegetation developed on them. Even more information is available when the radar is variably polarized or different wavelengths of radar are used to illuminate the scene. Although 'invisible' spectral properties still do not give us the information that a field geologist uses when mapping and are still affected by all the different veneers over unaltered rock, nevertheless the scope for geological interpretation is expanded dramatically. As will become apparent in the next chapter, it is the unusual features of rocks that are associated with economic amounts of physical resources, and the processes that produced them which are most readily detected by remote sensing, particularly when sophisticated methods of enhancing the data are used.

4 Using the information

In Chapter 1 I outlined the main areas of social, economic, and military needs that can be served in some way by the products of remote sensing. Chapter 3 covered the kinds of information which scientists can obtain about the different parts of the Earth that can be monitored, in order to broaden and deepen our knowledge about how the Earth functions, or has functioned in the past. In this chapter the focus is almost entirely on practical applications aimed at serving society as a whole.

Summarized extremely briefly, all forms of society depend on exploiting the natural world for water, food and fibre, and the energy and material resources that can only be supplied from the inorganic part of our planet—the rock systems that make up its outer few kilometres. In these three areas concerned with the supply of commodities remote sensing has exploded over the last two decades. The breadth of the topic is so wide that it is possible to give only a few examples illustrating what is possible now and what may be possible in the fairly near future.

The global economy is not just the winning, processing, and use of commodities, but also involves their transport. In order to do this efficiently means that lines of communication must be rendered as safe as possible, and in the case of remote or underdeveloped areas new routes must be opened up. Remote sensing can provide important inputs to this as well as to the siting of new cities and ports.

The world is not an entirely safe place, nor is it possible to make it so. Pressures of population or the economic benefits of certain areas force increasingly large populations to live in naturally hazardous regions, where flooding, volcanic and earthquake activity, windstorms, and forest fires are a constant threat. Modern industrial society depends increasingly on sensitive installations such as dams, oil refineries, and nuclear power plants, and also needs to dispose of dangerous waste-materials safely. Natural phenomena that are not in themselves life-threatening can put such facilities at considerable risk. Remote sensing can identify areas under threat by natural hazards, and in some cases can provide a measure of early warning.

Partly through natural causes, but increasingly due to human activity, the environment upon which our survival depends is constantly changing. These changes affect natural plant and animal life and also our exploitation of the natural world by agriculture, cropping forests, and managing livestock and marine resources. As well as economic disruption the changing environment can pose threats to human life, particularly when pollution is involved. The most obvious effect is the decline in the quality of life when aesthetically pleasing scenery is disrupted by the spread of cities, industry, and large-scale agriculture whose design takes no account of human sensibilities.

There is no avoiding the fact that remote sensing was primarily invented and developed for military intelligence purposes. Though current activities are naturally shrouded in secrecy, this book would be incomplete and naïve if it did not address this, quantitatively the most important use of the data that are available. Our reaction to the different methods of surveillance depends very much on individual political views. Some aim at delivering deadly force as efficiently as technology allows, others monitor the intentions of potential opponents to do the same. Some intelligence applications seem at first sight to gather whatever information reveals itself, as a means of knowing just what is going on and what might emerge as a threat to what are perceived as justifiable national and international interests. The same systems that can be used to predict crop failure and make ready its economic exploitation can also detect and monitor crops of narcotic plants, such as opium poppies, coca, and marijuana. A final issue is that while 'Big Brothers' may be able to watch, rudimentary knowledge of the technology of remote sensing also enables the less well-heeled successfully to hide and confuse.

4.1 Water resources

In arid or semi-arid areas, where rainfall is either unpredictable or falls in clearly defined seasons, provision of a dependable water supply is a matter of life or death. Human survival without water is limited to a matter of a few days, while that of livestock is measured in weeks. Any agriculture, other than that which depends opportunistically on rainfall, requires pumping of ground-water or storage of seasonal surface flow in reservoirs. In such areas, and also in those with consistent surface streams, the quality of the water supply is important. Water-borne disease is the single greatest health hazard globally, and there are also clearly defined limits on the inorganic dissolved content of water, beyond which it becomes undrinkable. Even irrigation water, if it has a high content of dissolved salts, eventually pollutes the soil with salts precipitated by evaporation which sterilize it for many crops.

Although humid areas have their own problems associated with water supply, this section concentrates on the role of remote sensing in solving the much greater problems of arid areas. It can serve the management of surface flow as well as assist in the exploration for subsurface or ground-water. By providing information on the global distribution of atmospheric moisture content and clouds it helps assess the likely distribution and, to a less precise degree, the amount of rainfall on a continuous basis.

The water regime in an area can be expressed by the interplay between four main factors: rainfall, surface flow, evaporation and transpiration by plants, and infiltration of water into rocks and soil. From the standpoint of water resources, surface flow is only useful when confined to channels of streams and rivers or ponded in lakes. Apart from the rainfall involved, the amount lost to the surface by evaporation and infiltration, the main controls over the volume of water in a channel are the area on which rain is collected to enter the channel—its catchment—and the nature of the drainage within the catchment. Where the drainage is in many small unconnected streams, each channel has only a small flow. The volume increases as tributaries coalesce, the run-off from each small catchment contributing to the larger streams. So, the larger the stream the greater is its total catchment. Because flow varies dramatically from season to season, and sometimes from year to year, an obvious strategy is to store water behind some kind of dam, or to divert it into a cistern. Without this provision, potentially useful rainfall flows unhindered away from the area where it could be used. Although simple dams can be constructed as long earth barrages across broad valleys, the best sites are where all the flow is constricted in a part of the valley that is narrow.

Figure 4.1 illustrates this principle well; here a number of constrictions or thresholds, marked by triangles in the larger valleys, are potential dam sites. Supply to each depends on upstream catchment. Whether such sites would prove to be useful depends on a number of factors. One is the nature of the materials comprising the valley floor and its sides. If these are highly permeable—able to transmit water easily beneath the surface—a dam would leak. As well as the anticipated volume of water carried by the stream during rainy periods, which indicates whether the reservoir could be filled, the velocity of run-off is important.

Fig. 4.1. Panchromatic photograph from the Large Format Camera carried by the Space Shuttle of part of the Red Sea Hills in the Sudan. It shows a number of ephemeral drainage courses connecting areas of deep superficial sands and gravels, seen as white. The small triangles indicate constrictions or thresholds in these water courses where slowing of water flowing both at and beneath the surface may allow ponding, which can be exploited by shallow wells. Width is 200 km.

Above a certain speed sediment carried by torrents would eventually fill the reservoir and make it useless. The safety of the dam is a prime consideration, and any signs of possible slope instability or earthquake activity mean that the site must be rejected.

The potential for water to be stored beneath the surface as ground-water requiring an intricate balance between many factors. To get there it must be able to infiltrate through the soil quickly. As a general rule, if penetration does not exceed a metre in a day, then water is lost by evaporation and little is stored. The speed of infiltration depends on the permeability of soil or rock, and the velocity of any surface flow. If slopes are high, infiltration is poor, but if flow is slowed by bouldery ground or by abundant plants then it can increase. The thresholds indicated on Fig. 4.1 acts as 'brakes' on stream flow during flood, with water ponding up above them. This gives water the opportunity to seep into the ground. How much water can be stored below ground depends on the thickness of permeable rock or soil. The best subsurface reservoirs, or aquifers, are poorly cemented coarse sediments which contain many interconnected pores. The only such sediments on Fig. 4.1 are those in the main valleys, which are highly reflective. However, the crystalline rocks forming the mountains are not totally devoid of potential. Some crystalline rocks are deeply affected by weathering to become crumbly, porous masses. The spectacular circular bodies on Fig. 4.2 are granitic intrusions that have been broken down in this way to form potential aquifers. Another natural means of preparing unfavourable ground to store water is through the fracturing action of earthquakes. Along the lines of geological faults even crystalline rocks are broken and ground up to form highly permeable zones into which water can infiltrate and be stored. Figure 4.3 shows the distinctive straight alignments of

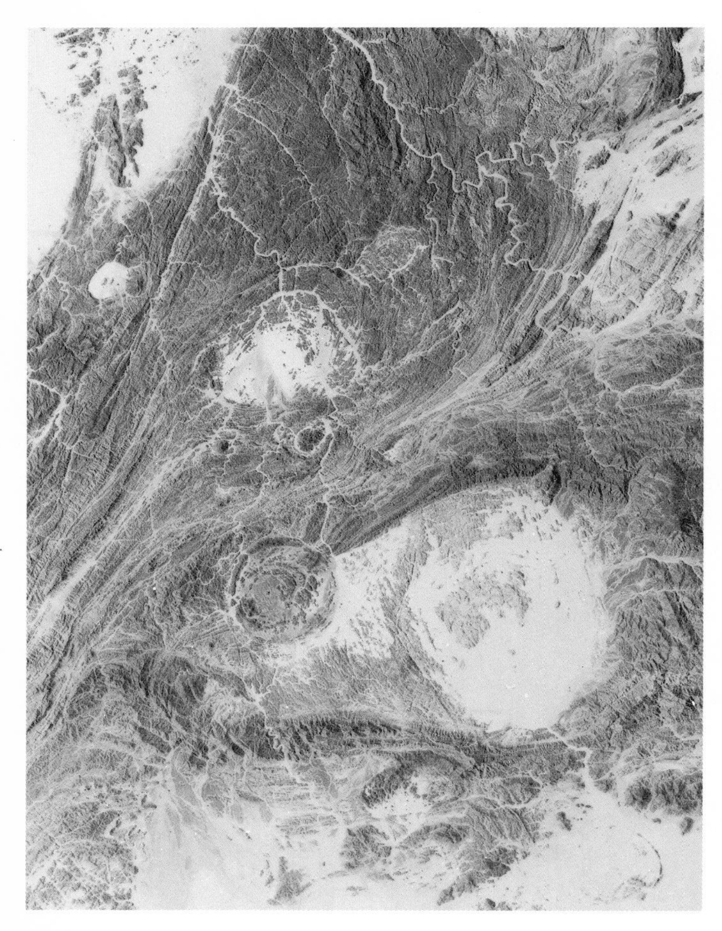

Fig. 4.2. This Large Format Camera image of part of the Red Sea Hills shows some clear circular features. Several are occupied by light sands and gravels, indicating that the granites which form them have been deeply weathered to form potential traps for ground-water. Width is 80 km.

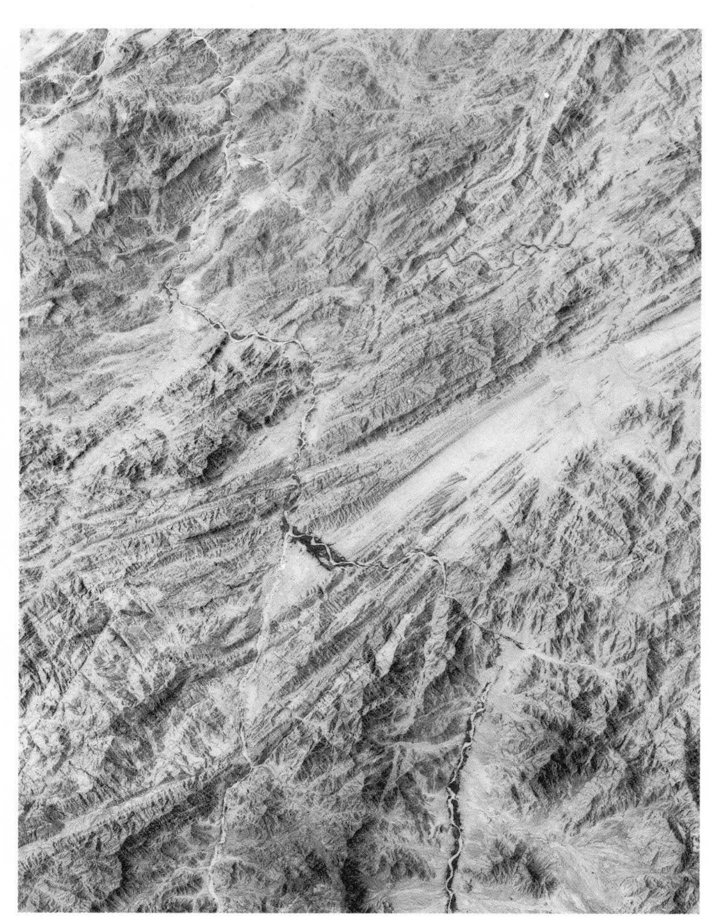

Fig. 4.3. Large Format Camera image of part of Eritrea showing clear linear features controlled by fracture systems into which surface water has seeped to provide water supplies. Width is 80 km.

several valleys that are controlled by zones of weak and fractured rock associated with old fault lines.

Like surface water, ground-water flows too, although the speeds involved are much lower due to the viscous drag imparted by narrow connected pore spaces. To all intents, the direction of flow is roughly parallel to that at the surface. Also, in an analogous manner to surface water, that beneath the surface can be ponded where the flow is slowed down or stopped. The provision of natural subsurface dams is due to reduction in the rock or soil permeability by some kind of barrier. This is usually due to a lateral change in the rock type. Sometimes the barrier is where impermeable rock rises to the surface downflow of permeable sediments or soil, as in the case of a flat sediment plain becoming constricted by rocky walls as it enters a canyon. The thresholds shown on Fig. 4.1 probably serve this function as well as providing good sites for surface dams. Wells sunk upstream of the threshold should encounter shallow ground-water ponded in this way. Another kind of subsurface dam is commonly produced where vertical sheets of impermeable igneous intrusions, known as dykes, cut across water bearing more permeable strata. Such dykes are often composed of rock that is very resistant to erosion, and they stand as long ridges that are easily seen on remotely sensed images. An excellent set of examples is shown in Fig. 4.4, a Landsat image of an area in the semi-arid central part of south India. The dykes form

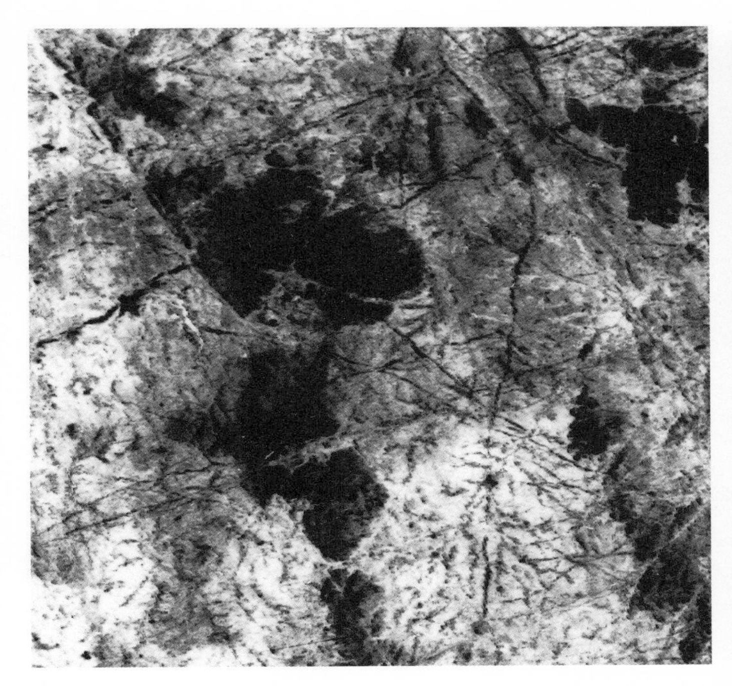

Fig. 4.4. Enhanced Landsat MSS band 7 image of part of Andhra Pradesh, India, showing abundant igneous dykes. Width is 40 km.

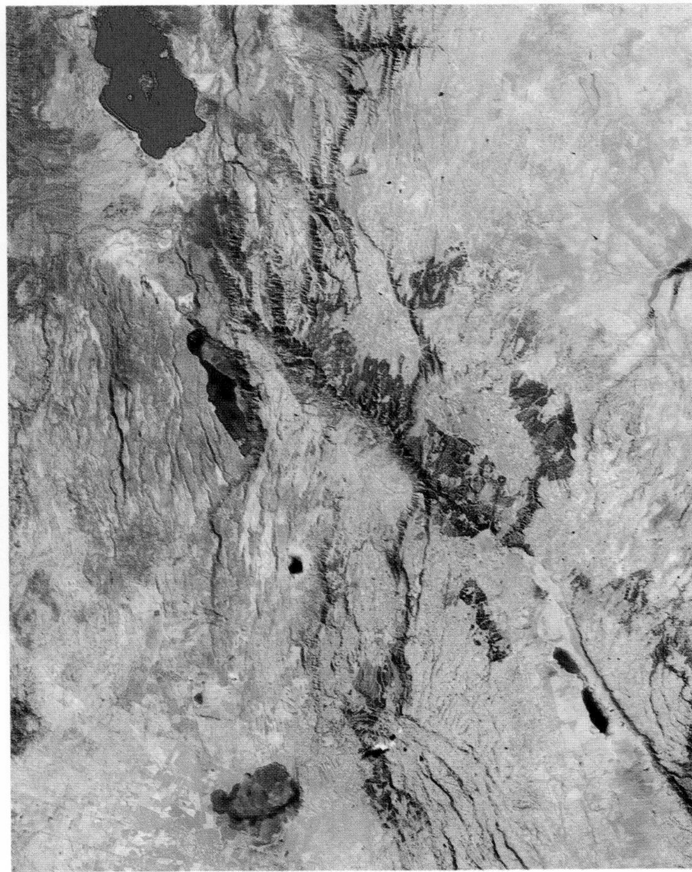

Fig. 4.5. Approximately natural-colour Landsat TM image of part of the Rift Valley of Kenya. Width is 40 km.

sharp black streaks, running east–west, across the general geological grain of highly sheared and fractured bedrock that runs north–south. Each of the dykes forms a potential dam beneath the surface.

The presence of ground-water at shallow depths is often indicated by natural vegetation. In arid areas, many hardy shrubs and trees have roots that penetrate to depths of up to 30 m. During the dry season they have access to shallow water reserves and so remain in leaf. After rains, even areas with no appreciable ground-water are covered with grasses that germinate in the short period while the soil is damp. They are no guide to dependable subsurface water supplies, and so dry-season images are most useful. Figure 4.5 is a dry-season Landsat Thematic Mapper image of part of the Rift Valley of Kenya, where band 4, which is characterized by high reflectance for living vegetation, controls green. Stands of leafed plants therefore show as various shades of green, and indicate areas where near-surface ground-water is present. Most of the intricate linear features on the image are small valleys controlled by the faults that formed when the Rift opened up. Where they are associated with green vegetation is a guide to ground-water that has seeped into the fracture zones formed by faulting.

In the desert areas of the world, such rain that does fall is concentrated in mountainous areas. Because of the steep unvegetated slopes there, much of the water runs off without appreciable infiltration. The ephemeral streams, highly charged with sediment, emerge in flat areas as muddy torrents that lay down wide tracts of sand and gravel into which the water that escapes evaporation may seep. Such plains are frequently featureless, and present a monotonous impression on images of visible and near-infrared radiation, with

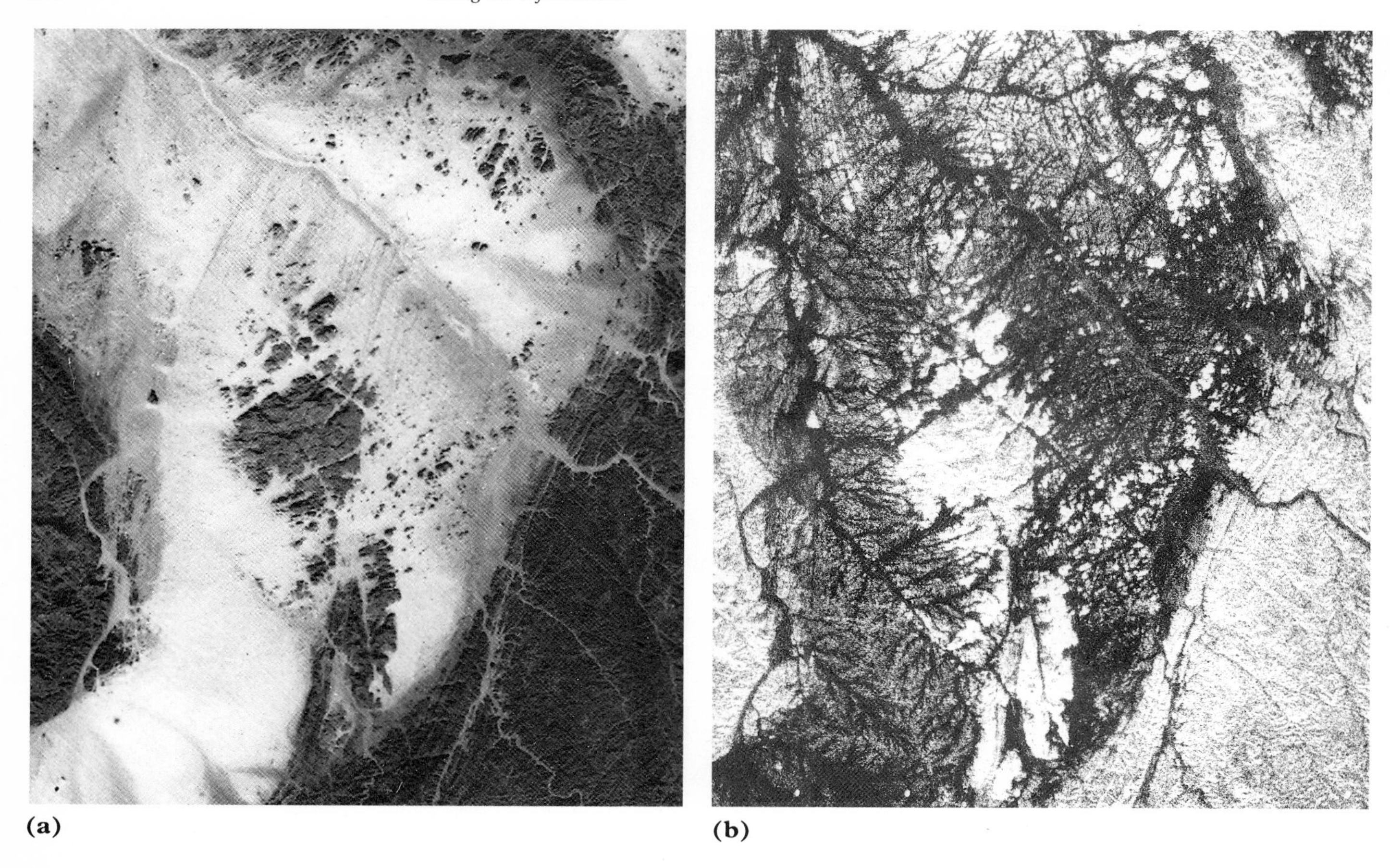

Fig. 4.6. (a) Landsat MSS band 7 image of part of NE Sudan showing a light sand and gravel plain rimmed by mountains; (b) SIR-A radar image of the same area as (a). Width is 40 km.

little guide to the places where the ground-water resides. Sometimes vegetation can give a clue, but often quite abundant reserves go completely unnoticed. Before the sediments were laid down, the torrents from the mountains would have carved now-hidden channels into the floor of the plains. It is into these fossil channels that later infiltration finds its way to form potential targets for well digging. Without a ready guide to their location, however, exploration is haphazard and expensive, as Fig. 4.6(a) graphically demonstrates. A large wadi in the mountains emerges on to a sand plain, but its former course is completely hidden among the highly reflective sediments into which its discharge will seep. As introduced in Chapter 2, in extremely dry conditions radar energy is capable of penetration for some metres beneath the surface, and where it is reflected by some buried feature of re-emergence shows details of the subsurface on a radar image. Figure 4.6(b) is a SIR-A radar image of the same area as Fig. 4.6(a). Because radar is diffusely scattered by rough surfaces the mountains appear bright, and the sand plain has a dark signature since it is smooth and most energy is reflected away from the platform. Nevertheless, considerable detail is visible in the area of flat sediments, due to penetration by a proportion of the radar beam. Surprisingly detailed information on the location and trajectories of many small buried channels is revealed, so helping pin-point likely targets for drilling with great accuracy.

4.2 Food and fibre

The clear distinction between the spectral properties of vegetation and those of rock and soil, covered in Chapter 2, makes remote sensing particularly useful in applications related to the biosphere. Indeed, many of the sensing systems were designed primarily for vegetation applications, ranging from the first development of infrared film for detecting camouflage to the launch of the first of the Landsat series of satellites that aimed at global monitoring of food production. Consequently, over the last 15 years a very large number of studies aimed at agriculture, forestry and pastoral farming have been undertaken, only a small proportion of which can be represented here.

As shown in Section 3.4, one of the best measures of biological productivity on land is some kind of vegetation index based on the contrast in very-near infrared and red reflectance of plant structures compared with soils and rocks. This is often applied to very large areas using low-resolution imagery, such as that from the NOAA AVHRR system. At very low cost and with relatively little effort these data enable the available green biomass, both agricultural and natural, to be assessed on up to a daily basis, although monthly or even seasonal data are probably all that is needed for most purposes. Figure 4.7 is a false-colour AVHRR image for a large area in the Middle East. Figure 4.8 is an image of the ratio between very-near infrared and red reflectances that has been simplified by showing areas with various ranges of the ratio in different colours. Blue is water, dark green is where green vegetation is most dense, and the range through emerald green, orange, and yellow expresses decreasing biomass in the semi-arid areas. Showing in red on Fig. 4.7 are the irrigated and rain-fed agricultural areas of the Lebanon and Israel, particularly along the Jordan Valley, the forests of the mountains bordering the Jordan Rift, the intensively farmed plains around Damascus in Syria, and irrigated parts of the Tigris and Euphrates valleys. These areas show as dark green in Fig. 4.8. Much of the rest of the area is rangeland with low rainfall, used by nomadic pastoralists, merging to the east with semi-arid to arid desert. However, in this large tract of marginal land there is considerable variation in the vegetation index, as shown in Fig. 4.8, indicating differences in the amount of spring vegetation available for browsing. Information of this kind is naturally of great interest to the nomadic herdsmen, as grazing varies in any one area from month to month, and from year to year. AVHRR data, being captured on a routine daily basis can therefore give advance information concerning where to drive herds to take advantage of available vegetation. Also, they can indicate to administrators which areas have been overgrazed, by comparison with archived scenes of previous seasons with similar rainfall patterns. Given this information, herdsmen can be induced to drive their livestock to less intensively used areas before the animals become too weak to be moved.

A step beyond this relatively local and simple managerial use of low-cost remotely sensed data, such as those from the NOAA AVHRR, is to apply similar image manipulation techniques to monitor the continent-wide seasonal progress of natural vegetation (Figs 3.40 and 3.41). Given several years of archived data it becomes possible to compare the current state of vegetation early in the growing cycle with cases at the same time of year from years of good biological productivity and those characterized by widespread drought and famine. This allows a crude estimate of eventual harvest production to be made. These estimates can be refined for smaller, critical areas by the use of timely, high-resolution imagery, such as that available from Landsat, combined with measurements made in the field. By itself, remote sensing aimed at vegetation alone gives an imprecise and sometimes misleading impression of what may happen to agriculture, especially in areas where farming relies on primitive methods in small plots of land. Another approach to drought forecasting and famine early warning is to concentrate on rainfall and evaporation over critical time periods. This is possible using information on cloud cover, from meterological satellites such as Meteosat, measurements of rainfall from ground stations and data on deviations in surface temperature from the normal, derived from imagery of thermal emission. Both NOAA and the UN Food and Agriculture Organization are developing means of systematically maintaining a watch on potential famine conditions in sub-Saharan Africa.

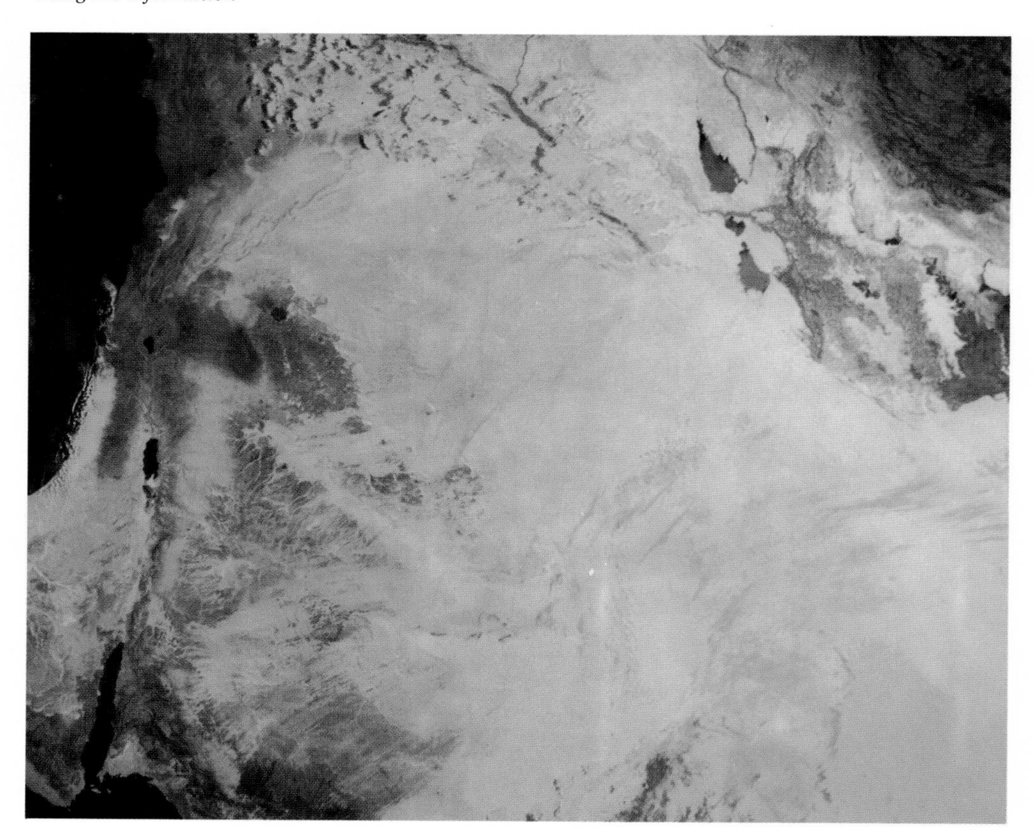

Fig. 4.7. False-colour AVHRR image of the Middle East, with well-vegetated areas showing in shades of red. Note that some clouds appear in the same yellow hues as the desert. Width is 1100 km.

Fig. 4.8. Colour-coded image of vegetation index derived from AVHRR data for the same area as Fig. 4.7. Dark green indicates abundant vegetation; light green, browsable grasses and shrubs in the desert; orange and yellow represent very low vegetation cover.

Other information required to assess the potential for disaster stemming from drought cannot be obtained from remote sensing. This includes information on the location of stockpiles of food, changing population densities, and market forces that trigger hoarding and rises in the price of basic food commodities. Moreover, although drought has played the major role in the Sahelian famines this decade, an important contributory factor has been politics, particularly the disruption of normal farming practice by wars in northern Ethiopia and southern Sudan. In spite of these unmeasurable factors, information from remote sensing has the potential to direct attention to possible famine areas in advance of disaster, allowing suitable measures to be taken to alleviate the worst of the attendant problems.

A far less risky environment for pastoral farming than sub-Saharan Africa is the grassland prairies of western North America, and the savannahs of Africa and South America. There, rainfall is more consistent and higher, and temperatures are not so extreme as to induce rapid evapo-transpiration. Consequently, grazing is always potentially abundant for most, if not all of the year. These areas are among the greatest producers of meat, dairy products, and hide in the world, capable of generating high profits with a small population of herders or supporting high populations of people dependent on their own relatively small herds of cattle and other livestock. These advantages, however, have limits that are easily exceeded by overgrazing. Moreover, the vegetation that grows there is not all of equal nutritional value for cattle. Figure 4.9 is an example of simulated Landsat TM data derived from an aircraft-mounted multispectral scanner over the rough-fescue grasslands of south-western Alberta in Canada. The red, green, and blue components of the image are controlled by TM bands 3, 4, and 5. Band 5 gives an estimate of vegetation moisture content, low values indicating high moisture, and bands 4 and 3 express the 'greenness' of the vegetation and its biomass. The green areas are pastures with a high proportion of forbs, favoured by cattle, and are heavily grazed. The lighter the green coloration the greater the new growth of grass. The reddish-brown areas are dominated by other species, such as bronne grass, that are not so nutritious. The mottling of these low-quality pastures is due to variations in biomass: the paler the shade of brown, the denser the grasses. Towards the right of the image can be seen a rectangular area slightly different in tone to those surrounding it. This is an area of well-grazed pasture enclosed by fencing, at the centre of which is a large, isolated meadow of high-quality grasses showing as green. On ranches as large as the one shown only in part here, remote sensing forms an important input to the efficient management of the rangeland, which would not otherwise be possible.

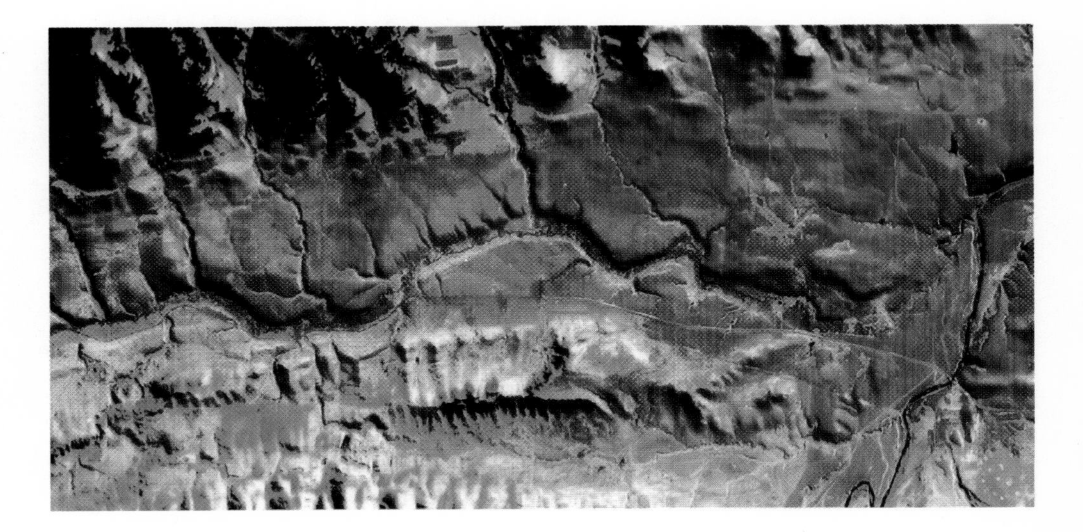

Fig. 4.9. Simulated Landsat TM false-colour image of part of the prairies of Alberta, Canada. Bands 4, 5, and 3 control red, green, and blue. Width is 10 km.

Provided the resolution dimensions of imagery of the farmland is smaller than that of individual plots, it is possible to see the division into fields. This is because different crops are spectrally contrasted due to their leaf-cell structure, their morphology, and to the different stages to which crops have grown at any time of year. Each field is often dominated by a single species and has some kind of boundary, such as a hedge or fence, and thus is contrasted with its neighbours. Another factor in highlighting fields is their shape, often angular or geometrically regular (Figs 3.54, 3.57, and 3.58), particularly in areas of modern farming. In less well-developed areas plots are often smaller than 20 m across, and because they depend on intricate local irrigation or on patches of naturally fertile soil they become much less easy to distinguish because of their irregularity (Figs 3.55, 3.56). This problem is compounded by the tendency to farm in gardens containing a variety of different crops. To overcome this problem, the design of the French SPOT system was oriented towards the most common field size in the Third World, having dual resolution of 20 and 10 metres.

While individual fields are usually quite visible on satellite images, it is not possible to suggest their crops simply by inspection, however well they may be discriminated (Fig. 4.10). Instead the power of a computer is used to erect different classes based on the varying reflectances in the different bands, themselves depending on the spectral properties of the surface in each field. As explained in Section 2.5, this classification must usually be based on some knowledge of field contents from ground studies, which the computer uses to identify different categories throughout the whole scene. Figure 4.11 is a classification based on the image data shown in Fig. 4.10. Yellow areas are maize; red, soya beans; orange, newly ploughed fields; and cream, barren or fallow land. Green areas are deciduous forest and small stands of trees along valleys, while purple areas are towns. The blue water is divided into light silty, and dark clear classes. The black areas remain unclassified. Although the precision of such a classification can never be exact, compiling

Fig. 4.10. Landsat MSS false-colour image of part of the eastern USA. Width is 30 km.

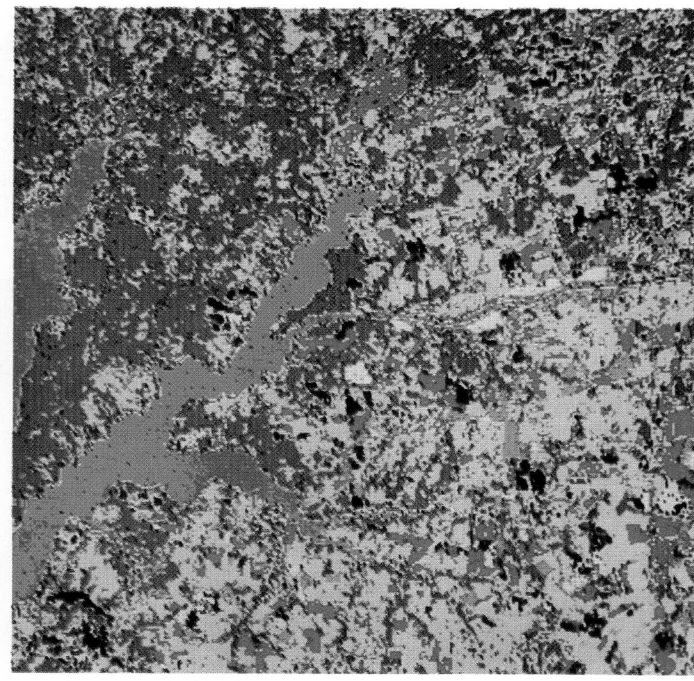

Fig. 4.11. Classification map of Fig. 4.12. (Courtesy of N. Short, GSFC, NASA.)

yearly crop maps by field work would involve a very large team whose task would be never-ending. The results shown for over 1000 square kilometres took perhaps a week to prepare, including several days in the field establishing training areas for the computer classification. The result is not merely a thematic map but enables areas of individual crops and even their eventual tonnage at harvest to be estimated. This is particularly important in Third World countries where food supplies need to be carefully managed and where information is almost totally absent. Where farmers' records of planting are used to calculate taxes or subsidies, as in the European Economic Community, such detailed classification can be used as a check on their honesty, with potential savings for the EEC alone in hundreds of millions of ECU.

Apart from the effects of drought, the greatest losses to world food production come from disease or infestation by parasites. Identifying and locating stricken crops as early as possible is vital in planning preventive measures. At the onset, effects are often merely slight discoloration of leaves, followed by reduction in canopy, and ultimately by death. The early stages are difficult to identify by eye and in large fields may go unnoticed until too late for effective remedies. Figure 4.12 shows a potato crop beginning to suffer from blight on an aerial colour-infrared photograph—a cheap and rapid method of agricultural survey. The large black area is a stand heavily affected by blight among otherwise healthy plants showing as bright reddish-pink. The field at centre has patches of well-established blight while that on the far right shows the first signs as green streaks. It is interesting to note that the blight seems to be related to vehicle tracks through the crop, earlier spraying or fertilizing probably having spread the blight spores.

Disease and insect infestation is also a major problem in forestry. The larvae of many species of moths and beetles feed on the leaves of trees, often one species depending on a

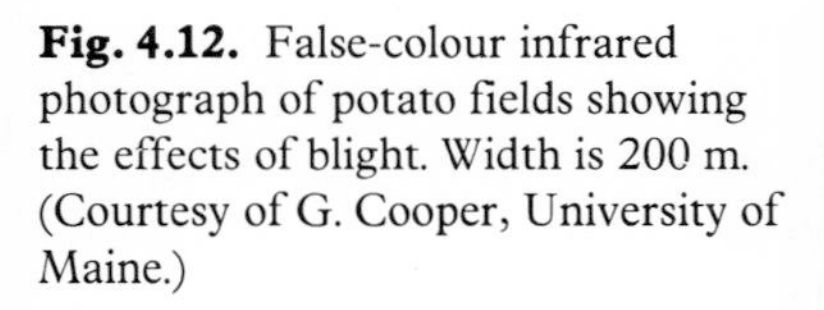

Fig. 4.12. False-colour infrared photograph of potato fields showing the effects of blight. Width is 200 m. (Courtesy of G. Cooper, University of Maine.)

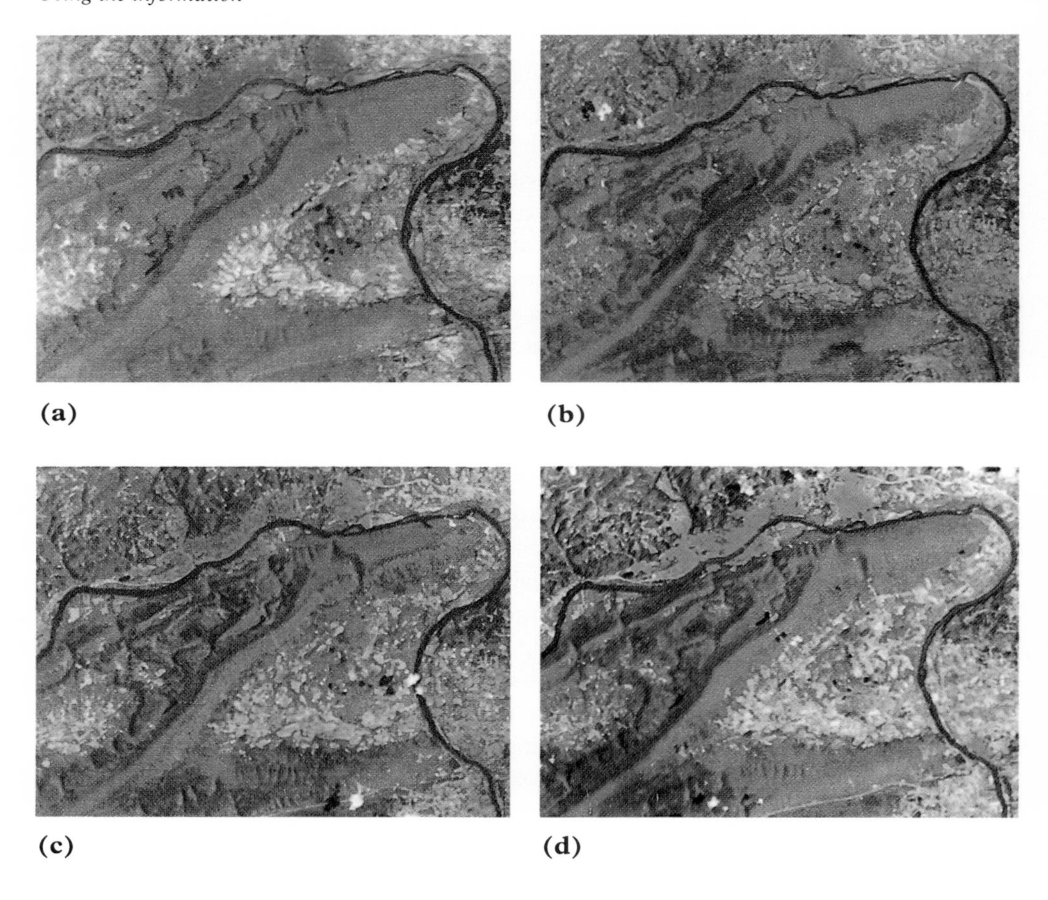

Fig. 4.13. Time sequence of Landsat MSS false-colour images of an area in the Appalachian Mountains of the USA, showing the defoliant effect of caterpillar infestation. The sequence for May, June, July, and August is shown by (a)–(d). Width is 35 km. (Courtesy of N. Short, GSFC, NASA.)

particular type of tree. In mixed forests this is not a great problem as individuals of the same tree species are relatively isolated, and so the larvae cannot spread very easily. Moreover their depredations are 'diluted' by the abundance of resistant species. The greatest problems are in forests dominated by a single species. Under favourable conditions the larvae can explode in numbers, laying waste very large areas in only a short period of time. Figure 4.13 is a time sequence taken in 1977 of Landsat MSS images of an area in the Appalachian Mountains of Pennsylvania, USA. Image (a) is from May, (b) June, (c) July, and (d) from August. In image (a) the upland area is covered by deciduous woodland, showing as red. Farmland is typically mottled in appearance, while built-up areas show as blue-grey. By June several linear swathes of forest appear as greyish-brown. This is due to stripping of leaves from oak trees by the larvae of gypsy moths. As the larvae grow their appetites increase, so that by July the oak trees are totally bare of leaves, the soil showing through the canopy of branches in much the same way as urban areas. At about this time the larvae pupate and oak trees are able to refoliate, so that by August the signs of damage are almost vanished. However, the stress on the trees is so high that two or three years of attack depletes the recuperative capacity of the trees and they die.

In semi-arid areas, the worst losses to crops come from the ravages of insects, particularly locusts. Since locust swarms can strip crops bare in a matter of hours, locating areas of damage after the event is of little use. It is vital to identify the location of conditions that allow locust populations to explode suddenly, and then predict their likely migration direction. Locusts normally occur in small swarms around areas of natural vegetation in deserts. A year of high rainfall allows the populations to increase dramatically. Soon all local vegetation is consumed and the swarms must migrate to new pastures, often in

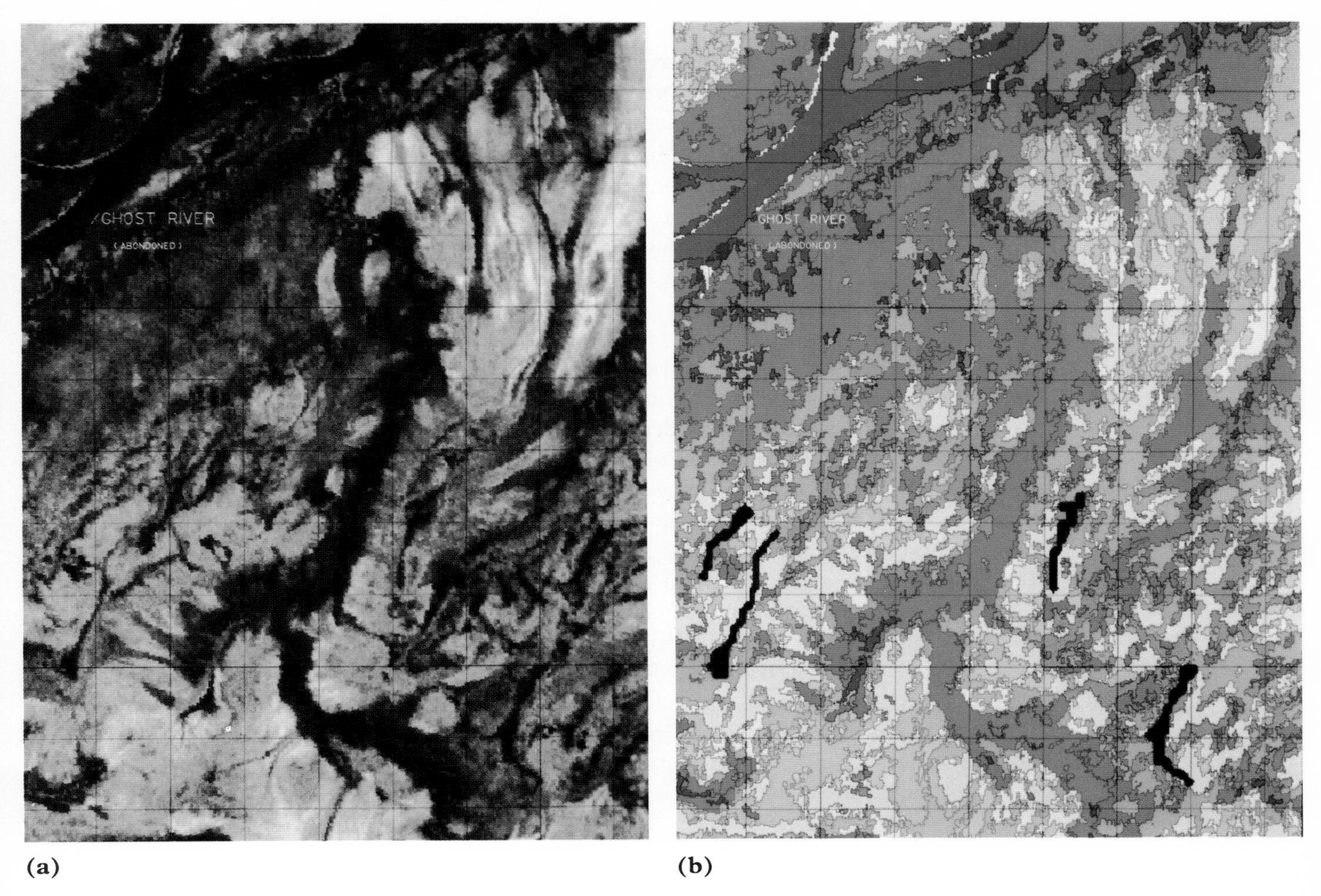

(a) **(b)**

Fig. 4.14. (a) False-colour Landsat MSS image of the coastal plains of Hudson's Bay, Canada. (b) Thematic map produced by supervised classification of (a). Width is 18 km. (Courtesy of the Ontario Centre for Remote Sensing.)

populated areas. Most of the breeding sites are well known, and remote sensing is able to monitor vegetation there, so enabling likely population explosions to be predicted in advance and control measures to be taken.

Using remotely sensed data to evaluate natural forest resources is one of its most challenging applications. The principal reason for this difficulty is the lack of homogeneity in the surface cover, in terms of tree type, density, and height. Many different species may be represented, and where their crown size is less than the resolution dimensions of the system employed each pixel represents a mixture. The problems are at their greatest in tropical rain forest where the number of important species may exceed a hundred. It is quite possible that operational evaluation will never be a viable proposition there. However, Fig. 4.14 shows the results from a pilot study in the simpler forest and swamp land of northern Canada. Image (a) is a Landsat MSS false-colour composite of part of the basin of the Albany River, near its outlet into Hudson's Bay. Relief is very low, and most of the soils, except by the river and in sandy valleys, are waterlogged. This is partly a result of the presence of a permanently frozen layer in the soil. These conditions restrict the growth of trees to the valleys, which are picked out in dark tones. The lighter areas are swamps covered with lichens, grasses, and sparse shrubs and trees. They are filled in some cases by appreciable thicknesses of peat.

Image (b) is the result of a multispectral classification of the Landsat data (Section 2.5), the different colours representing various classes of surface cover. The classes were separated by a computer, using the multispectral signatures in small training areas that represent each chosen class. Bright red and green are black spruce and deciduous forest, blue being open water. Dark pine areas are stands of tamarack on fens. The yellow and pastel shades are all various classes that are crossed by old beach ridges, shown in black. As well as showing the distribution of forest and wetland types, the classification allows the potential for harvesting timber and for peat reserves to be assessed accurately.

In order to banish the risk of famine and to develop the potential of many Third World countries, predicting drought and monitoring agriculture and natural vegetation is not enough. Prevention is far better than piecemeal attempts to curve an endemic problem. Experience from areas in North America and the Middle East of what in the developing world are regarded as barren or marginal land suggests that a far greater proportion of the Earth's surface can be productively and continuously farmed than is under cultivation at present. Planning new agricultural ventures requires the analysis of many different factors. Suitable soils need to be delimited, adequate rainfall or irrigation is necessary, slopes need to be such that water can be distributed over a wide area and risk of erosion minimized, and the socio-economic factors of proximity and access to markets and conflicts of interest must be evaluated. Weighing these many factors is a difficult task for the planners. A solution that incorporates input from remote sensing and from other sources, and renders them in a common form as maps is the function of a geographic information system (Section 2.5). Figure 4.15 shows part of such a GIS aimed at managing an area of high agricultural potential in Peru. The base is a map of drainage and roads on which is overlain the results of various kinds of survey. Image (a) outlines the area and colour codes the topographic slopes derived from digitized elevation data. Lilac and pale blue indicate low slope angles suitable for irrigation; greens, yellows, and reds indicate slopes that are too steep for efficient use of available water. Image (b) shows the results of a soil survey based on aerial photographs, the different soil types being colour coded. Those in magenta and pink are the most fertile; orange is adequate; yellow, green, and salmon are of poor quality. Image (c) shows the results of comparing the information on (a) and (b) alone as a guide to the best areas for intensive cultivation. Red, orange, and magenta are the prime areas identified, pale green being of marginal potential, while the dark green, blue, and grey areas are of little use. A refinement of this scheme would be to incorporate details of soil geochemistry and water quality linked to the requirements of different kinds of crops so that a comprehensive plan could be developed.

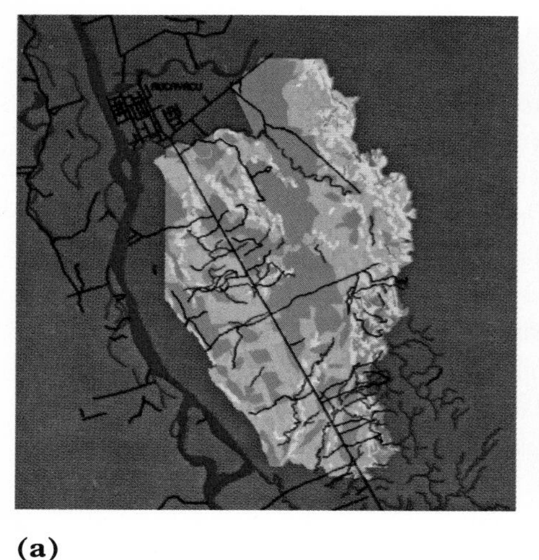

(a)

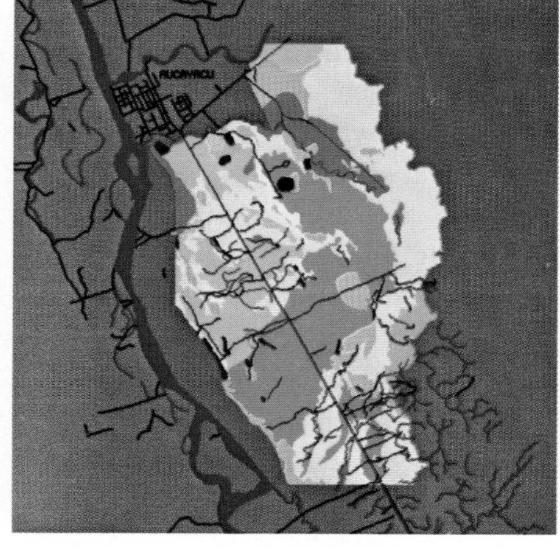

(b)

(c)

Fig. 4.15. Part of an agriculturally oriented geographic information system in Peru, showing (a) slope angles, (b) soil type, and (c) suitability for cultivation. Width is 10 km.

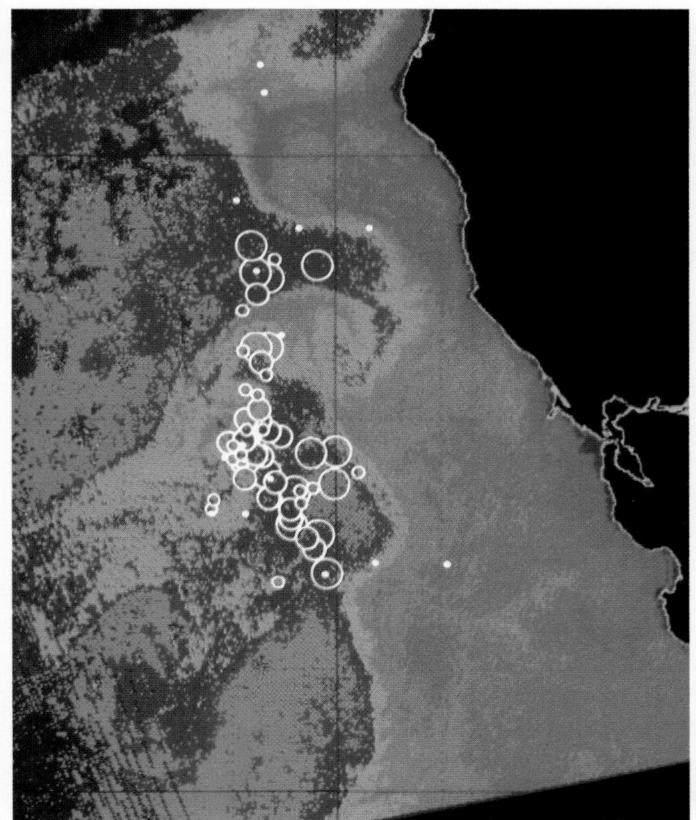

Fig. 4.16. Colour-coded Nimbus-7 CZCS data for the Pacific Ocean off San Francisco, showing variation in phytoplankton pigments for 21 September 1981 and circles whose size reflects the number of tuna caught in a four-day period. Width is 500 km.

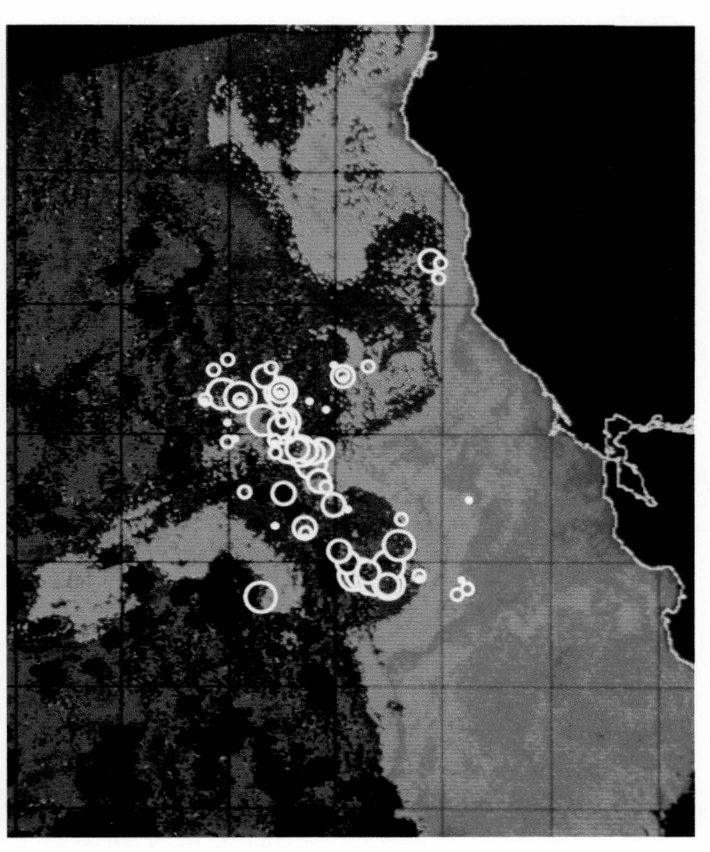

Fig. 4.17. Similar data to Fig. 4.16 for 29 September 1981, indicating location and size of tuna catch in the associated four-day period. (Courtesy of M. Laurs, NOAA.)

Exploiting and managing the food resources of the oceans is a goal of supreme importance in boosting the protein consumption of a majority of the world's population. The vast areas involved and the ephemeral nature of offshore conditions makes it a task of great complexity. Fishermen are obsessed by finding ways to catch more fish at least cost in the shortest time and at least risk to their boats and crews. Even among the most well-equipped fleets, subjective lore about fish-finding dominates their activities. In the case of the deep-water tuna fleets operating from California, part of the lore is that water colour is a good guide to productive fisheries. Tuna are pelagic predators high in the food chain that has its base in the marine productivity of phytoplankton. They locate their prey using sight, and so remain in clear water. However, prey species feed among the murky plankton-rich waters made fertile by upwellings of cold bottom water and the oceanward drift of coastal waters. The CZCS carried by the Nimbus-7, as discussed several times earlier, was designed in part to measure ocean colour. Figures 4.16 and 4.17 are CZCS images of the area of the Pacific off San Francisco—at mid-right—for two days, a week apart in September 1981. Water colour is measured as the ratio of blue to green reflectance and colour coded from red (high) to blue (low) in terms of the concentration of planktonic pigments. The location and size of weekly catches from the tuna fleet is shown as a series of circles representing number of fish caught. Even over such a short period of time, the patterns in the ocean water have changed markedly. One thing however is clear: the majority of fish have been taken from close to the boundary between clear and plankton-rich water, as the fisherman's lore predicted. Finding the boundary is, however, extremely difficult in such a huge area of ocean. Timely satellite images in the form shown obviously help improve the efficiency of the fleet considerably. Unfortunately, the Nimbus satellite was experimental and no longer operates. Until it is replaced with an operational system, such applications depend to a great extent on results from meterological satellites, such as the TIROS/NOAA series. They do not carry such an accurate means of estimating marine biological productivity, and estimates depend on using the weak red contrast associated with near-surface plankton together with indications of cold upwellings of nutrient-rich water from the thermal scanner.

4.3 Physical resources

Physical resources are generally taken to be commodities won from the rocks that make up the Earth's crust. They include building and other constructional materials, inorganic feedstock for chemical manufacture, the ores of metals, fuels formed by the preservation and change of ancient organic materials, such as oil and coal, and nuclear fuels such as uranium. Such resources form the primary input to many of the basic industries of highly developed society. If the justifiable aspirations of the underprivileged majority to standards akin to those available in the developed world are to be satisfied, then a very large increase in the discovery of these resources and their exploitation is required. Remote sensing has the ability to assist in this because of its low cost and high efficiency, compared with the conventional methods of field survey and prospecting.

Locating physical resources relies primarily on knowledge of the general geological make up of an area, to which remote sensing can contribute a great deal about the disposition of rocks by geological structures, and the distinctions between common rock types, which was partly covered in Section 3.6. Because of the factors mentioned there that mask a true picture of rocks exposed at the surface, and because the bulk of resources are hidden below surface, it is important to note that remote sensing is unable to locate them directly, except in rare cases. Instead, it aims at efficiently outlining geological conditions and highlighting unusual features that are commonly associated with economic deposits. Basic geological mapping using conventional field techniques is a slow task, requiring boundaries and distinctive outcrops of rock to be followed painstakingly. Exploration focused on the eventual discovery of a mine of oilfield involves extremely costly methods. The role of remote sensing in this context is twofold. First, by allowing observations from a limited number of field locations to be extrapolated over large areas, it makes the work of geologists more efficient and frees them to make more detailed observations on areas of great interest. Secondly, by helping identify unusual geological features to which resources may be related it cuts down the area over which high-cost exploration methods need to be deployed. Both can lead to considerable savings in exploration, thereby reducing the inevitably high economic risk associated with seeking the unknown. Decisions about withdrawal from probably unproductive areas, or committing to higher cost, more intensive ground exploration, can be made much earlier—a goal of most exploration managers.

Figures 4.18 to 4.21 demonstrate contributions of remote sensing to geological mapping, in addition to those already covered in Section 3.6. Figure 4.18 combines high-resolution Landsat MSS data in the very-near infrared with coarse-resolution day and night thermal infrared from the experimental HCMM system, as a false-colour image. This combination allows the albedo of the rocks together with the rate at which they heat and cool—dependent on thermal properties and density—to contribute to rock discrimination in this

Fig. 4.18. False-colour image of part of Morocco, combining Landsat MSS and HCMM data. Width is 160 km. (Courtesy of R. Haydn, GAF.)

remote and arid area of Morocco. The brownish, orange, and pink strata are sandstones and carbonates resting on more brightly coloured, ancient igneous and metamorphic rocks. The area is one of potential for phosphate deposits.

Figure 4.19 displays the first, second, and third principal components of Landsat MSS data (Section 2.5), whose brightness is modulated by Seasat radar data, in red, green, and blue. The radar data improves the resolution from 80 m to 20 m and adds geologically important information about surface roughness to that on albedo and iron staining carried by the MSS data. The broad structure and details of the different sedimentary units folded by the San Raphael Swell are clearly shown. Of particular interest are the bright yellow

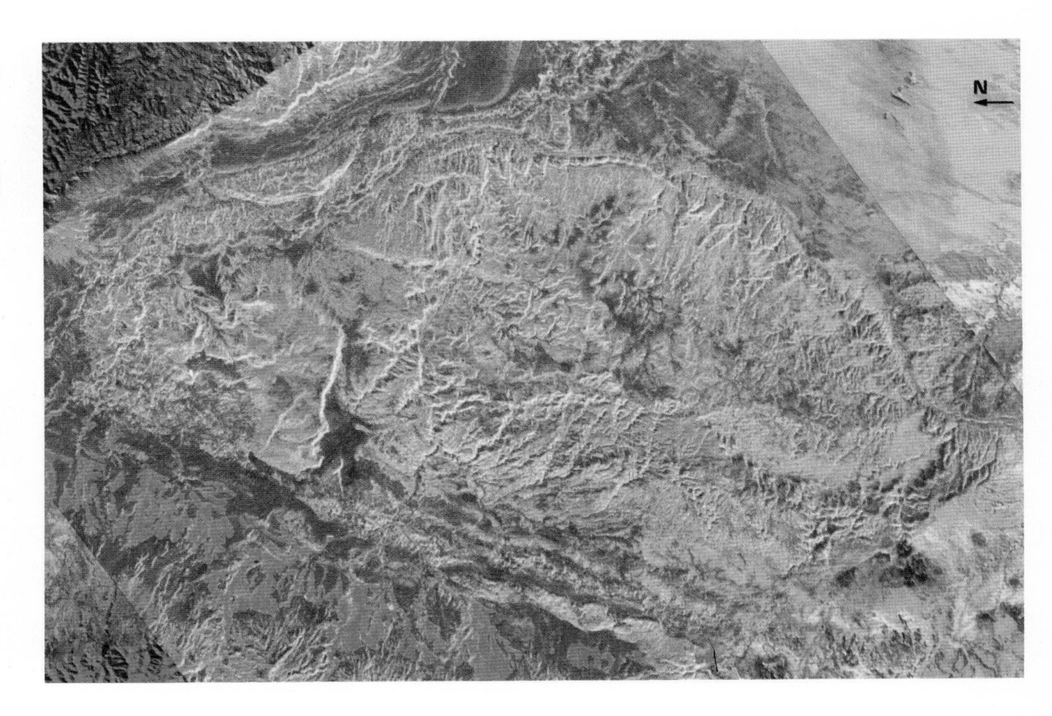

Fig. 4.19. False-colour image of the San-Raphael Swell, Utah, USA, combining Seasat radar and Landsat MSS data. Width is 100 km. (Courtesy of A. Kahle, JPL.)

Fig. 4.20. Approximately natural-colour image of airborne multispectral-scanner data of an area in Western Australia. Width is 14 km.

coloured rocks on the left at the northern part of the dome and in its core at the southern end. Both are sandstones of different ages which have associated uranium reserves.

One of the most difficult areas in the world for geological mapping is the centre of Australia. Because of the widespread development of deep, iron-rich soils, the monotonous natural red coloration of the surface masks all but the most meagre details of the geology, as shown by Fig. 4.20. However, the soil contains a number of minerals such as opaline silica, clays, and serpentine to which different parts of the near-infrared spectrum are sensitive. The different rocks break down to produce different proportions of these minerals. In Fig. 4.21 the three reflected infrared bands deployed by the Landsat Thematic

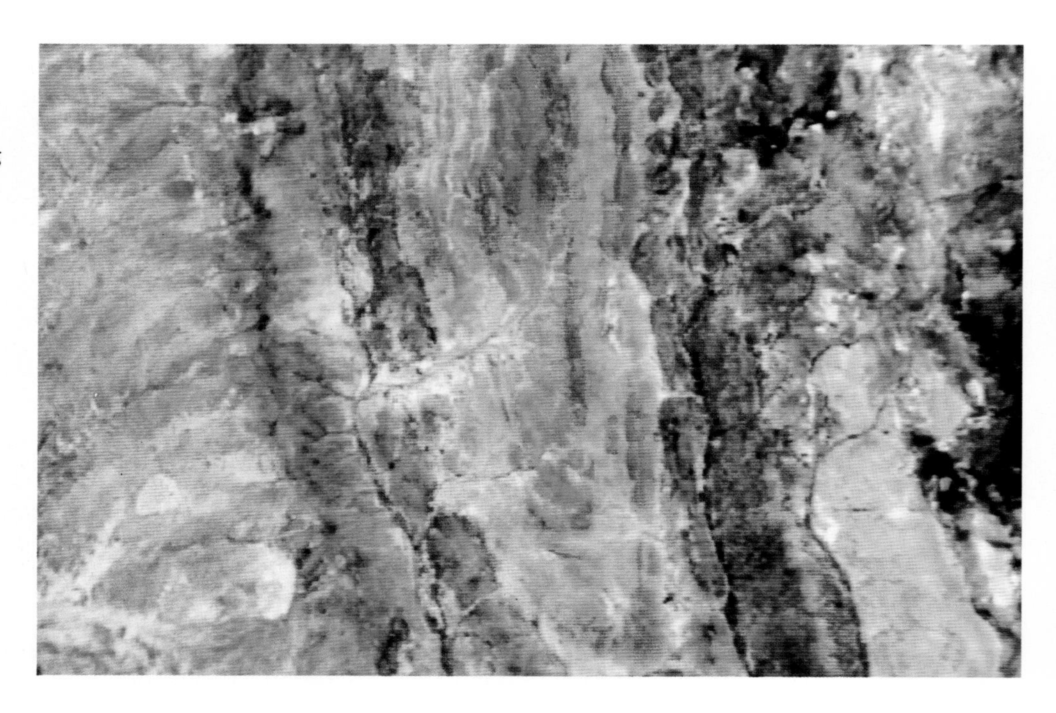

Fig. 4.21. False-colour image of the same area as shown in Fig. 4.20, using the equivalents of Landsat TM bands 7, 5, and 2 as red, green, and blue components.

Mapper are used as red, green, and blue components. The result is a dramatic demonstration of what is easily feasible in such intractible terrain. The different colours relate to a number of subtly different basic and ultrabasic lava flows, some of which host gold mineralization and others contain sub-economic concentrations of nickel.

Exploration for metal resources carries the highest of all economic risks that expenditure may not be matched by returns in the shape of mineable deposits. Because they are associated most frequently with very small but highly anomalous areas where a great many processes have all acted together to concentrate the metals involved above their normal abundances, it is very easy to miss even very high value deposits in the field. However, the anomalous processes involved produce unusual rocks and minerals associated with ores. It is towards these that remote sensing is directed. One feature common to many deposits, particularly those carrying economic gold and copper is that they contain large amounts of worthless iron sulphide. When weathered this material is converted to red, orange, and brown iron oxides and hydroxides. Although these minerals help identify brightly coloured altered zones to the field geologist, they frequently reside among other rocks with very much the same colour. As introduced in Section 2.2, these iron minerals have very distinct spectral signatures in the very-near infrared. These, together with their coloration, can be expressed with much more contrast to surrounding unmineralized rocks, using ratios of the bands involved in the spectral features. Figure 4.22 is a Landsat MSS false-colour composite image of part of the Chilean Andes, famous for its copper deposits. The lack of vegetation and the general monotony of the volcanic rocks in which these deposits formed leads to an almost featureless image. There are a few small white patches that are active mines. Figure 4.23 shows an image where the primary colours red, green, and blue are controlled by the ratios between MSS bands 6:7, 5:6, and 4:5. This is much more geologically informative, with iron-rich areas showing as orange tinges on the left of the image. The more the iron minerals present, the brighter the coloration. This is because the first ratio is always high for iron minerals with an absorption feature in MSS band 7, the 5:6 ratio is less than one, but still relatively high, and the 4:5 (green:red) ratio is always very low for reddish materials. The result is that all the active copper mines are identified, but so are many other previously unknown altered areas. Field studies oriented by this image did in fact locate important new showings of copper in several of the targets identified.

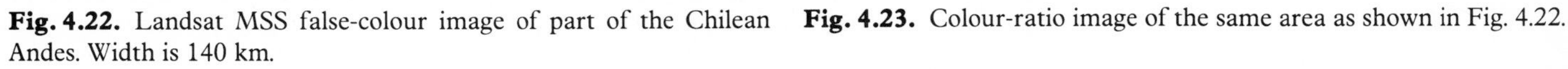

Fig. 4.22. Landsat MSS false-colour image of part of the Chilean Andes. Width is 140 km.

Fig. 4.23. Colour-ratio image of the same area as shown in Fig. 4.22.

Iron minerals are not always present above copper mineralization, and in many cases, iron-rich surface alteration has no mineable ores beneath it. Another set of minerals that are commonly associated with mineralization, due in their case to chemical alteration of rocks by the fluids that carry and precipitate metals in a very common category of ore deposits, are clay minerals. In the visible region they are very highly reflective white minerals, but so indeed are many other minerals, so this feature is no help. Where they differ from other minerals is in the 2.1–2.3 μm part of the near infrared. Due to their content of hydroxyl (OH) ions bonded with aluminium, clays have strong but sharp absorption features in this region (Section 2.2). The Landsat Thematic Mapper, at the behest of exploration geologists, carries a broad-band scanner (TM band 7) in this region to detect clays. Figure 4.24 is an airborne simulation of a TM colour-ratio image over a well-studied mining area in Nevada, USA. It combines TM band ratios 5:6, 5:2, and 3:4 as red, green, and blue. The clay-rich parts of the altered zone show red, due to the strong absorption in band 7, and low values of the other ratios for clays. The greenish areas are iron-rich. One of the most important characteristics of the mineralization is that silica has been added in certain areas. This reduces the clay content, but the areas of silicification are only vaguely shown in dark brown and bluish hues. One means of resolving the alteration patterns in more detail is to use the discriminating power of the varied thermal emission spectra of common minerals (Section 2.2). Figure 4.25 is for exactly the same area, based on three bands from the thermal spectrum gathered by a Thermal Infrared Multispectral Scanner (TIMS) on board an aircraft. The most highly silicified rocks show in oranges, and places where clays and silica are mixed are well discriminated as magenta patches. Although the study area is well known and no new deposits are located by the images, the potential of combining both methods is obvious. Many highly metalliferous deposits formed by alteration of common rocks deep in the crust are heterogeneous, and it is only the presence of a few of the zones that is connected with paying metal ores.

Research on mineralization involving the alteration of primary minerals to clays reveals that being able to distinguish between the many different species of clay minerals can be extremely important in assessing the mineral potential of an area. In the 2.1–2.3 μm region, the different clays have quite markedly different spectral features. Unfortunately these

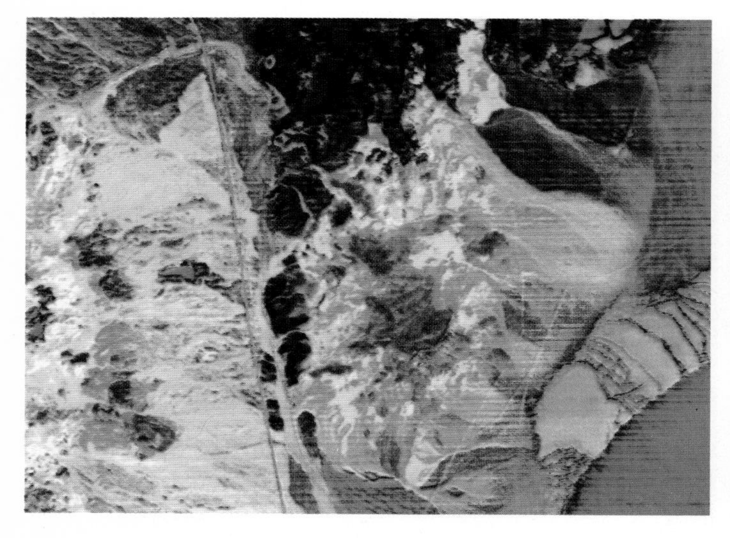

Fig. 4.24. Airborne simulation of Landsat colour-ratio image of an area of copper mineralization in the western USA. Width is 6 km. (Courtesy of A. Kahle, JPL.)

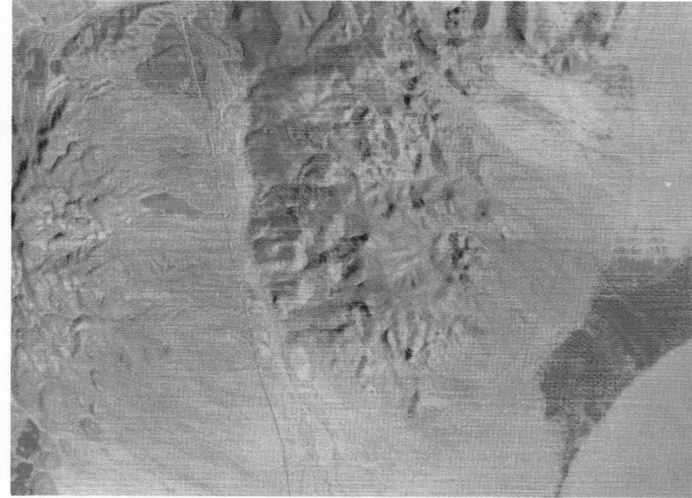

Fig. 4.25. False-colour image using data from narrow bands in the thermal infrared, for the same area as Fig. 4.24. Width is 6 km. (Courtesy of A. Kahle, JPL.)

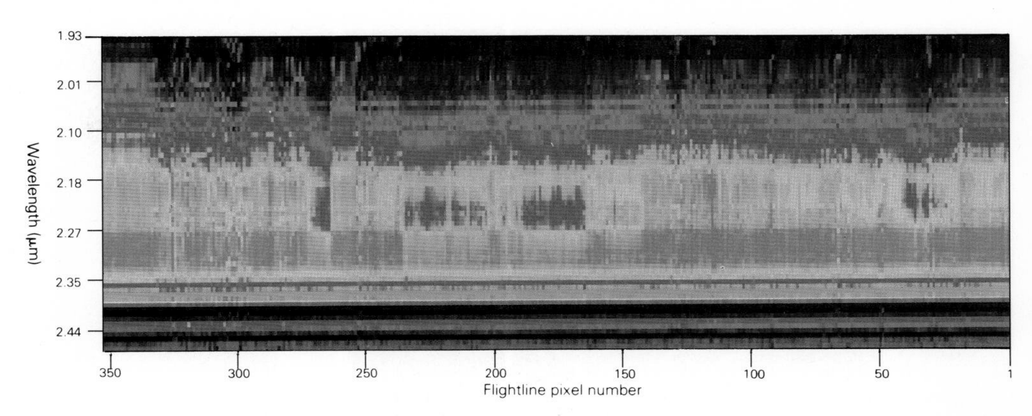

Fig. 4.26. Colour-coded spectral data from an airborne spectroradiometer. (Courtesy of S. Marsh, Sun Oil.)

cannot be discriminated by the wavebands employed by general-purpose remote-sensing systems, such as Landsat TM, because the bands imaged are too broad. Moreover, the spectral effects in this region due to other minerals are confused with those of clays. To overcome this difficulty new types of device have been built, which gather data from a large number of very narrow wavebands. Effectively, they measure spectra remotely rather than directly using a laboratory or field spectrometer. Using CCD technology (Section 2.3) and special diffraction gratings to split the spectrum now means that spectral resolutions down to a few nanometres are quite feasible. Figure 4.26 is a display produced by such an airborne spectrometer. Each vertical line represents a spectrum from 1.9 to 2.45 μm from a pixel along the flight-line. The colours from red through yellow, green, and blue to magenta are from high reflectance to low. There are several patches where reflectance is low in the 2.1–2.2 μm region. These are due to clay minerals at the surface. The broader troughs are due to one clay species, the narrower ones to another. Although involving very large volumes of digital data, this kind of spectrometry is now possible for swathes of ground rather than along a single line. One of the ways in which such imaging spectrometer data can be used is illustrated by Fig. 4.27. Each pixel on the image has its own unique spectrum relating to the mixture of surface minerals that it encompasses. For different parts of the spectrum different minerals will dominate this spectrum, if they are present in appreciable amounts. By using a computer to compare the actual spectra with the ideal laboratory spectra of different minerals, pixels that probably contain the mineral in question can be highlighted on the image and their spectra displayed for comparison with the standards. In Fig. 4.30 this has been done for two important clay minerals in a traverse across the mineralized zone shown in Figs 4.26 and 4.27. On the image, kaolinite-rich areas show up as magenta.

Most physical resources are anomalies compared with common rocks, that is, they contain unusually high concentrations of the sought-after elements and compounds. This results in chemically anomalous soils and ground-water, to which field exploration is often directed. An important result of these environmental anomalies is that they may result in changes in the vegetation growing on them. Sometimes peculiar species of plants grow, but more usually the excessive amounts of some elements have an adverse effect on common plants, which are then said to be stressed. This takes the form of stunting and changes in the leaf pigmentation, commonly yellowing the normally green leaves. Detecting such geobotanical anomalies has long been a field exploration tool. Because many remote-sensing systems were deliberately oriented towards vegetation monitoring they lend themselves admirably to large-scale geobotanical reconnaissance.

All the chromium and nickel used in the world's steel industries is won from igneous rocks with low silicon and high magnesium content. These major element deviations from

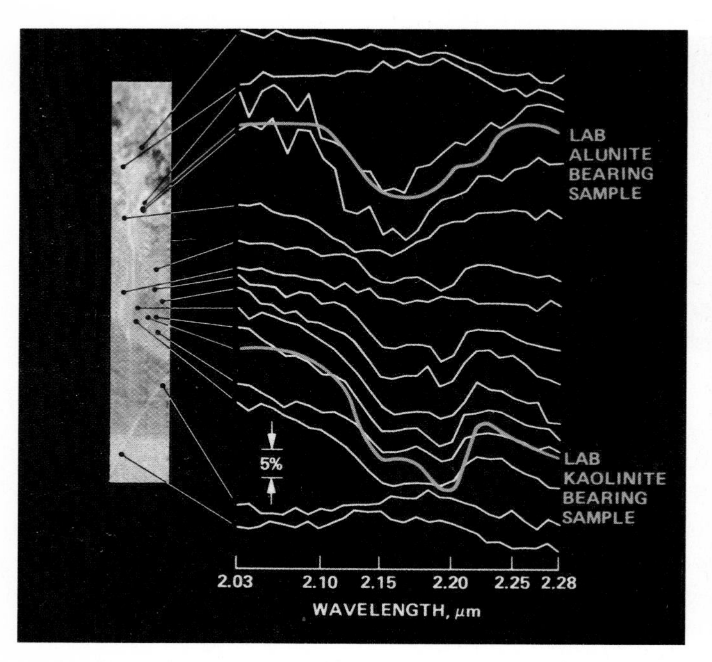

Fig. 4.27. Image and spectra extracted from airborne imaging spectrometer data. Note that the yellow spectra are from a reference set of laboratory samples. (Courtesy of M. Abrams, JPL.)

Fig. 4.28. Colour-ratio image of Landsat MSS data of part of northern California, USA. (Courtesy of G. Raines, USGS.)

the norm do not in themselves stunt vegetation. However, the high concentrations of several metals, including chromium and nickel, do have a toxic effect. Figure 4.28 is a colour-ratio Landsat MSS image of the humid coastal region of northern California, which normally supports a dense cloak of coniferous forest. Where such ultrabasic rocks outcrop, conifer growth is strongly inhibited. The image combines ratios of MSS bands 4:5, 4:6, and 6:7, which are sensitive to vegetation density, as red, green, and blue. Forested areas are red, but the poorly wooded areas in blue correspond almost exactly with outcrops of ultrabasic rocks, shown as black outlines.

In oilfields, one of the fluids trapped together with liquid hydrocarbons is natural gas. No hydrocarbon trap is completely secure from leakage, and the gas tends to seep out slowly, eventually to reach the surface. Whereas there are now extremely sophisticated and highly confidential means of 'sniffing' the presence of such natural seepages, the gas itself has some unusual results. It sustains various species of bacteria, which themselves produce highly acid conditions in ground-water. These conditions inhibit the symbiotic organisms that are essential to the uptake of nutrients by plant rootlets. If trees' roots penetrate below the depth affected by these gas-consuming bacteria, they begin to die. Consequently only shallow-rooted trees survive over gas seeps. Figure 4.29 shows how such unusual stands of trees, in this case maples showing on this colour photograph in their yellow autumn colours, form an excellent indicator of seeps, and possibly the reservoirs from which they emanate, in areas thought to be underlain by oilfields.

As explained in Section 2.5, remote sensing is not the only way in which images relating to an area can be produced. By converting the results of various kinds of non-imaging survey, such as data from points or along lines, to a series of interpolated values of the measured variable for square cells in a grid, it is possible to display them using an image-processing computer in exactly the same way as a true image. Moreover, these artificial images can also be enhanced in many ways and combined together with remotely sensed

Fig. 4.29. Stand of prematurely senescent maple trees (orange) in Virginia, related to gas seepage. (Courtesy of H. Lang, JPL.)

data and other kinds of geographic information. There are great advantages in doing this. First, the conventional way of displaying non-image data is in the form of contour maps of the variable, to which the human eye is certainly not attuned, because they are completely unnatural. An image version, especially when it has clues to depth and is coloured (Section 2.6) is far more easily interpreted. Second, contouring actually loses part of the information available, because it consists of a simplification of the data into a smaller number of steps than there are recorded values. Third is the ability of images of very different kinds of data in a common format to be combined together. This allows unsuspected correlations to be seen very easily. In exploration, where efficiency and extracting the full economic benefit from often very costly or useful data is of primary importance, these methods of data display and combination are assuming increasing importance.

Figures 4.30 and 4.31 are taken from a study of a small, potentially hydrocarbon-bearing, sedimentary basin in northern England. Figure 4.30 is a conventional geological map, showing very old unproductive rocks in purples and greyish hues. The potentially oil-bearing rocks are in bright blues, greens, and yellows, while younger unproductive rocks are coloured in orange. The area of interest runs NE from the Solway Firth (the inlet of sea) to the top-right corner of the map. Although much is known about the surface geology, oil occurs at considerable depth. Using seismic methods of subsurface exploration is high in cost, and to drill for information would be prohibitively expensive without have zeroed in on the areas most likely to contain hydrocarbons. Since geological structures beneath the surface cause the depths to high- and low-density rocks to vary, the gravitational field varies slightly according to the subsurface structure. Gravity can only be measured accurately at well-surveyed points on the surface, and its variation is usually displayed on contour maps. Figure 4.31 shows the gravity data in image form, after computer processing to render the variations in colour and to highlight the gravitational 'topography' by illuminating it from one side. Blues and greens indicate lower than average gravity, and therefore deeply buried low-density rocks—mainly the sediments in which oil may have accumulated. Showing yellow through red to magenta are the areas of high gravity where the high-density deep and unproductive rocks are closer to the surface. Quite a lot of the patterns confirm the surface mapping, such as the high over the exposed dense, old rocks and the marked straight line running NNW at the bottom right parallel to a major fault. In

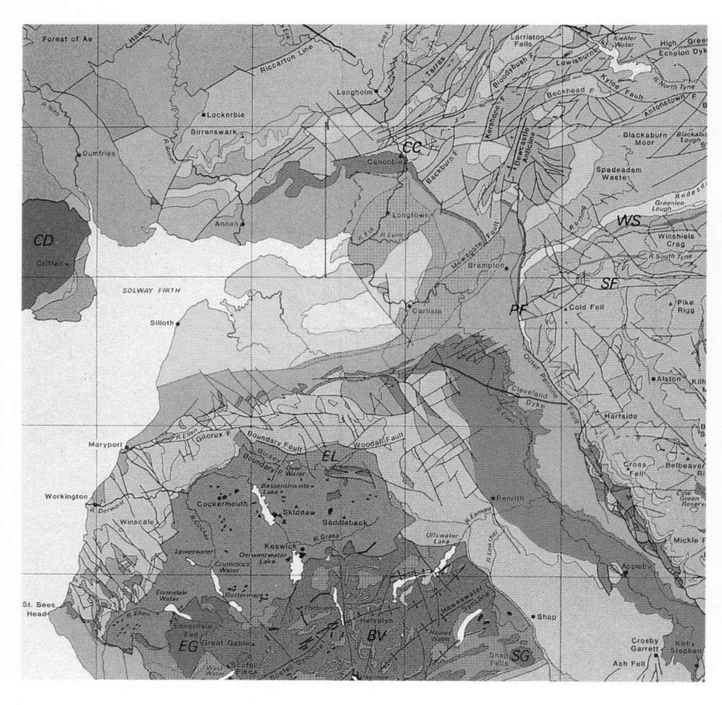

Fig. 4.30. Geological map of part of northern England. Width is 100 km. (Courtesy of the Northumberland Natural History Society.)

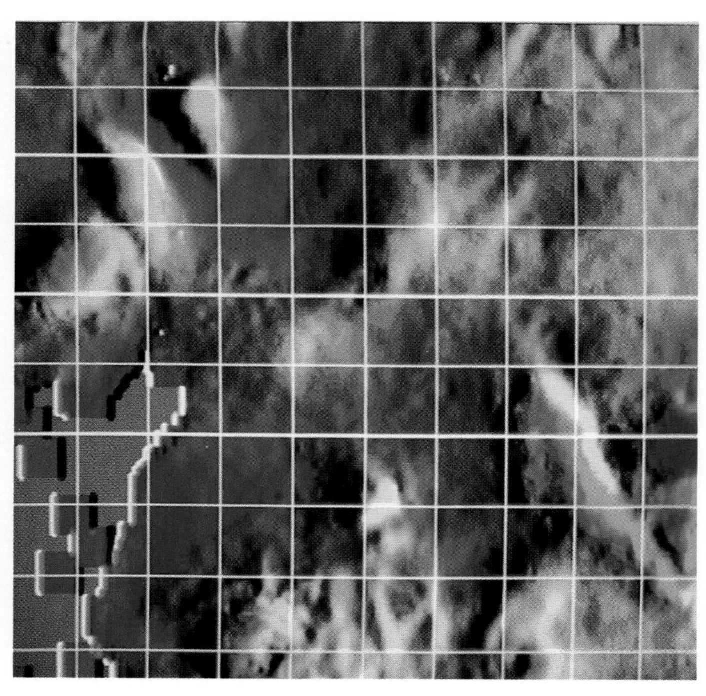

Fig. 4.31. Coloured image of gravity data illuminated from the side, corresponding to the area shown in Fig. 4.30. Width is 100 km.

the basin itself, however, the gravitational image indicates many subsurface structures that are unsuspected from field work. They are not apparent on the contoured gravity map, and would help guide seismic surveys far more cost-effectively than the geological map.

Figures 4.32 and 4.33 illustrate a rather different set of data for an area in New South Wales, Australia. There the target is uranium and other metal resources. Figure 4.32 is a geological map, where ancient igneous and metamorphic rocks are shown in red, blue,

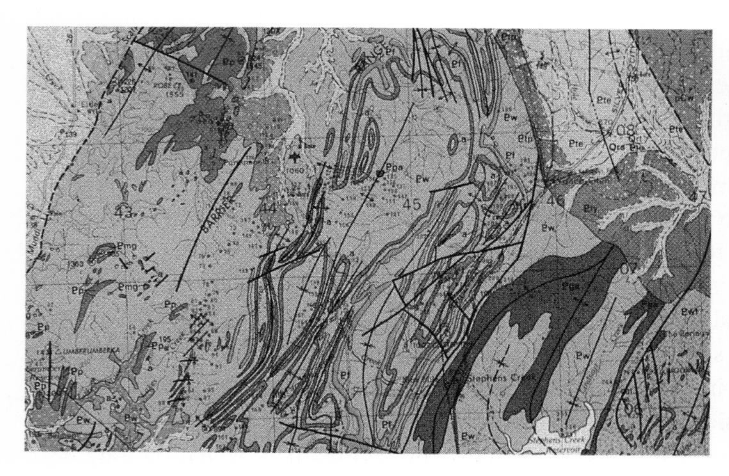

Fig. 4.32. Geological map of part of New South Wales. Width is 30 km. (Courtesy of the Geological Survey of New South Wales.)

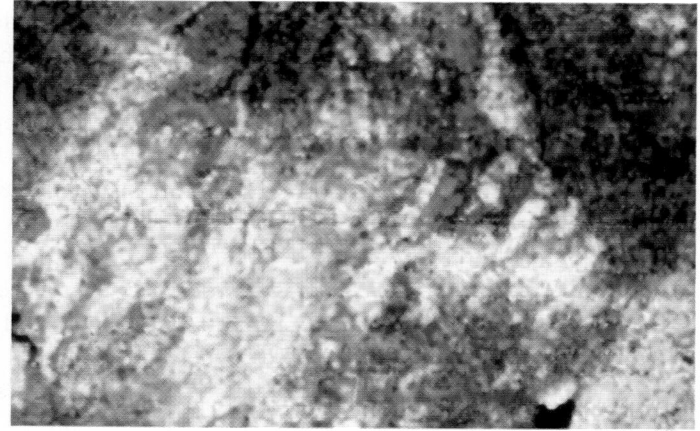

Fig. 4.33. Image of airborne gamma-ray spectrometry data for the same area as shown in Fig. 4.32. Red is controlled by energy emitted by uranium, green by thorium, and blue by potassium. (Courtesy of BMR and CSIRO.)

green, and pinkish-grey. A series of old sediments laid down during a 750 million-year-old glacial epoch are in orange and brown, and younger, mainly superficial sediments are in yellows. Figure 4.33 is an image whose red component is controlled by the variation in gamma-ray emission caused by uranium, green by thorium-emitted gamma rays, and blue by those due to potassium. The simple rules of additive colours show rocks rich in each of the three elements and neither of the others as primary colours, and those where two or all three are present in similar amounts as yellows, magenta, and creams. There are two important features to note. First the radiometric image shows much of the variation in surface geology quite well, the old glacial deposits showing as a dark patch because they are depleted in all three elements, and the young sedimentary cover as greens and blues due to their low uranium. Of economic interest are the red areas, where uranium is very much higher than in the surrounding rocks. Different proportions of the three elements are a good guide to the processes that formed the rocks at the surface, so images of this kind are unique in directly identifying rock types rather than a few spectrally unusual minerals.

4.4 Communications

If development is to spread to the underprivileged parts of the globe, probably the primary requirement is the provision of road and rail links. Very rarely is the most efficient potential route between two points a straight line. There are many important considerations. Hazards due to potential flooding, subsidence, and landslips must be avoided or at least identified. Heavy vehicles demand gradients below a limiting value, and, of course, suitable materials for construction must be as close as possible to the new route. This is by no means guaranteed, as modern roads require ballast and wearing course aggregates with stringent specifications. To a large extent, much of this information is available from images such as those shown in Sections 3.3, 3.6, and 4.3. Indeed, remotely sensed images provide a great deal more information about the terrain to be crossed by a new route than do conventional maps. One important set of information is generally missing from them, however. In order to estimate the appropriate gradients involved in a route and the volume of debris that must be shifted requires accurate information about the topography. Until recently this was only available from contoured maps, themselves often unavailable in remote and poorly developed areas, or by painstaking and lengthy photogrammetric measurements from stereoscopic aerial photographs.

The SPOT system, being capable of oblique views of the Earth's surface, has the potential for overlapping stereoscopic coverage of the whole land area of the globe. Because the image data is provided in digital form it is now possible, using complex computer software to derive elevations for any point automatically within an area covered by overlapping oblique scenes. This is based on the apparent shift in the positions of objects at different elevations from one scene to the other, due to the oblique viewing. The higher the surface, the greater the apparent shift. Despite this, much the same patterns are present on both images. Put extremely simply, the computer seeks the same patterns on both scenes and measures the relative shift of each pixel constituting the pattern. The shift is then converted to an elevation. The great advantage of this method over conventional contouring is that an elevation is calculated for every 10 m^2 pixel in the area to produce a digital elevation model. Given an array of elevations it is then a simple matter to calculate slopes over the whole area. Moreover, the wealth of information about the nature of the land surface stemming from the imagery itself can be combined with the elevation data to produce a three-dimensional model of the surface within the computer. Figure 4.34 shows a dramatic and extremely useful means of manipulating such a model. Instead of just appearing as a map, the terrain is displayed in perspective as if viewed from a high vantage

Fig. 4.34. Simulated perspective view of part of the Alps based on a digital elevation model and Landsat TM data. (Courtesy of ESA.)

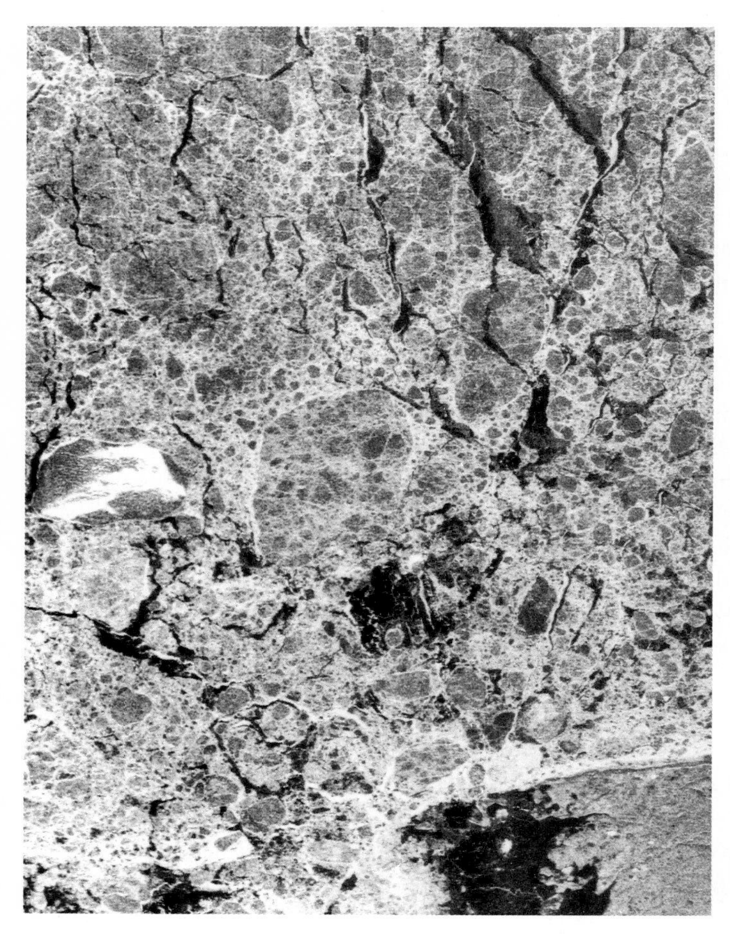

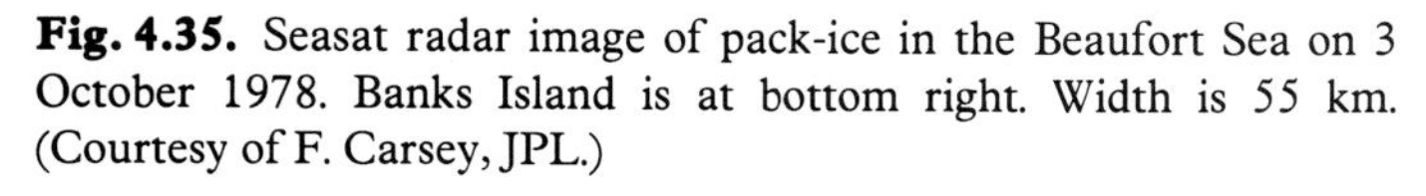

Fig. 4.35. Seasat radar image of pack-ice in the Beaufort Sea on 3 October 1978. Banks Island is at bottom right. Width is 55 km. (Courtesy of F. Carsey, JPL.)

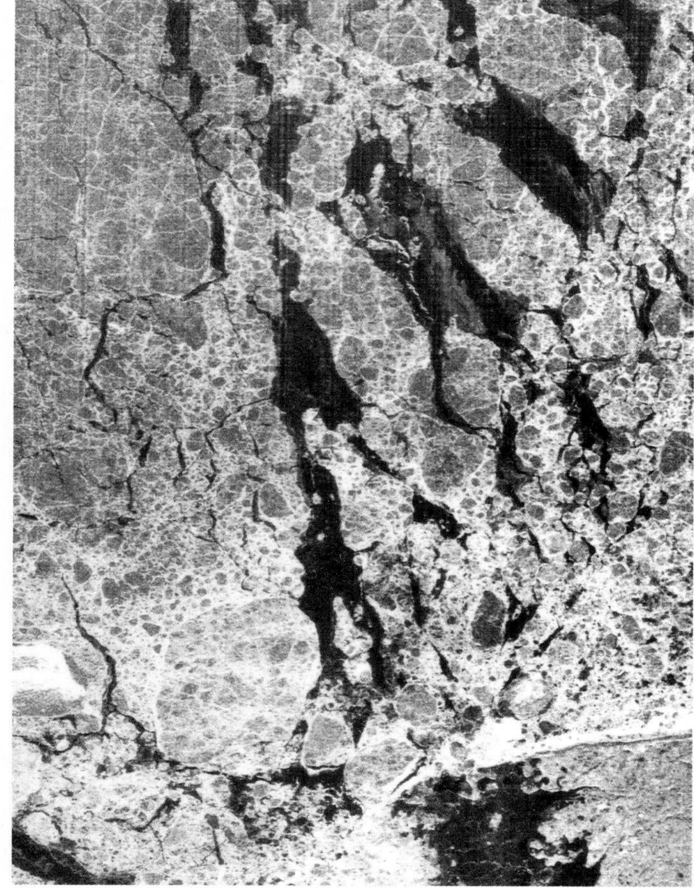

Fig. 4.36. Seasat radar image of pack-ice in the same area of the Beaufort Sea as Fig. 4.37 on 6 October 1978. (Courtesy of F. Carsey, JPL.)

point. Every critical part of a proposed route can be viewed from any angle to assess its feasibility and any apparent risks or difficulties. Although field studies are still essential, most of the routing and planning can be done at the work station with a high level of efficiency.

Normally, marine communications pose no problems in terms of routes, although if timely information is available diversions can be planned to avoid storms. Such information is available now on a routine basis from the various meteorological satellites in geostationary orbit (Sections 3.1 and 3.2). As indicated in Section 3.2, experiments with satellite radar remote sensing can supplement meterological observations with information on wave patterns and sea-surface roughness to further refine modern navigation. This is one of the intended purposes of the ERS-1 and Radarsat systems (Section 5.4).

The search for energy resources has partly focused on the potential wealth of the Arctic Ocean north of both Canada and the Soviet Union. Development is currently hindered because of the great navigational problems for survey vessels and tankers posed by the constantly shifting polar ice-pack. During the brief summer, the Arctic is not as locked by ice as might be imagined, but its movement by wind and currents is extremely complex and unpredictable. Figures 4.35 and 4.36 are radar images from Seasat of the same part of the polar pack in the eastern Beaufort Sea, in the Canadian Arctic. They were taken three days apart in October 1978. At the bottom right of each is part of Banks Island. Even over such a short time there have been considerable changes in the ice state. In the later image many large leads have developed, but of great interest is the movement of two large adjacent floes, just below centre on Fig. 4.35. Three days later they have moved about 20 km southwards. Using the floes as points of reference enables both stable features in the ice and new ones to be detected, as well as calculation of the directions and rates of movement in the pack-ice. Information of this kind would be invaluable for navigation in such waters.

Navigation is not only a problem at sea, but also along major rivers traversing plains comprised of soft sediments. The natural tendency of such rivers to meander, and for shoals and sandbanks to develop as a result of sedimentation means that their courses change quickly, unless enormous expenditure is invested in stabilizing their banks. Figure

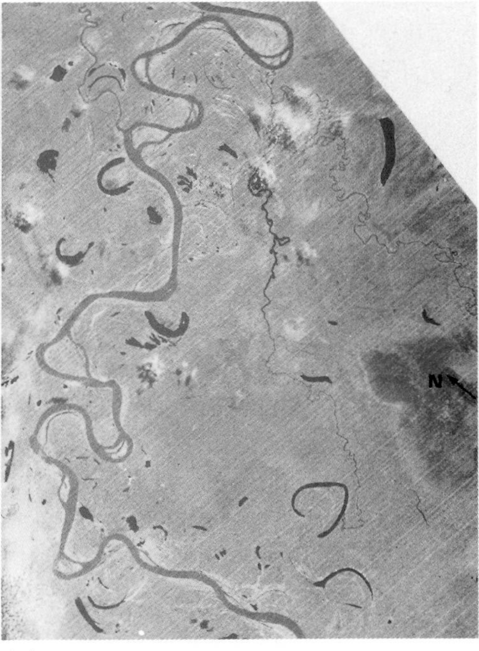

(a)

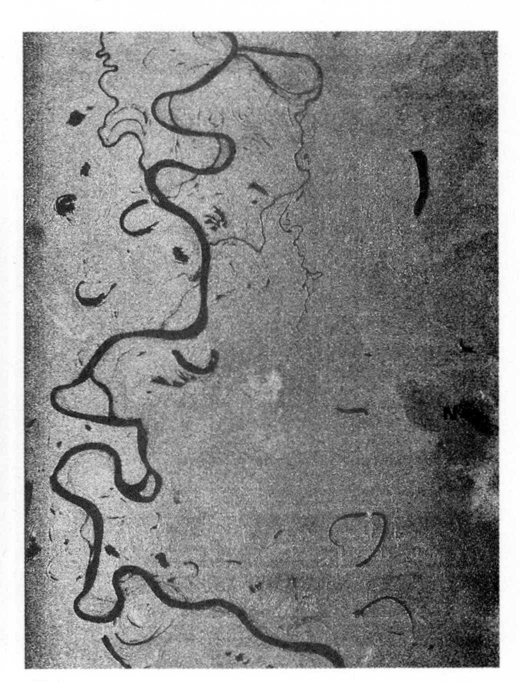

(b)

Fig. 4.37. (a) Landsat MSS band 5 image of part of the Rio Ucayali, Peru for October 1972; (b) SIR-A radar image for the same area as (a) for November 1981. Width is 50 km. North is at top left.

4.37 illustrates the problem nicely. The left image is from Landsat in 1972, that on the right from the SIR-A radar experiment in 1981. A few moments' comparison of the two images shows considerable changes in the course of the Rio Ucayali and its tributary over the nine-year period. Repeated coverage, by radar because this area is generally swathed in cloud, can provide the only economic means of updating navigational charts for this and other inland waterways in such remote areas.

4.5 Disasters

There are two aspects to the use of remote sensing in the context of life-threatening or economically disastrous natural hazards. One is to give forewarning so that measures to blunt the worst effects can be taken, the other is to monitor the effects of a disaster soon after it occurs so that relief can be as efficient as possible.

Floods possibly constitute the most common and destructive of all natural hazards, and the threat from them is very widely distributed. The areas most prone to flooding are along the courses of great rivers, particularly in their deltas. Floods distribute silts and clays over wide areas to produce deep and highly fertile soils that are easily irrigated. The agricultural advantages outweigh the threat of floods, and deltas in temperate and tropical areas are often the most densely populated regions on the globe. Floods can rise due to unusually

Fig. 4.38. Landsat MSS false-colour image of SW Queensland, during floods. Width is 140 km.

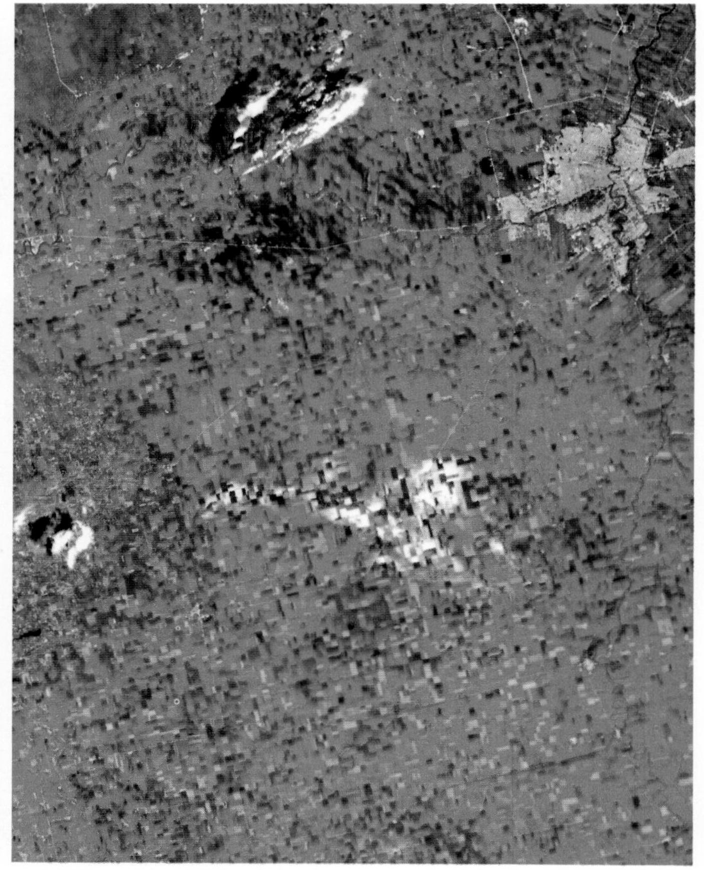

Fig. 4.39. Landsat MSS false-colour image of part of Manitoba, Canada, for 18 July 1984, with Winnipeg at top right and tornado damage at centre. Width is 85 km. (Courtesy of CCRS.)

high rainfall, and can also result from melting of deep snow-masses that accumulate high in the river's catchment. In this case estimates of snow cover and depth can be obtained from satellite imagery. Where retrospective information relating to snowfall associated with earlier floods is available it becomes possible to assess annual risk in the lower reaches of the drainage system.

Figure 4.38 shows a huge and unpredicted flood that occurred in south-western Queensland, Australia, in early 1974. The flooded area is in various shades of blue. As it happens, the area is normally relatively arid and unproductive. The monsoonal rains were between four and six times the average, up to a metre of precipitation having fallen in the headwaters of what is part of a drainage system towards the centre of the continent. The extent of the flood is a measure of the very low relief in the area, so that water was able to flow outside the usual channel system to occupy virtually the whole of a structurally controlled topographic basin. Had this event taken place in a populated area the image would have served several purposes. First, the areas of unflooded ground standing a few metres above the water level show as orange 'islands'. It is there that survivors and their livestock would have been most likely to congregate and where relief would be most beneficial. The low relief of the area would have made these dry patches difficult, if not impossible to predict from topographic maps. Second, the stream lines picked out in darker blues show where current flowed most strongly and where soil would have been stripped to devastate the land for future replanting. The lighter-blue areas are sluggish silt-laden water where the soil may even have been improved by flooding.

Natural disasters caused by windstorms are generally restricted to areas of high climatic instability. For instance hurricanes and typhoons are generated by the mixing of air masses of different temperature and density at the intertropical convergence. Meteorological satellites are extremely efficient at detecting such cyclonic systems in their infancy, and enable them to be tracked in detail so that adequate warning can be given. Tornadoes or whirlwinds are far less predictable as individual entities, although areas prone to them are well known from historical records. They spring up suddenly, move rapidly, and die away equally quickly. Although small they are intensely destructive, and because they are so unpredictable, the best that remote sensing can do is to assess the location and extent of damage. Figure 4.39 shows a graphic example of the devastation that they can cause where they touch the surface. At top left is the Canadian city of Winnipeg, surrounded by a red patchwork of cereal fields in the prairies of Manitoba. The pale streak just below centre is the track of an isolated tornado that has completely stripped the fields of vegetation.

As shown in Fig. 2.12 the wavelength at which maximum energy is emitted by a body is proportional to the fourth power of its absolute temperature. The emission spectrum of a body also has a characteristic shape. Because of these factors, normal temperature variations at the surface, in bodies of water, and in the atmosphere can only be monitored in the mid-infrared between about 8.0 and 14.0 μm and in the microwave region, virtually no energy being emitted in the visible and near-infrared. However, where temperatures are much higher than normal there is significant emission in the near-infrared, although temperatures over 1000 °C are needed for any radiation to be detectable in the visible part of the spectrum.

One of the greatest hazards in forested or scrubland areas is fire, ignited either by lightning strikes or accidentally. The hazard is especially high in hot, dry climates and where the trees incorporate highly flammable resins, as in softwoods and eucalyptus. In fact, some kinds of woodland, such as eucalyptus forest, require regular fires in order to regenerate naturally. As well as destroying valuable timber, forest fires pose threats to life, property, and air traffic, and the smokes which they generate can have significant effects even on global climate. Figure 4.40 shows a forest fire on a natural-colour image. Although

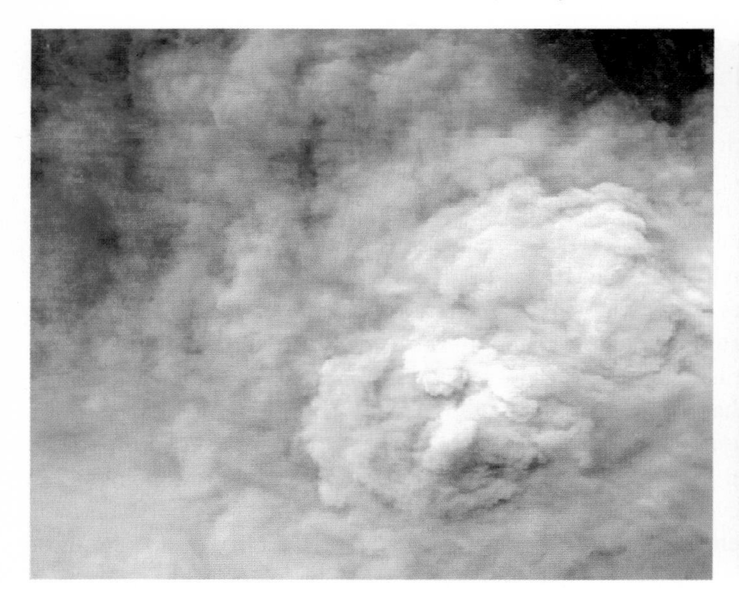

Fig. 4.40. Natural-colour airborne scanner image of forest fire in California. Width is 5 km. (Courtesy of J. Myers, Ames Research Center, NASA.)

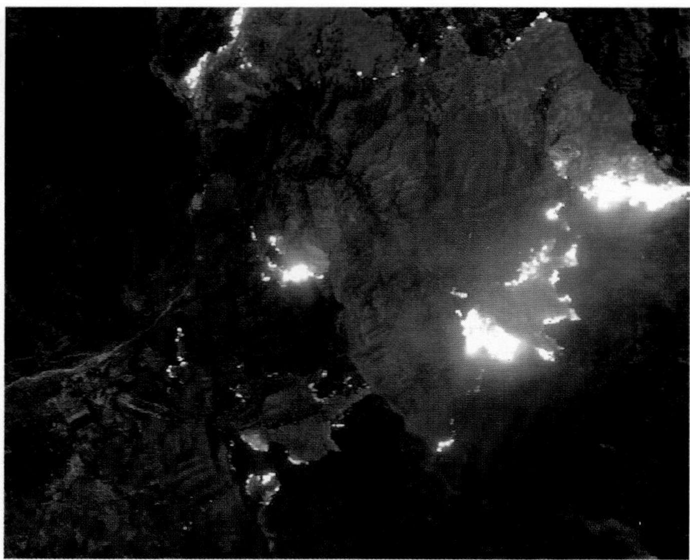

Fig. 4.41. Airborne scanner image incorporating the equivalents of Landsat TM bands 6, 7, and 5 in red, green, and blue of a forest fire in California. (Courtesy of J. Myers, Ames Research Center, NASA.)

it is quite clear that something is burning, the scattering effect of smoke is to prevent visible radiation from the fire emerging. The location of the blaze itself, needed for fire fighting, is hidden. Figure 4.41 combines the equivalents of Landsat TM bands 6, 7, and 5, in the mid- and near-infrared, as red, green, and blue. Because scattering by fine smoke particles and steam is reduced at longer wavelengths, radiation is able to emerge to reveal many details of which fire-fighters would be unaware. Because of the high emission in all bands the active fires show as yellow and cream, the main blazes being obvious, as are several minor fires. The burnt-out areas are clearly outlined, parts showing as bright blue, possibly due to ashes at the surface. At bottom centre the fire has jumped a stream course and a satellite fire is beginning to develop.

Although active volcanoes are very restricted in their global distribution, volcanic rocks provide extremely fertile soils and the high relief associated with the volcanoes encourages rainfall. Consequently, despite the risks, volcanic areas are often highly populated. The hazards posed by volcanic activity are numerous. Despite their high temperatures, lava flows themselves do not constitute the greatest danger, for they generally flow slowly enough for evacuation to be timely, and they flow along well-known depressions in the land surface. The main threats are from sudden explosive ejections of ash and debris and the triggering of huge mudflows formed from a mixture of ash with both rainfall and melted snow and ice.

Figure 4.42 is a spectacular image of the erupting Augustine volcano in Cook Inlet Alaska, in 1986. It combines Landsat TM bands 6 and 7 in red, 4 and 5 in green, and band 3 in blue. Snow and ice cover on the flanks of the cone show as blue, and the vegetation around the coast as greenish. The apex of the cone, from which the large ash plume is emerging is coloured orange, recording intense heat emission. Hot ash and lava are flowing in streams down the flanks of the volcano, the most active being picked out in bright red while ones that are cooling are in darker shades. Beneath the plume is a large grey-brown area of fallen ash.

Predicting volcanic eruptions is obviously of greater social use than monitoring their aftermath. Many volcanoes remain dormant for tens or even hundreds of years, and these

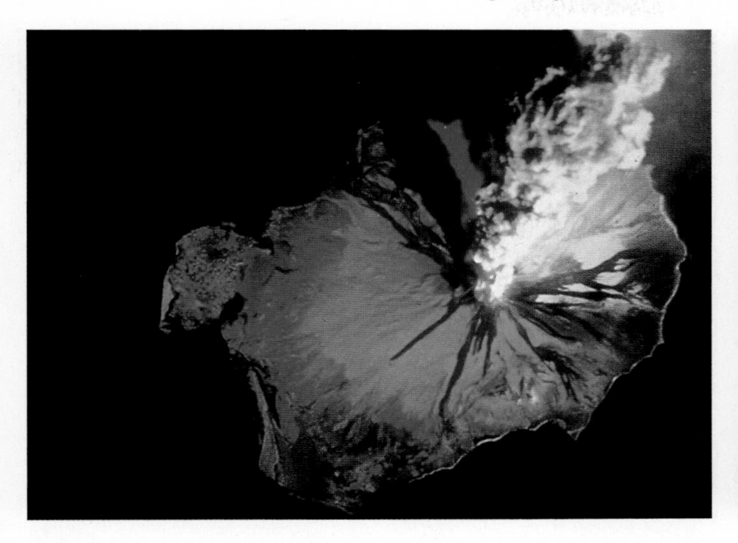

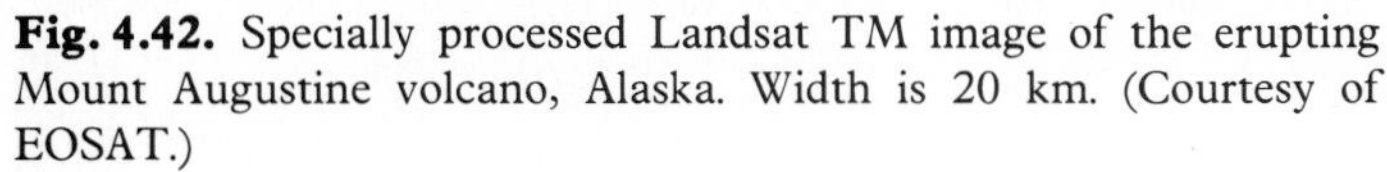

Fig. 4.42. Specially processed Landsat TM image of the erupting Mount Augustine volcano, Alaska. Width is 20 km. (Courtesy of EOSAT.)

Fig. 4.43. Landsat TM image, using bands 7, 5, and 4 as red, green, and blue, of Mount Erebus, Antarctica. Width is 15 km. (Courtesy of D. A. Rothery, Open University.)

pose the greatest problems. Regularly active cones have been intensively studied and are sometime literally bristling with monitors. For instance, an eruption is often signalled well in advance by small changes in ground elevation and by chains of small earthquakes, both easily monitored if the instruments are in place. Disastrous events are most usually unexpected, simply because it is not feasible to monitor directly all volcanoes. However, it is relatively simple to use satellite remote sensing, with its regular cycle of overpasses, to keep a watch on the temperature structure of the craters of volcanoes. Eruptions are often presaged by small outpourings of lava inside the crater, which are invisible from lower elevations. Because lavas have temperatures in the region of 1000–1200 °C, these events emit near-infrared radiation. Figure 4.43 shows a Landsat TM image of the volcanic Mount Erebus in Antarctica, just prior to an eruption. Significant emission in the very-near infrared within the crater imparts a yellowish signature. Theoretically it would be possible to monitor all the life-threatening volcanoes with Landsat data; however the manpower and facilities are not available at present. Using much coarser resolution AVHRR data would, in most cases, fail to signal activity covering areas smaller than 0.5 km across.

Barely a year goes by without awful news about the devastation caused by major earthquakes. Like volcanoes, the distribution of seismically very active zones is globally quite restricted. However, the effects of a major earthquake are not restricted to the fault zone where the primary movement takes place but are transmitted by the Earth's crust over thousands of square kilometres, placing at risk urban communities whose location seems, at first sight, to be safe.

Predicting seismic events is still a much less precise matter than forecasting the weather or volcanic eruptions, and depends on deploying complex arrays of sensitive instruments. Remote sensing helps locate zones that have been active in the past and which may break again. As Fig. 4.44 from California shows, many recently active faults have a distinct signature. They often form straight lines at the surface, and are thereby easily seen on images, such as this high-altitude aerial photograph. Careful inspection reveals the active nature of the fault line from the numerous examples of offset small streams, the truncation of spurs and the partial blocking of valleys by laterally transported ridges.

Whereas major seismic zones are restricted in their location, sensitive modern instruments reveal that there are very few parts of the land surface that are not occasionally

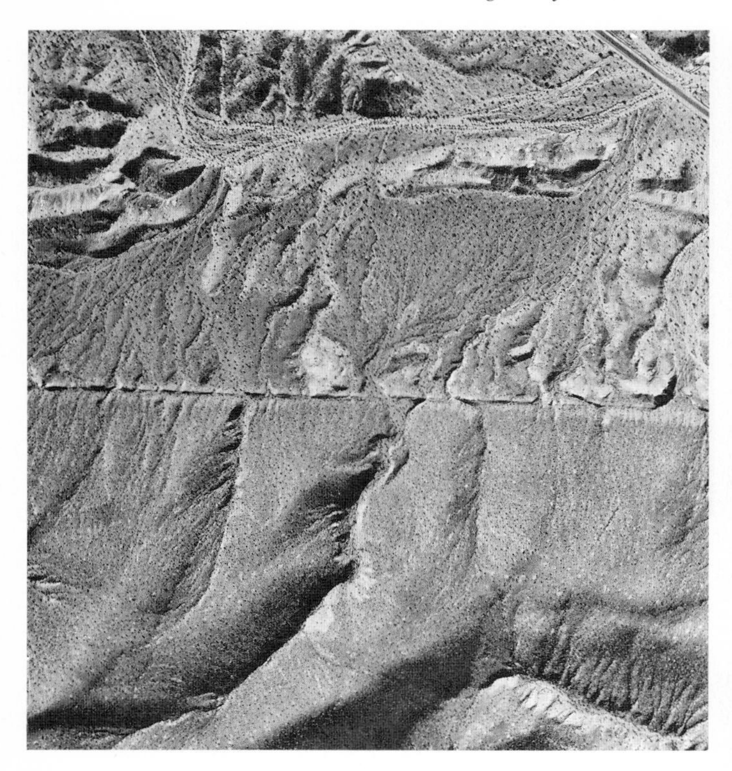

Fig. 4.44. Aerial photograph of an active fault zone in California. Width is 1 km.

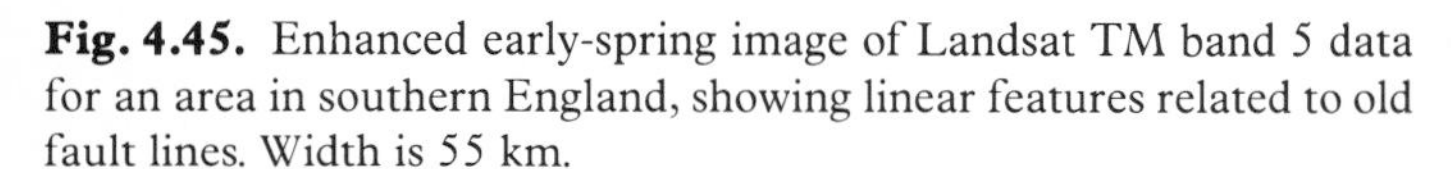

Fig. 4.45. Enhanced early-spring image of Landsat TM band 5 data for an area in southern England, showing linear features related to old fault lines. Width is 55 km.

affected by minor tremors. Although such events often pass unnoticed by the population, even small displacements pose a threat to installations such as nuclear power plants where secondary effects due to malfunction or leakage could be potentially devastating. Such small earthquakes are frequently focused on lines of very old faults which form crustal weaknesses. Detecting these is extremely difficult, as any topography associated with them is generally worn away by protracted erosion. Under special conditions remote sensing can allow them to be detected, as Fig. 4.45 shows. Here the usual camouflaging effects of intensive agriculture on the low relief of part of southern England (Fig. 3.57), where several nuclear installations are located, has been removed by the effect of spring frost on a Landsat TM band 5 image. For the first time a whole number of subtle linear features running roughly East–West, related to deep crustal weaknesses, are revealed by the topography illuminated by a low sun in the South-east.

4.6 Environmental change

As the human population and its economic activity increase inexorably, so its effects on the natural world grow too. As seen in Sections 3.5 and 4.2, the effects of agriculture are distinctive from afar, and are inevitable if we are to survive and global standards of nutrition are to increase. However, humanity's fate is inextricably linked with the health of the environment. Our relationship is one of both unity and conflict, in that survival depends on exploiting natural wealth and changing the environment in many unavoidable ways, yet our activities must maintain some kind of balance so that change does not feed back to our physical or psychological detriment.

The bulk of this section focuses on detecting and monitoring the changes imposed on our surroundings by human activities. However, it must not be forgotten that there are vast areas of the planet that remain in very much their natural state. As well as using remote sensing to keep a watch over any changes that may take place there, it is vitally important that such wild places are analysed and understood, the better to protect them from accidental change. One of the largest areas of natural habitat in the northern hemisphere is Alaska. Figure 4.46 is a land-cover map of part of north-western Alaska derived by classification of Landsat MSS data. Each colour represents a specific category of vegetation in the complex boreal environment. White represents cloud cover. The black areas are recently burned forest, the greens are various forest types and the pale blue, yellow, and magenta represent different types of tundra. There are smaller classified areas in other colours. Each colour is not only a class of vegetation cover, but also a specific habitat for the animal life of the area, and presents different problems for conservation.

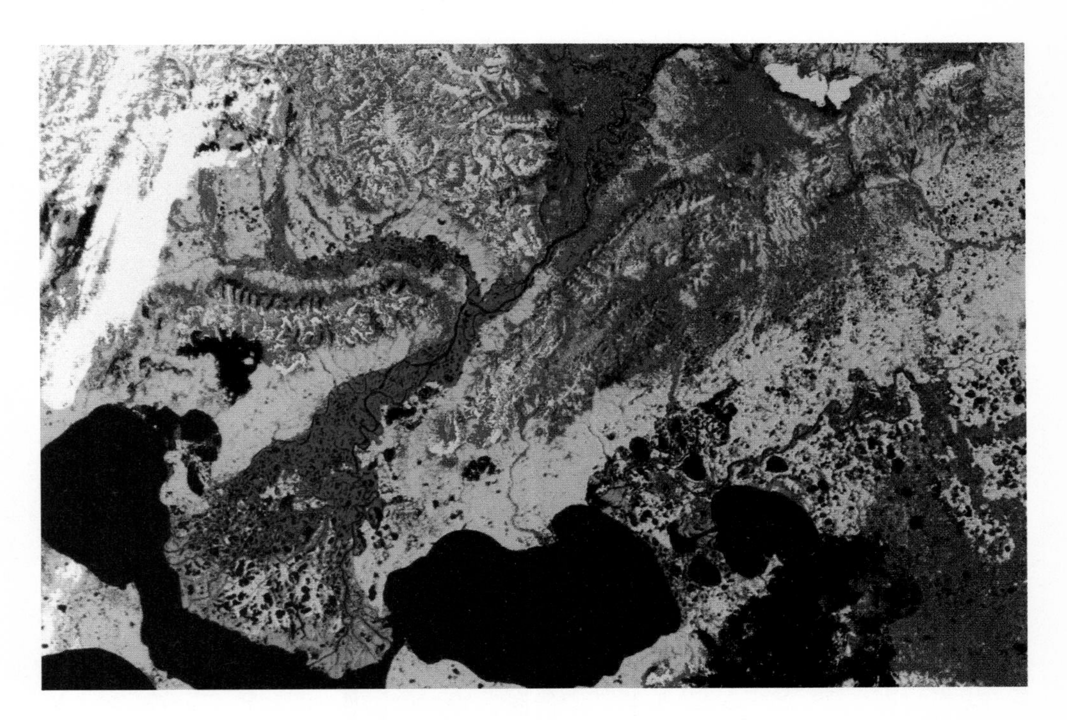

Fig. 4.46. Classified Landsat MSS image of part of NW Alaska. Width is 150 km. (Courtesy of D. Craighead, Wildlife–Wildlands Institute, Montana.)

In Section 3.1 we saw how seemingly trivial amounts of non-toxic materials used in aerosols have begun to attack the atmosphere's content of ozone over the polar regions with potentially disastrous results. Increased burning of fossil fuels elevates the carbon dioxide content of the atmosphere, and as this absorbs energy originating from the Sun but re-emitted by the Earth as thermal radiation, atmospheric temperature increases by the 'greenhouse effect'. Smokes from fuel-burning and forest fires enter the upper atmosphere to have the opposite effect. By increasing the planet's albedo, incoming energy is reflected back to space to give a cooling effect. Another product of the burning of fossil fuels is that the sulphides contained within them form the gas sulphur dioxide, which when dissolved in rainfall forms an acid. One potentially harmful effect is the dissolving of toxic metals from the soil, resulting in their transportation into water courses, where they kill aquatic life. The most visible effect of acid rain is on forests. Figure 4.47 is part of a Landsat TM image of forested mountains in New Hampshire, USA. The blue, green, red order of colours used in the image is controlled by 0.56, 1.65, and the ratio of 1.65 to 0.83 μm reflectances. The 0.56 μm waveband is a measure of the green-ness of the canopies, 1.65 μm is in the region sensitive to moisture in vegetation, and the 0.83 μm region is where healthy plants reflect most solar radiation. Low moisture and chlorophyll content in leaves, indicating unhealthy trees, should produce a red coloration, while healthy vegetation should appear blue to cyan. The image graphically shows that very large areas of the natural woodland, especially on the peaks of the hills, is under severe stress, thought in this case to stem from acid rain produced by industrial activity in the north-eastern USA. Very similar effects are now being monitored in forests throughout central Europe and Scandinavia.

Water pollution is a growing problem, to which attention is being drawn by the well-publicized activities of such environmental action groups as Greenpeace and Friends of the Earth. Among the most polluted rivers in Europe is the Humber in eastern England, along which are a great variety of industrial developments, several major cities, and a number of harbours. Figure 4.48 shows a particularly noxious view of the river near its mouth. It is in approximate natural-colour and was produced by an airborne multispectral scanner. The cream-coloured plume is effluent of titanium dioxide, a base for paint manufacture, through which the wake of a ship can clearly be seen. The dark streak above the entrance to the harbour is a major oil slick discharged from a coastal tanker. Other murky features are

Fig. 4.47. Landsat TM image, using the ratio of bands 5:4, band 5, and band 2 as red, green, and blue, of an area of damaged forest in the eastern USA. Width is 30 km. (Courtesy of B. Rock, University of New Hampshire.)

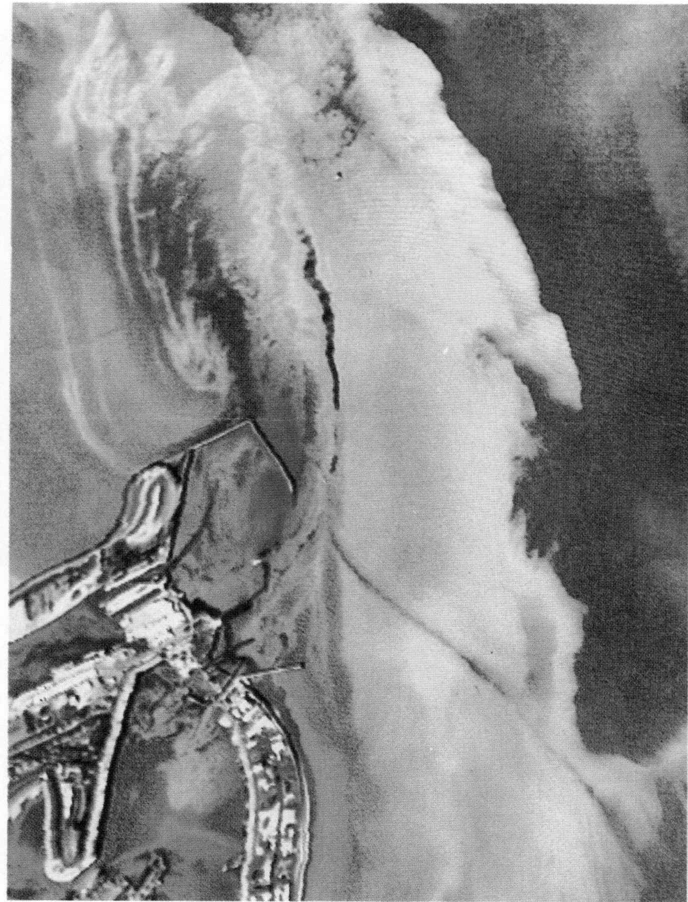

Fig. 4.48. Approximate natural-colour image, from an airborne scanning system, showing pollution in part of the Humber estuary, NE England. Width is 1 km. (Courtesy of Hunting Technical Services.)

domestic sewage as well as the natural sediment load of the river. The patterns in the various suspended contents of the water are due to tidal flow, together with alignments caused by waves.

The Humber is one of many rivers contributing pollution of various kinds to the North Sea, others being the Thames, Rhine, and Elbe, as well as sources that flow into the Baltic Sea. Unchecked discharges over a long period have undoubtedly been building up to a dangerous level of pollution in this restricted body of tidal water. In May 1988, this culminated in a major event that assumed headline proportions for several months. It took the tangible form of a so-called 'red tide', where a sudden increase in phytoplankton, in this case red algae, enveloped the coasts of Denmark and southern Norway. The decay of the algae reduced oxygen levels dramatically, leading to the destruction of near-shore fish stocks. At the same time and over the following months, Common and Grey seals became infected with a virus, now known to be related to canine distemper, and thousands died in a very short period on both eastern and western shores of the North Sea. It was this last problem that occupied the media throughout the summer of 1988, rather than the less emotive algal bloom.

It is difficult to establish a definitive connection between the two events, linked only by the time of their occurrence, but a common cause is widely suspected. Algal blooms are connected to sudden increases in nutrients in the sea, coinciding with favourable temperature conditions. Figure 4.49 shows the progress of the algal bloom over a few days in May 1988. The two images are colour-coded renditions of NOAA AVHRR thermal data, where the sequence of blues, greens, yellows, red, and magenta indicates increasing surface temperature. The land areas of Denmark and southern Scandinavia are picked out clearly in 'warm' colours, white being cloud cover. On 15 May (Fig. 4.49(a)) the algal bloom appears as a green or warm mass surrounded by blue water at normal temperatures in the seaway between Denmark and Scandinavia. The Kattegat, as this channel is known, is where the tidal current systems of the North and Baltic Seas interact. By 19 May (Fig. 4.49(b)) the bloom had grown and spread to surround Denmark and extended westward as a plume into the North Sea. Tidal currents in the North Sea flow in an anti-clockwise

(a)

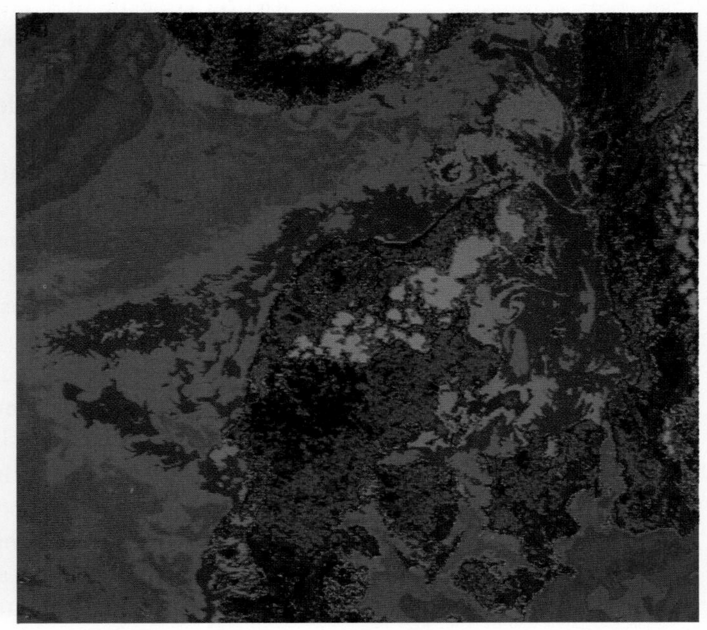

(b)

Fig. 4.49. Colour-coded images of NOAA AVHRR thermal data for the North Sea around Denmark on (a) 15 May 1988 and (b) 19 May 1988. (Courtesy of Peter Fend, Ocean Earth Inc.)

circulation, so the waters involved in the algal bloom would soon be carried to the shores of eastern England. Although a red tide was not reported there, the epidemic among seals that began in the same general area as the algal growth did manifest itself some weeks later. Though not yet fully investigated, it may be that disease organisms, like those which devastated the seal population, proliferate in the same marine conditions that encourage algal blooms. In this case both were transmitted to a wide area from their original culmination in the Kattegat by tidal circulation. This site itself seems to have acted as a focus for the accumulation of pollutants from a number of rivers, driven there by complex current systems. Without remote sensing, the spread of the problems could not have been charted, linked, and analysed. Given the sort of information in Fig. 4.49, early warning of similar events can be made, and more importantly, the sources of the pollutants that are involved can be traced.

Probably the most alarming environmental change of the post-war period was the release of highly radioactive gases and dust by the accident at the Chernobyl nuclear power plant in the Soviet Union on 26 April 1986. The plume swept rapidly across Northern Europe, carried by highly variable winds into Scandinavia and northern Britain. Wherever rain fell from the plume during the weeks following the accident, radioactive fallout polluted the soil and vegetation, thereby entering the human food chain via livestock. The plume itself was invisible but, by chance, Landsat-5 passed over its source only three days after the event. The 30 m resolution of the Thematic Mapper is sufficient for the layout of the Chernobyl complex to be identified as a power plant. On Figs 4.50(a) and (b) it shows as a grid of roads and buildings at top left, next to an artificial lake. The colours over land are produced by TM bands 7, 5, and 3 in red, green, and blue. Over the lake and the River Pripyat, from which the plant draws its cooling water, the colours are derived by density slicing of data from the TM band 6 thermal channel, the warmer the colour the higher the water temperature. The purpose of the man-made lake, ironically, is to prevent thermal pollution of the natural water courses and possible environmental damage. The clue to what happened, if we did not already know, lies in the patterns within the cooling lake. A year before the accident (Fig. 4.50(a)) a plume of hot water emanating from the plant can be seen being forced by an artificial dyke into a circulating pattern in the lake. Immediately after the accident, this pattern has broken down, showing that the plant is inoperative (Fig. 4.50(b)). The only direct sign that something untoward is happening within the plant itself is a tiny red dot near the centre of the grid pattern (see inset). This is located directly over the reactor that the accident struck, and indicates unusually high energy in the 2.2 μm waveband controlling the image's red component. This could be due to the fire in the reactor's graphite core emitting short-wavelength infrared energy due to temperatures above 1000 °C.

As well as giving information on the disaster at the reactor, the images also provide clues to why the failure might have occurred. The site is within the flood plain of the Pripyat River—notice the patterns of curved lines that are scars left by former river meanders. Such areas are underlain by muds and silts that are unstable, particularly when loaded with such a large mass as a nuclear reactor. On the inset to Fig. 4.50(b), the red 'hot spot' lies directly on a dark line running NE. Some analysts pronounced this to be a smoke plume. However, the feature persists on images before and after the failure. Another view is that this is a zone of failure in the foundation to the plant, along which moisture has percolated.

A common feature to both developed and underdeveloped areas of the world over the last two to three decades has been major changes in the size of urban areas. In the under-developed world this has taken the form of an uncontrolled and poorly monitored flow of people from rural areas into cities. This is generated by growth in population together with the increasing effects of famine and the number of destitute and landless farmers. In

Fig. 4.50. (a) Approximate natural-colour, Landsat TM image of the area around the Chernobyl nuclear plant in the Soviet Union, during normal operation. Colours in the cooling reservoir are derived from TM band 6. (b) Landsat TM image of Chernobyl three days after the nuclear accident in April 1986. The inset is a close-up of the nuclear installation itself. Width is 15 km. (Courtesy of GAF and Ocean-Earth Inc.)

developed countries a similar phenomenon has occurred, due to the decline in agricultural employment, but it has been accompanied by a shift of urban population from the once-crowded inner-city areas to suburbs, driven by increasing wealth and a desire for a better quality of life. In cities of the Third World, the migration has resulted in the occupation of areas unsuitable for housing and the building of shanty towns, generally near city centres. In Europe and North America the result has been an inexorable outward expansion of the area of towns and cities, encroaching on what were formerly rural areas.

The change in the urban–rural interface is most marked in the growth of entirely new cities, such as Milton Keynes, some 80 km north-west of London (Fig. 4.51). This, the largest urban development of the post-war period in Europe, aimed at providing rehousing for people from the inner-city slums of London, and was initiated in 1970. Figure 4.51(a) is a Landsat MSS false-colour image of the area designated for the new city in 1976. Clearly visible are the existing lines of communication, including the Leeds to London M1 motorway towards top right and the main London to Birmingham railway running from bottom centre towards top left. Existing small centres of population around the periphery of the new city appear as dense blue-grey patches. Development in the city at this time appears mainly as signs of the network of wide boulevards between which the new housing and industrial areas were planned. Some of these had already been constructed by 1976. Figure 4.51(b) illustrates a sophisticated method of change detection that uses Landsat data from three years (1976, 1980, and 1983) and principal component analysis. Areas

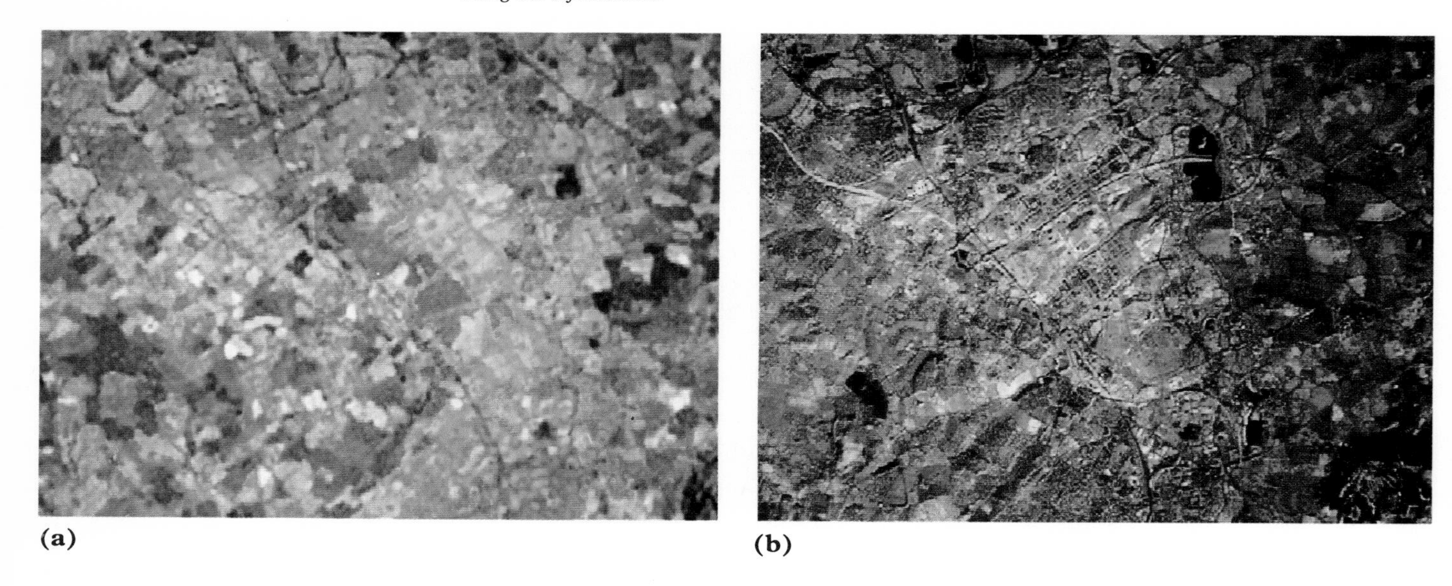
(a) (b)

Fig. 4.51. (a) Landsat MSS false-colour image of Milton Keynes new city in June 1976. (b) Image of the same area using principal components based on Landsat data from 1976, 1980, and 1983. Width is 15 km.

showing little change are in greens and blues. Drastic changes in land use between 1976 and 1980 show as orange and yellow, whereas those produced between 1980 and 1983 are in red. As well as depicting changes in the urban area from 1976 to 1983, many more city roads are apparent than in Fig. 4.51(a), including a high-speed through route running north-westwards from the southern city limits. The large system of rectangular yellow and red plots near the centre is a major shopping centre, other areas, mainly red, being factory estates. Since a small, flood-prone river flows through the city, attempts have been made to reduce the risk of flooding by the construction of black balancing reservoirs, already started in 1976, but complete by 1980. The different kinds of new construction have subtly distinct spectral properties, so it would be possible to classify the developments according to their use, as well as locating them and measuring their area.

4.7 Military uses

All the images in this book are derived from instruments whose technology was developed initially for military purposes, whether they be simple aerial cameras using black and white film, or infrared film that enables camouflage to be distinguished from natural vegetation, multispectral digital scanners that enable sophisticated image processing to detect and recognize patterns, thermal devices that detect warm-blooded animals or recently disturbed ground at night, and radar imaging capable of operation under all climatic conditions and of detecting tanks and military installations. The repetitive coverage by orbiting satellites also helps military intelligence detect changes due to building of new missile installations and redeployment of forces.

The first requirement in land-based conventional warfare is a thorough knowledge of the battlefield, both to plan the deployment of forces with maximum efficiency and to position artillery and train gunfire over hills and ridges. The threat of invasion of Britain by Napoleon Bonaparte's forces early in the nineteenth century prompted the British government of the day to institute the systematic production of accurate maps by the appropriately named Ordnance Survey, ordnance being artillery. Today, the largest volume of sales of even civilian remote sensing products is to the world's military establishments. The French SPOT satellite produces digital images at 10 m resolution with the capacity for stereoscopic rendition of surface topography. Accurate models of ground elevation with great detail about natural and cultural features can be produced from these for anywhere on the Earth's surface, as a perspective view if required (Fig. 4.34), both as input to battle plans and as an essential element in the guidance systems of cruise missiles. Similarly, they can be used in flight simulators to give pilots almost first-hand experience of potentially hostile terrain.

All forms of warfare, even today, are at the whims of vagaries in the weather. Accurate forecasting, based on satellite data, can give considerable advantage in warfare, as the Falklands War in 1982 demonstrated, by preparing personnel for likely conditions and selecting the most favourable time window for operations.

A great deal of low-grade military intelligence can be gathered from civilian satellite remote sensing, as indeed it is. However, an unpublicized convention among all the institutions acquiring and releasing data from systems in the public domain regards their resolution. Below about 5 m resolution it is possible not only to detect objects of military significance, but to begin to identify them. Currently, the finest resolution of images available to all-corners is 5 m, from the recently declassified Soviet KFA-1000 system.

Figure 4.52 shows that publically accessible remote-sensing data have sufficiently good spatial resolution to allow changes of great military significance to be detected and analysed, provided they are large enough. The two Landsat images are from an area on the disputed border zone of Iraq and Iran, that formed the battle zone during the 1980–88 war. The conflict was partly over disputed territory near to the Shatt el Arab, a waterway through which the waters of the Tigris and Euphrates flow to the Gulf, and which defines the frontier. Because it flows over flat, low-lying ground, the Shatt el Arab has changed its course in historic times, though it has been stabilized since the 1940s. The images show a major, undeclared changed in the period 1977 (Fig. 4.52(a)) to 1986 (Fig. 4.52(b)), taking the form of a complex series of geometric lakes and channels north of the Shatt el Arab.

Fig. 4.52. Images of the Iran–Iraq war zone near the Shatt el Arab: (a) Landsat MSS data from 1977, (b) Landsat TM data from 1986. Copyright 1986 Ocean Earth/EOSAT. Width is 50 km.

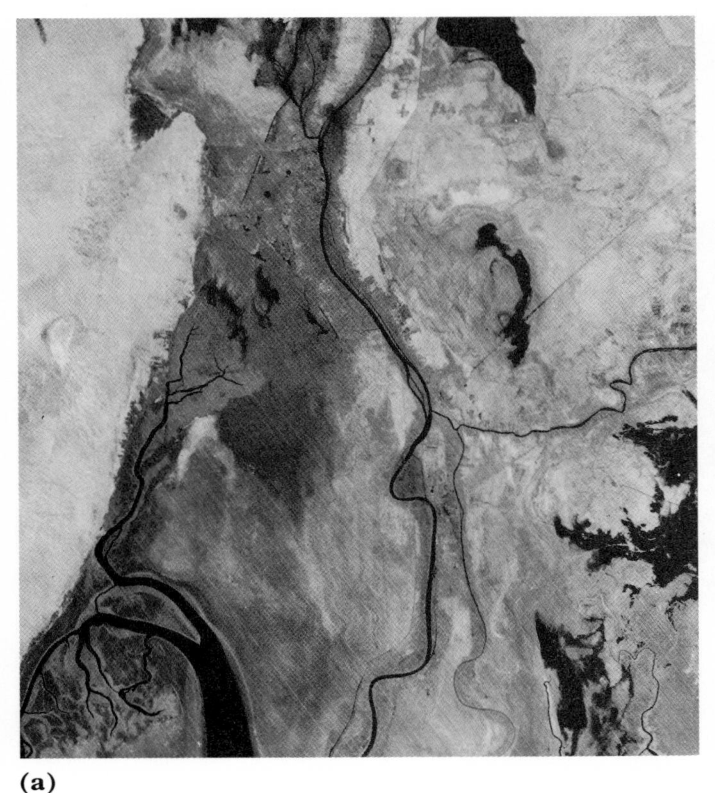

(a)

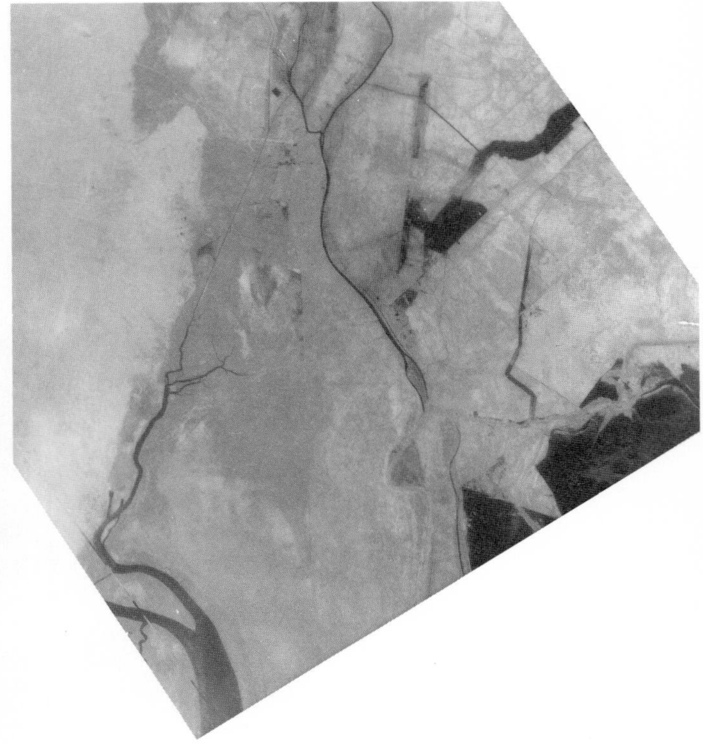

(b)

Though much of the construction took place during the war under fire from Iranian forces, the first signs that Iraq was undertaking major hydrological engineering appeared before the Iranian revolution, in 1978. Understanding the purpose of the excavations and their filling with water is not easy, especially when it is clear that they continued and accelerated under war conditions. However, it now seems clear that, whatever its original function, the plumbing system served as a major barrier to the advance of Iranian forces on Basra, the second largest Iraqi city, and its extension aimed at improving it as a defensive line. To the south-east of the area shown, a further and even more bizarre series of parallel channels were constructed during the war, also detected by remote sensing. These were widely publicized by the world media in 1987, with much speculation centring on whether they were made by the Iranians, with the intention of draining the barrier shown in Fig. 4.52(b), or by the Iraqis as a means of extending the defence system. Neither regime has admitted building the parallel channels, but if they were completed and did drain the lake system to the north, such a flow of water would ensue that the outlet of the Tigris and Euphrates waters would change to one of the old courses of the Shatt el Arab. This would isolate the present Iranian territory and oilfields around Abadan, and establish them as *de facto* Iraqi territory. Such strategic speculation may never be confirmed. However, such a regional diversion would have important effects on the ecology of that region, perhaps flushing out the salt built up in the soils at the head of the Gulf that destroyed it as a major agricultural resource several centuries ago, and decreasing salinity in the Gulf itself. Either would go some way to repairing the enormous economic and social damage that resulted from the war in which the engineering works seem to have figured so highly.

Civilian remote sensing, though useful to military analysts in detecting changes, is incapable of resolving the details that allow an analysis of what the changes really are. The United States intelligence community has developed unmanned digital imaging systems based on charge-coupled devices (Section 2.3) operating in the visible to near infrared, which are deployed on its KH-11 and KH-12 intelligence satellites. Theoretically they are capable of 5 cm resolution from a 250 km orbit—fine enough to read banner headlines, if not the date on a newspaper. Because of atmospheric effects, a more likely limit is about 15 cm from space. Leaked KH-11 images suggest a resolution from this older system of about 1 m. Even finer resolution is possible using fine-grained photographic film and long, high-precision lens systems deployed on manned aircraft, such as the U-2, and also on

Fig. 4.53. Panchromatic oblique image of a Soviet naval dockyard taken by a KH-11 system deployed by the US intelligence community. The original appeared in *Jane's Defence Weekly* of 14 November 1984.

intelligence satellites, such as the US Big Bird series, that either develop and digitize photographs on board for telemetry to the ground or release exposed film canisters to be recovered on re-entry. The USA does not have a monopoly on satellite intelligence, their capabilities probably being duplicated by systems deployed by the Soviet Union. The latter, however, are carried by satellites that only stay in orbit for a matter of weeks compared with the extremely robust US platforms. France plans to launch a military version of SPOT shortly, called Satellite Militaire de Reconnaissance Optique (SAMRO) that is reported to employ a CCD-based system capable of resolution of around 1 m. Figure 4.53 gives a clear idea from a leaked US KH-11 oblique image of what sub-metre resolution can show, for a Soviet naval dockyard. The resolution can be judged from the detail shown in buildings. The intelligence interest lies in the partly completed vessel on the slipway, thought to be half of a nuclear-powered aircraft carrier.

Radar, with its all-weather capability, and ability to detect angular man-made objects, as well as variations in the surface roughness of the ground, is the military surveillance system *par excellence*. The development of synthetic-aperture radar, where resolution is a function of electronics and power output rather than the physical dimensions of the antenna, has meant that all the advantages of radar are now able to be deployed from orbit. An often overlooked advantage for even better resolution airborne radar coverage is due to the need for radar to 'look' to the side in order to produce images (Section 2.3). This means that an aircraft need not enter potentially hostile airspace to obtain images, and its mission can go

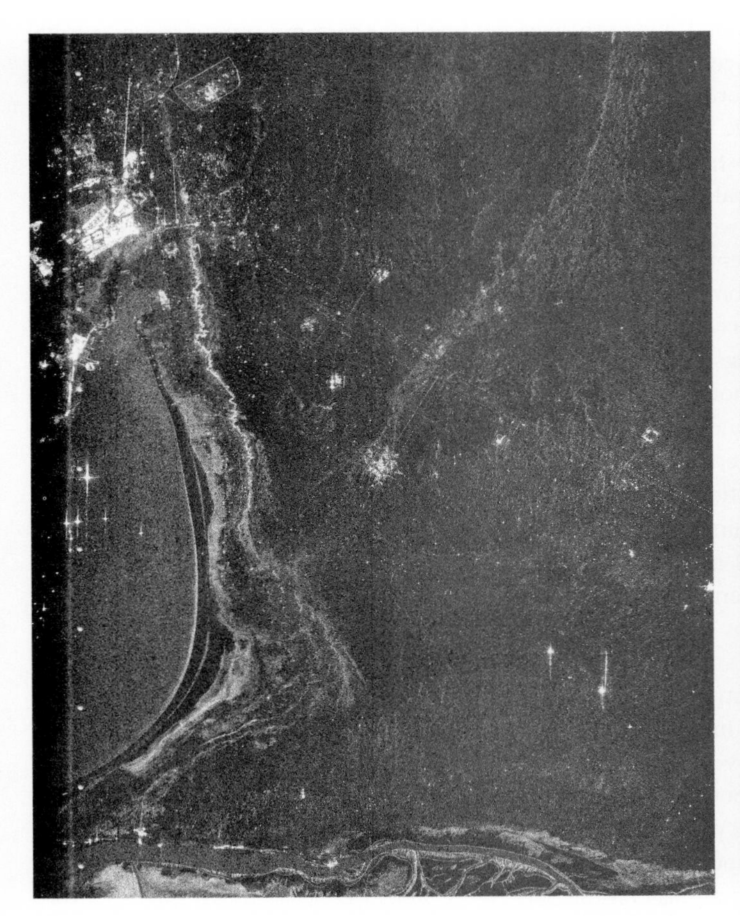

Fig. 4.54. SIR-A radar image of Kuwait, showing specular reflections from small cultural features. Width is 55 km. North is at bottom right.

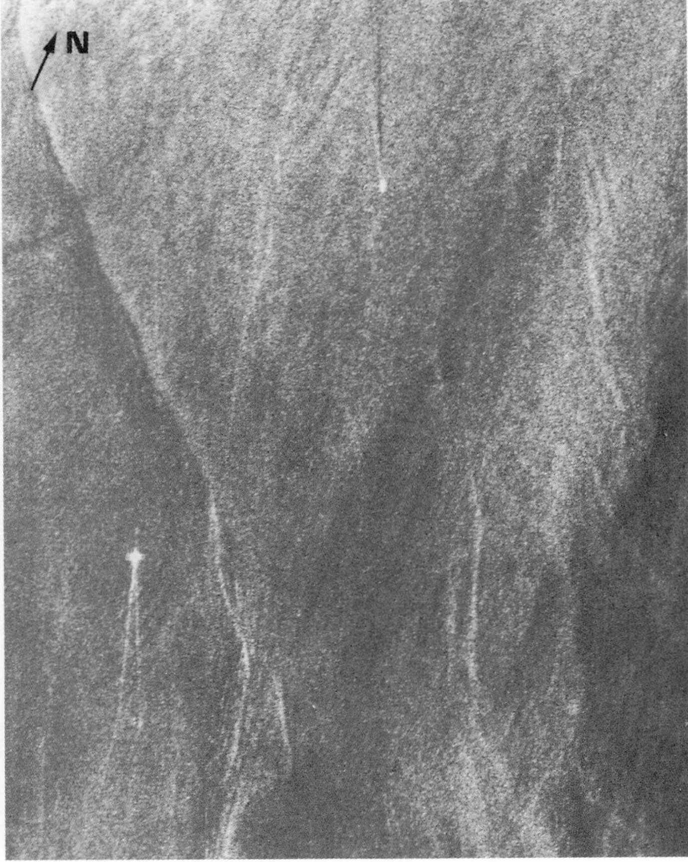

Fig. 4.55. Seasat radar image of the Atlantic off Florida showing ships' wakes. Width is 25 km.

undetected from the ground. Figure 4.54 shows a host of objects on a SIR-A radar image of Kuwait, that are detected only because they consist of metal and contain angles that act as corner reflectors. Some are lines of power pylons, other small buildings and some installations in an oilfield. Figure 4.55 is a Seasat image of part of the Atlantic Ocean off Florida, where ships are easily recognized by their typical star-like blooms due to reflection from their angular metallic superstructure. More interesting are the traces of their wakes, which show their heading and give some indication of their size. Radar is able to show this information uniquely because it detects small variations in the roughness of the sea, and ships' wakes consist of regular interferences between natural waves and those formed in the wake. The USA has plans to deploy several high-resolution imaging-radar intelligence satellites aimed at detecting armour movements in the northern European area and marine traffic.

The thermally emitted part of the spectrum has some interest for the military analyst, largely because heat dissipated from machines, buildings, and personnel is nearly impossible to hide, unlike their surface characteristics in the shorter wavelength part of the spectrum. The development of various organic compounds for coating aircraft, and designs that include no angular features can render them almost invisible to radar, but nevertheless they form heat anomalies. Figure 4.56 gives a mundane example of what is possible. The left-hand image (a) is a nearly natural-colour image of part of Heathrow Airport, captured by an airborne simulator of the Landsat TM system. The general layout of runways and terminal buildings are very clear, as are aircraft in parking bays and making their way to the runways. The right-hand image (b) was obtained at the same time using a thermal scanner. At first sight, it seems to be less informative. However, if the area to the left of the distinctive shape of Concorde—on the second terminal spur from the top—is examined, the thermal image reveals the apparent presence of two aircraft that are not present on the colour image. These ghosts are the cool shadows of aircraft left behind after they have departed. There are several other instances of this useful piece of intelligence in the image. Digital image processing would enable the subtle differences in darkness of these ghosts to be related to the actual temperature of the surface, enabling some estimate of the time of departure to be made. Some thermal infrared systems are so sensitive that it is possible from images of aircraft to determine whether they contain fuel—full tanks generally give a cool signature on the wings—or if their engines are warmed-up ready for take-off.

Except in extremely hot climates, the body temperatures of warm-blooded animals, including humans, are generally higher than those of their surroundings. Each individual therefore constitutes a thermal anomaly which a mid-infrared imaging system can detect, given suitably fine resolution. This is more readily achieved at night, as the ground cools quickly while bodies remain at a constant temperature. Figure 4.57 is a night-time airborne thermal image of a rural area. On it can clearly be seen details of roads and a stream, showing white or warm, and fields in different tones—those recently ploughed are warm, those with sprouting crops show as striped, while pasture is cool. Trees are shown as grey irregular blobs. Houses, and the access roads to them are clearly shown as geometric patterns. At about the centre of the image, just above the road intersection, is a large open area of asphalt on which numerous cars are parked—this is the car-park of a bar, so the image was taken in mid-evening with a large clientele. Of most interest are the clusters of small bright specks just above the car park and below the road on the right. These are small herds of cows, the first in a pasture and the latter in an orchard with a clear pattern of trees, into which they have escaped, presumable through an open gate. The cows could easily be squads of infantry making a foray on the bar.

Thermal imaging is now so sensitive that not only living animals can be detected, but also those that are dead and which are decomposing. The heat generated by decomposition

(a)

(b)

Fig. 4.56. (a) Approximate natural-colour image, from an airborne scanning system, of Heathrow airport, London. (b) Thermal image of same area as (a). Width is 2 km. (Courtesy of Hunting Technical Services.)

allows the location of murder victims, even in shallow graves, up to several months after their demise. On a less lugubrious note, thermal imagery is also capable of detecting the small thermal anomalies formed by the change in density of soil after it has been dug. This has permitted the British Army to detect buried arms caches in Northern Ireland.

The bulk of military remote sensing requires several elements, other than the provision of data at a suitable resolution and of appropriate wavelengths, that are not critical for most civilian applications. The data must be available for intepretation as soon as possible, and in many cases they need to be of a repetitive nature to detect change. Much of the analysis still needs to be done by human skills, and one operation may require many interpreters. Countermeasures no longer aim primarily to hide objects and personnel—this is almost futile given modern technology—but to confuse the interpreters and waste their

Fig. 4.57. Thermal image of rural area in England. Width is 700 m.

time. There are many possibilities, a few examples serving to illustrate the point. Since radar images rely for detection of vehicles and installations on their content of corner reflectors, a countermeasure is to construct many simple corner reflectors and scatter them about the terrain, perhaps moving them periodically in intriguing patterns. A very small group using minimum resources—a few sheets of metal—can create sufficient confusion for real deployments to be hidden in it. A plough can be used to dig geometric patterns in open country to give the illusion of a sizeable camp on a radar image. Similarly, a black plastic bag filled with water absorbs heat during the day and remains warm enough during the night to create a convincing illusion of a soldier, especially if several are deployed in loose formation, which is moved from day to day. Given countermeasures such as these, even ill-equipped guerrillas can force remote sensing from the agenda at a tactical level. On the other hand, developments in relating the spectral properties of the surface and the atmosphere to their composition, largely for a change in the civilian sphere, open up novel means of intelligence gathering. For instance, it is theoretically possible to use high-spectral resolution data in various parts of the spectrum to analyse the composition of gases emitted by industrial plants. This could be used to detect factories that are producing chemical warfare agents, that otherwise masquerade as fertilizer or pharmaceutical plants.

Perhaps the most comforting aspect of the increase in sophistication of military remote sensing is that it could enable any agreements on limitations of missiles or conventional hardware to be verifiable without the need for large teams of observers in the field. Indeed, the US Big Bird and KH satellites are intended primarily for monitoring deployment and proliferation of nuclear missiles. However, since the real capabilities, particularly the resolution of the most advanced military systems are the most closely guarded secrets of all, it has proved possible to use the excuse of difficult verification to dissimulate this vital issue in arms control. At the same time as the 1980–88 Reagan administration avoided extending treaties on nuclear weapons limitation by claiming the impossibility of verification, they used images from intelligence satellites to accuse the Soviet Union of violating existing missile treaties. The contradiction implies that all that is needed for verifiable treaties is political will, not increased technological capabilities. Until the results from such systems become available for universal public scrutiny, the intelligence community and its masters can say and do vitually what they please. Figure 4.53 gives a clear indication of just what can be verified. Moreover, if images with such resolution were publically available, many of the applications described in this book would become even more socially and economically useful.

5 Operational issues

The first four Chapters have provided a copiously but not comprehensively illustrated summary of the important problems that can be addressed by remote sensing: what the data are and how they are collected; what they show in terms of the various parts of the terrestrial environment; and some examples of how they have been applied. It remains to summarize what they can and, equally important, what they cannot do; what kinds of costs remote sensing entails compared with more conventional means of environmental analysis; what the requirements for skilled personnel are; how the acquisition and distribution of data are managed; and the prospects for the future.

5.1 Benefits and costs

Remote sensing has several unique attributes. Foremost among these is the ability, through judicious use of the interactions between electromagnetic radiation and matter, to detect and in some cases to measure accurately the presence of a wide variety of phenomena, elements, compounds, and organisms in the oceans and atmosphere and at the land surface. Most of these functions can be achieved by other means, but they are restricted to analysis at widely spearated points, whereas remote sensing allows large areas to be examined at once. However, while the whole atmosphere can be monitored, only the top 40 metres of the seas and a few micrometres of the land surface and its vegetation cover are accessible. Deeper understanding in these cases relies on informed interpretation of surface evidence and on data provided by more deeply penetrating methods that are not strictly remote sensing, such as deep-ocean probes and geophysical surveys.

Depending on the altitude from which data are acquired and on the optics of the systems used, the great majority of remote-sensing data are in the form of map-like images covering large areas. Since they are acquired almost as 'snap-shots' they give a synopsis of environmental conditions over a very short period of time, whereas other methods of analysis must be built up by lengthy and painstaking ground surveys. The view of a single image ranges from a kilometre or less with low-level aerial surveys, through tens and hundreds to a few thousand kilometres from polar orbiting systems, to views of an entire hemisphere from geostationary satellites. The larger the field of view, the coarser is the resolution of the images, from tens of kilometres achieved by geostationary metsats, through around 1 kilometre, 100 metres to 10 metres from polar orbit, to less than a metre from airborne surveys. The spatial resolution of any data forms one limit to potential use and benefit. Climatologists are served well even by very coarse resolution, oceanographers and meterologists in the main by that around one kilometre, as are users concerned with regional vegetation patterns and global geology. It might seem that other users would be happiest with the finest resolution that is available, but that is not necessarily the case. This is partly because of the limit posed by the recording medium, and associated problems.

Consider a standard Landsat MSS scene, which covers 185 × 185 km with a resolution of 80 m. This contains some 7 million pixels, so all four available bands amount to 28 million. The recording of the data in digital form occupies one large reel of magnetic tape. To record the same area with a resolution of 8 m would need 100 tapes for a ten fold improvement, and 10 000 for an 80 cm resolution, equivalent to that of high-level aerial photography or detailed images from military satellites. Looked at another way, the 80 m Landsat MSS data can be enlarged to a scale around 1:100 000 before the 'grain' of the image becomes obtrusive and detail is lost to the interpreter, who would then be dealing with an image almost 2 m across. Even this is too large for the synoptic advantage of the scene size to be exploited. Enlarging an 80 cm resolution scene to the limit would give a scale of 1:1000 and an image 200 m across—clearly absurd. Users decide on the area within which they wish to work and select a resolution appropriate for displaying it at a size where everything recorded is visible on a conveniently sized image. If more detail is required, then enlargement to the resolution limits is at the expense of the synoptic view. Greater detail still demands finer resolution, larger scale, and focusing attention on yet smaller areas. This is called a hierarchical approach, with many advantages, that are only provided by remote sensing. At each level different attributes of the scene are interpretable, both locally and in the context of the whole area of study.

Any remotely sensed image displays far more information than a map at the same scale, no matter how detailed the map is. Because digital images can be registered accurately to cartographic map projections, they might seem ideal candidates for supplanting conventional maps completely. It all depends on the use. The vast bulk of maps are for navigation of one kind or another. In a well-developed area with a complex network of roads and urban areas, the navigator would be bewildered by the wealth of detail on an image. Moreover, at a useful scale of 1:250 000 most of the roads would not show on an image, no matter what its resolution. A useful road map displays routes at more than ten times their real width, and ignores most topographic detail and that relating to surface cover. For the geologist or agriculturalist, the image contains a wealth of intriguing and interpretable detail. However, left in the original form, even with lines drawn on to mark important boundaries and other interpreted features, the image would not convey its 'message' to another geologist or agriculturalist, without them having to perform some interpretation too. Transfer of knowledge obtained from images is therefore dependent on producing simplified maps from them, by extracting only those features that are relevant to the intended user. Without this much of the efficiency of remote sensing is wasted. Remote sensing does not mean the demise of the map, but its improvement, and also heralds the appearance of new kinds of map.

The most familiar of these new maps is seen daily on televized weather forecasts, showing the changes in cloud cover. Because cloud height can be monitored, and this controls precipitation, detailed maps of rainfall are replacing those based on widely scattered gauging stations. Moreover, the details of cloud formations allow the measurements of pressure and air temperature from weather stations to be extrapolated to give far more detailed maps on which weather prediction can be based. Although not as changeable as weather, the rotation of crops and yearly variations in times of planting and harvest meant that, before remote sensing, records of agricultural practice, if recorded at all, could not be expressed in map form in time for any action to be taken, because they needed to be compiled on a field-by-field basis. In the European Community this permits all kinds of malpractice, such as declaring that excessive areas are planted with crops carrying subsidies. Remote sensing theoretically now allows the contents of each field to be assessed and dishonest farmers to be warned and penalized, thereby saving huge sums. More importantly, it enables the agriculture of any area to be monitored, even to the extent of allowing crude predictions of yield, with great benefits in managing critical resources in Third World areas and planning relief measures to overcome shortages.

Natural vegetation and the wildlife and stock that it supports has never before been examined in detail, except for very small areas. As well as assessing the gross cover of green plants over huge areas using vegetation indices, remote sensing holds out the possibility of an almost complete inventory of habitat on a global scale. This allows critical areas where unique conditions remain to be identified, as well as areas undergoing human-induced change. In an economic context, this allows the movement of stock to be planned and managed, in order to take advantage of good grazing and relieve areas of overuse.

The repeated coverage on a regular basis by satellite remote-sensing systems provides perhaps the most important new opportunity—that of monitoring, comparison, and retrospective analysis. Geostationary metsats provide data on a near real-time basis, while polar orbiters revisit each point on the globe from between half daily to periods of 20 days or so. For the first time we are able to detect, document, and analyse environmental change. The progressive growth of image archives lays down a resource on which models of climate, oceanography, vegetation growth, and human activity can be developed and refined. Given sufficient data covering every aspect of the Earth accessible to remote sensing, together with that gathered by other methods, means that we can discover how its different parts interact as part of a closed system. This is the first step to being fully in control of our environment and its destiny. In the short term, repetitive remote sensing allows sudden events to be recognized and pin-pointed with great accuracy, instead of having to await the often fragmentary reports of people on the ground.

To avoid giving the mistaken impression that remote sensing is some kind of panacea, it is necessary to temper an account of what it can do with information about its shortcomings. Exactly what does an image reveal? No matter how data are acquired, they represent parcels of the surface with more or less the dimensions of the resolution cell. Irrespective of the wealth of information relating to that parcel, and this can extend into several hundred narrow wavebands, it is the content of this limiting area that induces the response in the system. Depending on the make-up of the surface and the resolution, the area comprises one or more different types of material, often many. Consider the most common cover type—vegetation. Unless the surface is covered over large areas by a single species of plant, the image is recording communities of several species. Moreover, the fundamental distinction between one species and another is genetic not spectral. They are classified in practice by leaf and canopy morphology, and in the case of some kinds of plants, such as Eucalypts, can only be divided because of subtle differences in the shapes of their seeds. Remote sensing alone cannot uniquely identify species. Where the same association of plants is characterized by different proportions, remote sensing may reveal a very different spectral signature that has no significance. Finally, only the uppermost level of a community's canopy is sensed. This raises an important point, which is of considerable comfort to the field scientist. The results of remote sensing, for vegetation at any rate, must be calibrated and validated in some way by detailed work on the ground. Without this they are misleading.

This is even more true for another application—the discrimination and mapping of different rocks. Except in the most arid deserts and areas from which glaciers have recently retreated, fresh rock does not occur at the surface. At the very least it is covered with a rind of the products of weathering together with lichens and mosses, but more usually by thick soils, and vegetation. Geologists subdivide rocks by their content of minerals formed at the same time as the rocks crystallized, or were cemented together from loose sediment. Very often this relies on microscopic examination, and some divisions are only possible on the basis of minute examination of the relationships between different minerals and the detection of small quantities of those that reflect fundamental chemical differences. Geological remote sensing depends for the most part on 'peeling back' these multiple layers and using field information and background knowledge to enable a true picture to be built up. Once again, the traditional geologist need not feel threatened by the 'black arts' of remote sensing. Although a combination of remote data, sophisticated image processing and skilled interpretation helps accelerate geological mapping and makes it more comprehensive and accurate, not a single mine or oilfield has been found by remote sensing alone, contrary to early expectations. None the less, it is able to highlight the unusual, through spectral and geobotanical anomalies, as well as to provide guidelines for mapping the mundane. Therein lies a real advantage of remote sensing in exploration for physical resources.

To a professional geologist such as myself, it never ceases to amaze that the vast majority of highly profitable mines for metals were not found, in the last analysis, as the result of profound insight into the doings of the Earth, although much intellectual energy has been expended in modelling the processes of mineralization. They were almost literally tripped over in the course of basic surveys and prospecting. By their very nature, economic deposits are small, rare, and thereby highly anomalous. To find them in the past meant traversing huge areas on the off-chance of locating what geological theory suggested ought to be present. This is extremely costly, as well as usually being frustrating. The great risk in exploration is that the cost of prospecting, which generates no profit whatever, will not be offset by the income from a mine. Just how much can be spent in a likely area must be matched against the chances of an economic discovery, as assessed by theoretical considerations. The lower the probability the less funds can be assigned per unit area. Resolving this dilemma hinges on reducing the area of search as soon as possible. Remote sensing allows this to be done, using two strategies. One is the identification of small areas of anomalous soils and vegetation, which past experience suggests may be related to resources. The other is outlining areas where a combination of common rocks and geological structures, conducive to mineralization or hydrocarbon generation and accumulation, are found together. Normal mapping may reveal these long after the initiation of a project—possibly too late for the economists. Remote sensing can reveal them at the outset, so that risk capital can be assigned only to small areas, and thus spent most efficiently.

Something like 75 per cent of the Earth's land surface remains to be mapped geologically at scales useful to the explorationist, and using modern ideas. There are undoubtedly more of the 'trip over' deposits to be found. But there are also those that are hidden to the worker on the ground because they are buried. Seeking them will become the major task of geologists if material standards are to be significantly raised worldwide. By helping speed up basic geological mapping and revealing otherwise invisible subtleties that can be analysed in the context of growing understanding, remote sensing has a crucial role to play in development.

All these opportunities and many more, conditioned by the realities of remote sensing, will bear some cost. The vital issue is whether these costs will be lower than those associated with conventional work. Where entirely new information can be provided remotely, the question is whether the advantages accruing justify the attendant costs. Ultimately, the answers lie in the sphere of global politics and economics—should we do things to generate profit, or to make life easier or indeed sustainable, irrespective of cost? This is not the place for a full discussion, but one issue clarifies at least my personal thinking. The single most pressing issue of our times, including nuclear disarmament, is the provision of safe, accessible, and abundant water supplies. Remote sensing is possibly the most efficient aid to locating and evaluating such needed supplies. Compared with the use of data in oil and mineral exploration, that directed at water exploration is almost trivial. There are no profits in water supplies. Be that as it may, it is useful to examine some of the costs associated with a programme of remote sensing compared with more conventional approaches to the same ends. The example is a simple one concerned with an inventory of land cover in an easily accessible part of the USA, covering about 2500 square kilometres.

The costs in 1974 of separate programmes for the same area, involving ground survey, aerial photograph interpretation, and visual interpretation of Landsat MSS photographic images bear close examination. For a ground survey, it would have taken eight years, but 40 different types of cover could have been distinguished. The overall cost would have been around US $100 000, or $40 per square kilometre. Air photograph interpretation would have taken only 18 months, but would have given only five categories at a total cost of about $40 000, or $16 per square kilometre. The Landsat analysis, involving some field work, took six months to give 16 categories at a cost of about $16 000, or $6.40 per square kilometre. At that time the Landsat interpretation was about one-fifth of the cost of the air survey and less than one-tenth of that using field methods. Moreover, it was dramatically quicker. To set against this is the lower number of categories mapped compared with the ground survey. However, the larger the area involved using remotely sensed data the lower the unit cost, and the use of computerized methods of classification brings costs down still further. Basic data costs have risen somewhat faster than those associated with computing and human input, in the time since this analysis. However, taking Landsat Thematic Mapper data as an 'industry standard', their cost for seven bands at 30 m resolution in digital form amounted in 1988 to only about US $0.11 per square kilometre or US $3300 per 185 × 185 kilometre scene (EOSAT prices), to which must be added the economic costs of performing an analysis.

Against costs should be set benefits that stem from the use of remote sensing. For most applications this is not easily calculated. Sometimes the benefits, such as provision of water supplies or early warning of famine, cannot be expressed in financial terms. A few simple examples will illustrate the point. Oil exploration companies use seismic surveying along lines as the principal method of locating oil traps. Each kilometre of onshore seismic line costs around US $4000. Orienting the trajectory of the lines can be improved and the total length required can be reduced by a better understanding of geological structure prior to the survey. Such an orientation using remote sensing for 10 000 square kilometres may cost US $20 000. If it saves 5 km of seismic survey out of several hundred, it is cost-effective. Even more spectacular is the routine use of remote sensing in poorly known areas to plan roads or pipelines. One example in Bolivia allowed a gas pipeline to be rerouted, shortening its length by 17 km to save US $12 million. Another, to which financial benefits cannot be attached, but which is of great importance concerns the effect of remote sensing analysis on water exploration. A major aid agency working in the arid parts of Sudan has had a success rate of 20 per cent during their extensive drilling programme. They did not orient their exploration using a geological interpretation of remotely sensed images. Where this input was employed by another agency working in a similar area in adjoining Eritrea, under war conditions, they had an 80 per cent rate of successful well drilling.

These few examples suggest that remote sensing is a sound economic proposition in many fields. It must become a matter of routine to compare its costs and benefits with those of other methods in every conceivable application. A proper evaluation, however depends very much on what it can and cannot achieve. Space and the diversity of possible applications prevents this being discussed further here. The onus is very much on the potential users, and the clarity of their expectations. This can only be optimized by some training beyond the familiarization provided by this book.

5.2 Training

The brevity of this book is a measure of the great breadth of the applications that remote sensing can serve, as well as the book's intended role in familiarization rather than as a teaching text. The theory involved in understanding the interactions between electromagnetic radiation and matter is a specialized subject in its own right, as too is digital-image processing. In each of the application areas there is a large and growing body of scientific literature and case studies of specific applications, as well as a number of specialized textbooks.

There can be no mistake in suggesting that remote sensing has the potential for revolutionizing many, if not all surveys of the environment and the resources that it holds. It has already transformed meteorology and climatology, and seems likely to do the same for oceanography. Its ability to monitor human activities and their effects opens up avenues for clearly directed planning, in many cases in those areas whose development, or lack of it, has hitherto been in a state of anarchy.

Truly vast amounts of data in the form of images of all parts of the spectrum, at many scales and with a range of resolutions, already reside in archives, and more are inevitable, at least in the short term (see Section 5.4). This includes almost global coverage of the land surface by the Landsat MSS system, with many areas covered by images from several dates since the launch of Landsat-1 in 1972. Archives of NOAA AVHRR data and those from other metsats have been global and multitemporal for somewhat longer. The SPOT, Landsat TM, and Japanese ERS-1 systems are rapidly gathering higher resolution imagery, also aimed at eventual global cover. Considerable areas have been covered by experimental systems, both mounted on satellites and aircraft, aimed at investigating the potential uses of data from less familiar parts of the spectrum than the visible and near infrared. Soon to be added to this are new or improved kinds of data from a variety of planned orbiting systems (Section 5.4). An idea of the volumes of available imagery is given by the fact that the world's land surface is covered by between 5000 and 6000 185 × 185 km Landsat images, and by about 50 000 60 × 60 km SPOT scenes. Given the range of available data and the repetition in many cases, the archives contain in the order of hundreds of thousands of images acquired from operational satellites alone. The number produced by experimental systems is difficult to estimate, but may be in tens of thousands. Turning to mundane aerial photographs gives millions of individual frames. The number of interpreters needed to exploit the potential of this mountain of information fully must be large. At the time of writing, the database was grossly underused. This is due partly to a widespread lack of understanding of what the subject is all about and its potential, and partly to delays in assigning funds to its use. The other reason is that the growth in the number of people capable of using the data increasingly lags behind the rapid growth of the database and the means of employing it effectively.

Consider a reasonable estimate of six months for the complete information content of a digital Landsat MSS or SPOT scene to be extracted and documented by one trained analyst using sophisticated image-processing techniques and a (probably unlikely) blend of interpretive skills across the board of the environmental sciences. This gives an idea of the number of trained personnel required to do full justice to the available data. Many applications, such as the detection of environmental hazards and the monitoring of crops require analysis of images on a regular short-term basis. A crude estimate of the required number of highly trained personnel on this basis would be around 10 000 people concerned with detailed image enhancement and analysis of global Landsat MSS data alone. Of course, this is a purely hypothetical situation. In reality the principal users of remotely sensed data can be divided into two groups—the specialists devoted to sophisticated computer handling of image data, and various groups of environmental scientists who interpret the images in the light of their own responsibilities and experience. The first require a thorough background knowledge of the physics of radiation, the theory of digital image processing, and the operation and programming of computers. The second need sufficient training in these fields to be able to interact efficiently with the first, and to understand the information held in image data that is relevant to their field. Because this information relates to all surface phenomena, interpretation in one context demands a familiarity across the board of the environmental sciences, as well as specialist knowledge in one field, such as geology, ecology, or meteorology. Even with a thorough training in one discipline and experience in applied remote sensing, it is difficult for a primary user to function efficiently in isolation. Experience at NASA and other research environments suggests that the best operational strategy revolves around multidisciplinary teams.

Except in a purely research-oriented programme, groups of environmental scientists do not work on problems in isolation, but advise economists, managers, planners, and politicians. In turn these personnel request information of various kinds from the scientists. Clearly there is a need for these decision makers to be at least familiar with the opportunities presented by remote sensing, the basics of how they are achieved, how much they cost and the time-scales involved. That is part of the intended function of this book.

Finally, taking the positive view that remote sensing is here to stay, to develop rapidly, and expand into a major influence over the way the world is managed, means that it has to be introduced to the present and future generations of school and undergraduate students. Preparing a whole generation oriented towards remote sensing is by no means the primary objective. Images of the planet are an extraordinarily graphic means of enriching and expanding the teaching of many subjects, principally those covered by the humanities and physical geography, as well as illustrating in practice several aspects of theoretical physics. At higher levels, remote sensing demonstrates the interconnectedness of all the static and dynamic phenomena characterizing the Earth's outer parts, as the basis for study of Earth System Science. It also provides much of the evidence allowing this all-sided approach to understanding the Earth to be deepened and widened.

Training in these various contexts demands a variety of approaches. At the level of school education the most easily understood images illustrating different aspects of the environment can be incorporated into existing curricula, together with brief explanations of how they are acquired and the kind of information that they contain. Remote sensing forms an excellent basis for teaching the practicalities of electromagnetic radiation and its interactions with matter—the only means by which it manifests itself to us. University science courses provide the first opportunity for beginning specific teaching about remote sensing and all its aspects, in the context of applications in the different disciplines.

Decision makers currently involved in environmental management and its exploitation are likely to have little time, nor the scientific background necessary for detailed training. Their immediate needs are served by familiarization, such as that provided here. By far the most urgent requirement is to draw practising environmental scientists into the operational use of image data, for which there are two options. The first is short residential courses of intensive study and practical experience, on release from their normal duties. Several institutions host such courses, ranging from a few days to several weeks duration, in a variety of application areas. There are three main drawbacks to this strategy. One is that important personnel have to be diverted from active tasks. Another is that the materials with which they will work during a general course may not be directly relevant to their own programmes. Residential courses mean a sudden break in routine and with familiar surroundings, so that full benefit from the training might not be achieved. Various studies have shown that the degree of long-term retention of skills from such short courses is low, particularly if they are not put to immediate practical use. A second option is in-post training based on distance-learning techniques, as pioneered and developed by the British Open University. As well as fitting study into daily routines, general aspects of the subject can be combined with practical use of data relevant to a trainee's immediate work programme. In this way practice reinforces theory and leads to a much higher degree of retention of skills in the long term. The highest level of training relates to experts in remote sensing and image processing in their own right. This inevitably must involve advanced graduate training on a full-time basis to Master's level, and even through independent research to a Doctoral degree.

Only the last of these categories of personnel is catered for in any substantial way. This has led to the curious situation of an eminently practical subject being dominated by theoreticians, and a plethora of ever more abstruse pilot studies on hitherto undiscovered potato fields and the monitoring of obscure species of penguin. Remote sensing has yet to come of age. Without a major educational effort, planned rather than haphazard, it will remain dominated by what the small community of practical remote sensers term 'Trekkies', after addicts to the television series 'Star Trek'. Some information on available courses is given in Appendix 1.

As the public at large has become familiar with views of the Earth from space, through metsat images in TV weather forecasts, so a simple step forward is to provide synoptic and, where possible, very detailed analyses of political/military hot spots and environmental crises based on satellite images. As well as adding a new dimension to media coverage of world events, such a venture would serve a real need for precise information in the news and begin the process of educating everyone in the opportunities presented by remote sensing. Only one agency (Ocean Earth in the USA), independent of the sources of image data, government, and the media themselves, at present generates timely images of newsworthy locations for newspapers and the television, together with informed analysis.

5.3 Data distribution

All the agencies involved in the acquisition and distribution of remotely sensed data have different policies regarding prices, access, copyright, and end use. To discuss them all would be confusing and take up a lot of space, so the relevant authorities are listed in Appendix 2 so that readers can obtain the information that they need. In this section a few important general issues are covered.

One of the conventions associated with the peaceful uses of outer space, passed by the United Nations, specifically discourages the use of remotely sensed data to gain military advantage. If ever there was a pious hope, this is it, for not only are the majority of orbiting craft filled with means of gathering high-quality military intelligence but a very large proportion of sales of public-domain remote-sensing data is to military users. Only sales of the recently available Soviet data carry an exclusion clause to this effect. How it is enforced is not clear.

The history of satellite remote sensing is one of altruism subsequently modified by commercial and military considerations. At one extreme, data from geostationary and polar orbiting metsats are available to all-comers. Even schools, suitably equipped with a cheap parabolic receiver tuned to the frequency of data telemetry and suitable recording and display facilities, can obtain images from the geostationary systems and from the NOAA AVHRR at the time of their capture. Archived data from metsats held by the operating agencies is obtainable on digital tape at the cost of materials and handling. An even more open-handed policy has been adopted by NASA with regard to the products of its non-military experimental systems—images and tapes of the Earth and other planets are available free of charge to anyone from NASA archives, provided they are used for education or research.

The most famous of all remote-sensing systems is the US Landsat series. Although the driving force behind the initial funding was an opportunity to monitor Soviet and other agriculture for low-grade strategic intelligence, it was launched under what was termed the 'open skies' policy. This meant that data were available at far below economic cost to all-comers, with no restrictions on use or further copying and distribution. The on-board recording capability of Landsats 1 to 3 meant that the US was hoped to be able eventually to build a complete global archive and act as the main distributor, through the EROS Data Centre of the US Geological Survey. Responsibility for the system resided first with NASA and then passed to another government agency—NOAA. Any country that was able to finance the necessary ground-receiving hardware and processing capability was allowed to take data from the satellite while it was above the horizon (Fig. 5.1), at an annual cost of $200 000 per station (now $600 000), and market them independently to users interested in the areas covered (see Appendix 2). It is hard to conceive of greater generosity, when the cost of developing the satellites and launching them was of the order of several hundred million dollars each. It was not so philanthropic as first sight would suggest. It was a foreign relations concession to satisfy Third World countries who objected to the idea of wealthy nations and multinational companies monitoring their resources with impunity.

Up to 1981 the cost of a digital Landsat MSS scene was held at $200, rising by 1985 to $730. Between 1973 and 1978 annual sales of Landsat data rose from $228 000 to $2 million, and levelled off to around $2.5–2.9 million in the period up to 1983. During this period the programme operated at a considerable loss, the annual operating cost having risen to $40 million by 1984. The $600 million cost of developing and launching Landsat 4 in 1982, carrying the new Thematic Mapper as well as an MSS, together with the changed policies of the Reagan administration made the US government seek to rid itself of what was then regarded as an intolerable financial burden. The system was privatized and since September 1985 has been operated by the Earth Observation Satellite Company (EOSAT). While the essence of the 'open skies' policy still remains, prices are at a more economic level, and EOSAT retains copyright over the data in the form in which they are distributed, but not derivatives from them. As a result, free exchange of Landsat data, which was a growing and useful adjunct of the previous policy, slowed considerably with privatization. EOSAT has the responsibility for developing and launching replacements for the current Landsats. As a commercial concern it is likely to attempt to recoup the high costs by increasing the prices of the data it supplies. If it is unable to do this, the company will probably relinquish its responsibilities. This issue will become critical around 1990. The US government has no plans to continue the Landsat programme itself, so the great advantages of continuous coverage may be lost, hopefully on a temporary basis only. Other operators may step in to fill the breach, such as SPOT-Image or the consortium of Japanese interests that operate MOS-1.

The distribution of SPOT data is on an even more 'businesslike' basis. Prices are around nine times those of Landsat Thematic Mapper data for an equivalent volume. Moreover, SPOT is covered by a tight policy of copyright. No data in any digital or photographic form can be passed to a third party, exchanged, marketed, or reproduced without payment of considerable royalties and display of the SPOT trademark—which is one reason why there are so few in this book. Because of its potential for stereoscopic viewing at 10 m resolution, a ready market for SPOT imagery has been the military community. There have been several controversial uses of SPOT data in the media since the launch of its platform—the Chernobyl disaster, a chemical munitions factory in Iraq, and a uranium enrichment plant in Pakistan. Such breaches of security are forcing a review of the 'open skies' policy.

It is now possible for commercial remote-sensing systems to produce images that rival the resolution of secret 'spy satellites'. There are several plans in the USA, Europe, and Japan for commercially confidential, high-resolution systems that will sell to the highest bidder, most notably to the media and possibly to the oil industry. New US regulations give

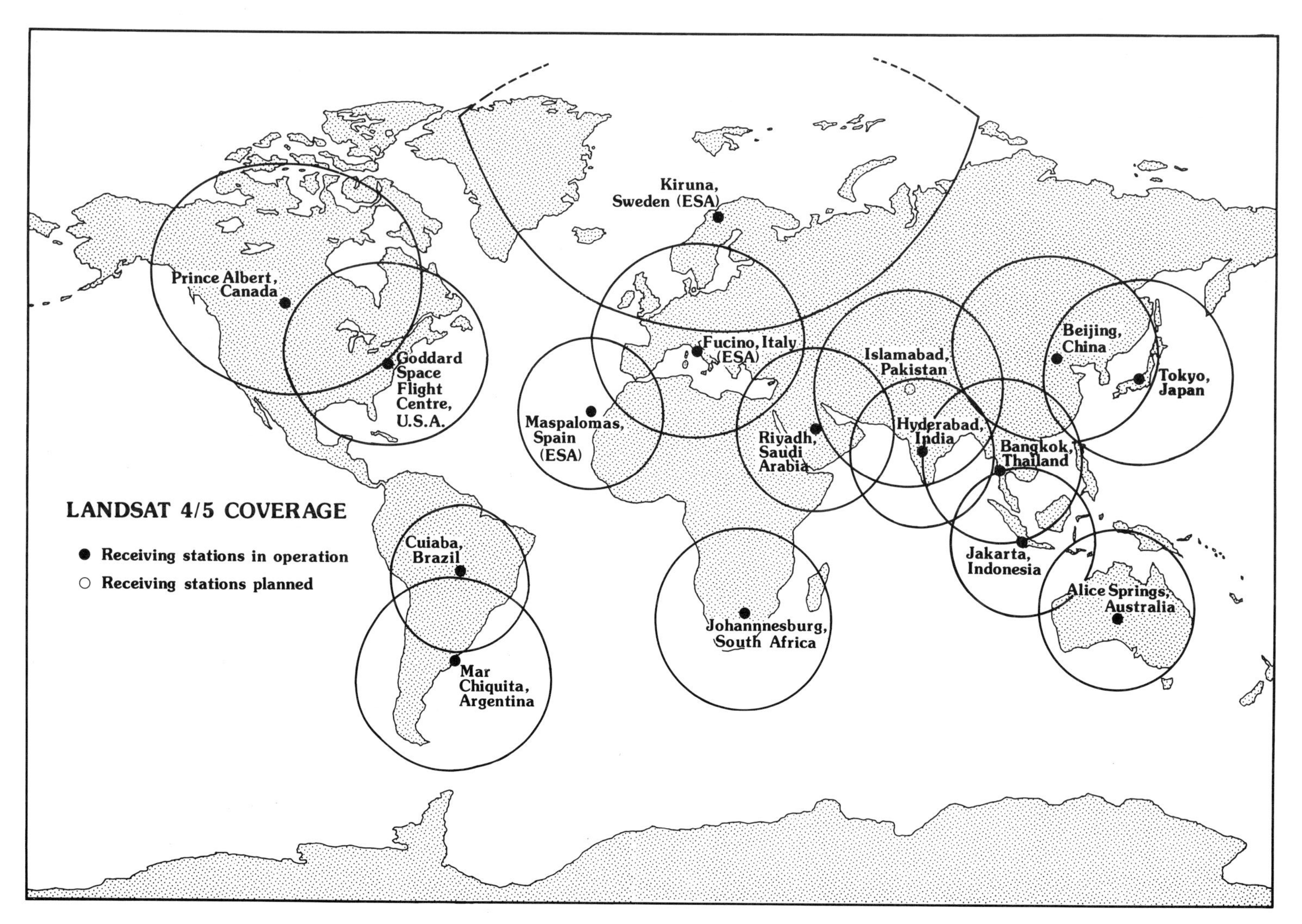

Fig. 5.1. Distribution of operational Landsat receiving stations. The outlined areas are those from which data can be received by line-of-sight telemetry. Many of the stations are also equipped for receipt of SPOT and other data.

the Departments of State and Defence the authority to delay, restrict, or veto the distribution of data considered to threaten the 'national interest' by companies run by US citizens. When the first Landsat TM images were processed by NASA in 1982, rumour has it that powerful interests in the US defence community attempted to have the whole system declared as restricted, because of what it revealed in the State of Michigan. This attempted veto was said to have been overridden at Presidential level. The mechanism for such restriction is now in place. If it is activated it leaves systems operated by European and Japanese agencies to capitalize on the opportunities for informing the public at large about issues of immediate concern, which manifest themselves in remotely sensed images.

Tempering the optimistic view of all the advantages associated with remote sensing are two important issues. Continuity of data supply from Landsat may cease, either for commercial reasons or through military 'over-ride'. Secondly, the cost of data may rise rapidly to match what can be afforded by wealthy users. This will leave most countries unable to use the advantages of remote sensing to help solve their immediate problems of development. Even now, many oil and mining companies, and certainly the intelligence communities of the USA, Soviet Union, France, West Germany, and Japan, together with those of countries with receiving systems, know considerably more about Third World countries than do their own governments. A large increase in costs, despite a greater ability to use data, will maintain this inequitable situation. India has had the foresight to avoid this potential stranglehold by launching its own high-resolution system in 1987, which produces data of almost as high a quality as Landsat MSS and TM. This may indeed become an important strategy for the non-aligned nations, particularly if they operate in consortia and distribute low-cost receiving stations independently of the main power blocs. Apart from these possible developments, there are several future programmes of orbital remote sensing that are already beyond the planning stage and will provide many new kinds of data through to the turn of the twentieth century.

5.4 Future prospects

There are many plans for the launching of civilian remote-sensing satellites in the period leading up to the end of the century, far more than can be covered exhaustively here. Some are pious hopes, others only in the planning stage. Here, only those for which funding is available and which have reached an advanced stage of planning are described. There are several categories.

First are the planned replacements for existing systems. The geostationary and sun-synchronous polar orbiting metsats, such as GOES, Meteosat, and the NOAA series, are of such importance and so cost-effective that replacements were assured shortly after the launch of current satellites. The Landsat programme's continuity has been approved and underwritten for the launch of the sixth and seventh versions in 1989 and 1991, possibly by the Space Shuttle, with some modifications to the Landsat-5 payload. These plans are, of course, subject to the continuing viability of Landsat's current commercial management by EOSAT. The MSS will be replaced by an instrument covering the same bands as at present, but with 60 m resolution. The TM will produce 15 m resolution panchromatic visible images, to rival SPOT, as well as the existing seven bands, and the resolution of the thermal band will be improved to 60 m. Landsat-7 may carry a pointable pushbroom instrument with high resolution and selectable bands in the VNIR and SWIR regions. This may serve the needs of both the media and oil industry. An identical system to the current SPOT was scheduled for launch in 1988, but no news had been received at the time of writing. Improvements to the pushbroom system and the addition of a sensor to monitor vegetation, and a radar altimeter are planned for SPOT-3, due for launch in 1990.

The second category covers new ventures with limited objectives and instrumentation. NASA plans to launch the Upper Atmosphere Research Satellite (UARS) in 1990, which will deploy a variety of spectrometers and radiometers to measure atmospheric chemistry and energy balance, and will use a lidar for measurement of wind velocities. The US Navy, in conjunction with NASA, NOAA, and the US Air Force will launch the Navy Remote Ocean Sensing System (NROSS) in 1990. This project will carry a radar altimeter aimed at measuring sea-surface elevation and a wind scatterometer to measure direction and speed of ocean winds, similar to those on Seasat, together with passive microwave imagers to give sea temperature measurements. The objective is to use well-tested methods to provide near real-time data for operational use by marine traffic. In parallel, NASA hopes, in conjunction with the French Centre National D'Études Spatiale (CNES), to launch a high-precision radar-altimeter system called the Ocean Topography Experiment (TOPEX) in 1991. This will have sufficient precision for ocean circulation to be monitored accurately. The Canadian Radarsat (1994) will carry a wind scatterometer and a synthetic aperture radar imaging system aimed primarily at sea ice. The European Space Agency (ESA) will launch Earth Remote Sensing Satellite 1 (ERS-1) in 1990, carrying a similar payload to Seasat and Radarsat, but with the addition of a radar altimeter and a scanning microwave radiometer to determine sea-surface temperature and measure atmospheric water vapour. Mainly land applications will be served by three planned systems, one using a multispectral pushbroom planned for launch by Brazil in 1991, another multispectral scanner (China 1989), and the Japanese Earth Remote Sensing Satellite (JERS-1), which will carry both radar and visible to near-infrared imaging systems from 1991.

For most applications of remote sensing, long-term continuity of data supply is essential, particularly when several different kinds of information are required. At present many users have to use data from several different satellites, some of which are continually operational but others have had a brief lifespan, such as Seasat and the Heat Capacity Mapping Mission. Unmanned satellites, when they malfunction, are generally beyond reach and cannot be serviced, although there are plans to use the Space Shuttle to retrieve, service, and relaunch some remote-sensing platforms. The third category of planned remote-sensing activity centres on using large satellites carring a great range of systems that can be serviced by astronauts. This is the objective of the International Space Station, consisting of three polar-orbiting, sun-synchronous platforms at an altitude of 824 km, one crossing the Equator in the morning (the European Columbus (1996?)), the other two in the afternoon (the US Earth Observing System (1995–8?)). Japan may contribute a fourth platform in 1998. The plan is not for manned space stations—the orbits required are too high and along the wrong paths for easy access by manned shuttles—but for each to descend to accessible altitudes every two to three years for maintenance, so giving an extended lifetime of perhaps 15 years. Before outlining the instruments that may be carried by these platforms it must be emphasized that the whole plan has not yet been assigned funds and there are several potentially disruptive issues to be resolved, not the least of which is the possible incorporation in the US segment of elements related to the Strategic Defence Initiative—ESA is not willing to collaborate in any mission that has a military content.

Should the International Space Station pass from its current status as several kilogrammes of documents outlining scientific plans and priorities to reality, it will please just about every environmental scientist with an interest in remote sensing. It will deploy five imaging systems: one similar to the AVHRR; two imaging spectrometers with narrow wavebands in the visible, VNIR, and SWIR, one with 30 m resolution for detailed studies, the other with a coarse 1 km resolution for studying gross features; a multifrequency, multipolarization, and multiple look-angle radar imager; and a 30 m resolution thermal-infrared multispectral scanner. There will be two kinds of altimeter, one using radar to measure sea- and ice-surface elevation to within 3 cm, the other based on a laser aimed at land topography. Eight instruments will be targeted on chemical constituents of the atmosphere, using information from virtually the whole electromagnetic spectrum up to microwave wavelengths. Atmospheric circulation and winds will be addressed by three systems, one based on lidar, one on Doppler shifts of oxygen absorption features in visible light, and the other on radar scatterometry. Three systems are oriented towards surface and atmospheric temperature, which, together with three monitoring atmospheric moisture content, rainfall, and snow cover complete that part of the plan aimed at climatology and meterology. Other instruments target on soil moisture, the physics of the upper atmosphere, data collection from ground stations, and very accurate laser ranging of ground reflectors for measurements of tectonic plate movements.

All in all, the Space Station offers breathtaking possibilities, that is, if it flies and if there are sufficient trained users to make the most of what it could provide.

Appendix 1 Training opportunities

Many universities and other institutions of higher education offer full-time undergraduate courses involving remote sensing, as well as opportunities for full-time graduate training, involving either advanced courses or research programmes. Such opportunities are unlikely to be of immediate interest to readers of this book. A more likely need is for information on part-time or short residential courses enabling the familiarization given here to be extended to more practical experience.

In the United Kingdom several institutions run short residential courses open to both UK and overseas trainees. All are adaptable to or designed for a variety of discipline interests, and range from one-day to month-long packages. Information can be obtained from:

The Short Course Secretary, Department of Agricultural Engineering, Silsoe College, Silsoe, Bedford MK45 4DT.

Remote Sensing Unit, Department of Civil Engineering, University of Aston, Gosta Green, Birmingham B4 7ET.

Carnegie Laboratory of Physics, University of Dundee, Dundee DD1 4HN.

Department of Earth Sciences, The Open University, Milton Keynes, MK7 6AA.

In Europe as a whole, the European Association of Remote Sensing Laboratories (EARSeL) helps coordinate information on training opportunities, which can be obtained from:

EARSeL General Secretariat, 292 rue St Martin, F-75003, Paris, France.

Apart from this it is worth noting the short courses organized in Europe by the following institutions:

Courses Secretariat, ISPRA, Centro Comune di Ricerca, 21020 ISPRA, Varese, Italy.

GDTA, 18 avenue Edouard Belin, 31055 Toulouse Cedex, France.

Student Affairs Office, ITC, 350 Boulevard 1945, PO Box 6, 7500AA Enschede, The Netherlands.

In the USA the best-known short-course programmes are run by:

Environmental Research Institute of Michigan, Ann Arbor, Michigan 48107, USA.

US Geological Survey, EROS Data Center, Mundt Federal Building, Sioux Falls, South Dakota 57198, USA.

Laboratory for Applications of Remote Sensing, Purdue University, 1220 Potter Drive, West Lafayette, Indiana 47906, USA.

Agencies with an international brief for training in remote sensing include the World Bank, The US Agency for International Development (USAID) and the British Overseas

Development Authority, each of which runs a number of occasional courses. The main provision of regular training programmes at a number of sites is by the United Nations, through the UN Development Programme and the Food and Agriculture Organisation. However, both agencies negotiate courses and training programmes with individual governments, usually centred around a specific problem, rather than opening their courses to all-comers.

An entirely new concept in remote-sensing training, involving home study of specially designed texts and visual aids, is currently being developed by the Dutch Open University in collaboration with the British Open University. This is a 50 hour package aimed at both familiarization with remote sensing and preparation for use of remotely sensed data in a variety of environmental sciences.

Appendix 2 Sources of images

Such a huge number of aerial photographs have been acquired for all parts of the Earth, by a large number of governmental and commercial agencies, that individual sources cannot be given here. However, an excellent guide to sources of preliminary information is given by the United States *Geological Survey Circular*, **834**:

US Geological Survey (1981). Worldwide directory of national earth science agencies and related international organizations. *Geological Surveys Circular*, **834**. (Obtainable from US Geological Survey, 507 National Center, Reston, Virginia 22092, USA.)

Information on satellite photographs of the Earth can be obtained from:

Skylab and Space Shuttle (Large Format Camera and hand-held cameras)

US Geological Survey, EROS Data Center, Mundt Federal Building, Sioux Falls, South Dakota 57198, USA.

Space Shuttle (Metric Camera)

DFVLR, Oberpfaffenhofen, 8031, Wessling, Federal Republic of Germany.

The various meteorological satellites deployed in geostationary and polar orbit have produced the most comprehensive cover of the Earth, for the longest uninterrupted period and with an unparalleled degree of repetition. Many are available from local reception and distribution centres, but here only the central administrative authorities are listed, as they can direct interest to national sources.

The National Oceanic and Atmospheric Administration of the US Department of Commerce is the main source of information and images from the GOES, Nimbus, TIROS-NOAA, and Seasat satellites. It can provide catalogues of imagery, photographic prints, and computer-compatible image tapes, obtainable from:

Satellite Data Services Division, NCC/NESDIS, World Weather Building, Room 10, Washington, DC 20233, USA.

Information on products from the geostationary Meteosat can be obtained from:

Meteosat Data Services, Meterological Data Management Department, European Space Operations Centre, Robert-Bosch-Strasse 5, Darmstadt, Federal Republic of Germany.

The repository of all declassified results from remote-sensing experiments carried by satellites launched by the US National Aeronautics and Space Administration, including images in various forms from the early years of space research, Skylab, Seasat, HCMM, Nimbus, and Shuttle missions, including Shuttle Imaging Radar is:

World Data Center A for Rockets and Satellites, Code 601, NASA, Goddard Space Flight Center, Greenbelt, Maryland 20771, USA.

As well as supplying free information on archives and various guides to the available data, images that are available at Goddard SFC are supplied free of charge to all non-commercial users on request.

The two most publicized and most useful sources of readily available remotely sensed imagery are the Landsat and SPOT systems. Both are distributed by the organizations directly responsible for their management:

EOSAT, 4300 Forbes Boulevard, Lanham, Maryland 20706, USA.

SPOT Image, 18 Avenue Edouard Belin, F31055 Toulouse Cedex, France.

Information on the availability of imagery held by both organizations can be obtained by specifying the geographic coordinates of any area of interest. Annotated computer printouts are supplied, together with instructions on the formalities of ordering and price lists. However, these organizations supply information only on the data that they hold, whereas licences have been issued for receipt and distribution of data from both systems to a number of ground-receiving stations in other countries (Fig. 5.1). This means that other imagery of areas covered by these stations may be available, particularly with respect to Landsat-4 and -5, which do not have onboard image-recording facilities. Because of this, telemetry to US Landsat receiving stations is limited by direct communication to relay satellites (TDRS) that enable images of the Americas, Europe, part of Asia and Africa to be recorded by EOSAT. Availability of the latest Landsat imagery for other areas is restricted to the line-of-sight communications shown in Fig. 5.1. Both EOSAT and SPOT Image will supply addresses of other receiving stations, together with those of country-by-country organizations that have negotiated franchises for purchase and resale of data.

Possibly the lowest cost option for obtaining large-area coverage based on Landsat imagery is through image maps comprising mosaics of several scenes. Some of these are documented in the *Remote sensing yearbook* by D. J. Carter (see Further reading).

At present, data from the Indian IRS-1 and Japanese MOS-1 satellites are available only for areas covered by their direct line-of-sight communications with parent institutions. Information on available data, prices, and ordering formalities should be directed to:

Earth Observation System, Indian Space Research Organization, K. G. Road, Bangalore 560009, India.

NASDA, 2-4-1 Hamamatsu-Cho, Minato-Ku, Tokyo 105, Japan.

In view of the many future opportunities to be provided by remotely sensed data from planned orbital platforms, a watching brief on developments can be maintained by asking to be placed on the mailing list of the following organizations:

Earth Observation System and Shuttle missions

World Data Center A for Rockets and Satellites, Code 601, NASA, Goddard Space Flight Center, Greenbelt, Maryland 20771, USA.

ERS-1, Columbus Polar Platform

European Space Agency, 8–10 rue Mario Nikis, 75738 Paris 15, France.

Japanese Polar Platform

Space Development Division, Research and Development Bureau, Science and Technology Agency, 2-2-1 Kasumugaseki, Chiyoda-ku, Tokyo 100, Japan.

Radarsat

Canadian Centre for Remote Sensing, 2464 Shefield Road, Ottawa, Ontario K1A 0Y7, Canada.

Further reading

This book aims at familiarization with the basic principles of remote sensing and at giving an overview of the various aspects of the terrestrial environment that can be addressed by examining and analysing images. The subject is large and growing rapidly, so this list of further reading is not exhaustive, but covers only those texts that are equally general but delve more deeply into the subjects related to remote sensing. Each contains its own list of references to more specialized books and papers published in various journals that are relevant to its subject area. Thus the reader can venture, with persistence and ingenuity, into an in-depth study of any aspects of the subject that are of immediate importance. Perhaps the most useful initial aids to further study are the two source references, that by Cracknell and Hayes including comprehensive lists of references on a subject-by-subject basis. An easy means of expanding experience with the interpretation of images of the Earth can be gained by browsing through the various image atlases that are available. Regular updates on the advance of the subject are best achieved by browsing in specialist journals, although important papers appear regularly in nearly all journals dealing with the environmental sciences. Remote sensing forms the topic of a number of international symposia and conferences, whose proceedings are generally published by the organizers.

Textbooks

Burrows, W. E. (1988). *Deep black: the secrets of space espionage*. Bantam Press, London.

Colwell, R. N. (ed.) (1983). *Manual of remote sensing (second edition)*, 2 Vols. American Society of Photogrammetry, Falls Church, VA.

Curran, P. J. (1985). *Principles of remote sensing*. Longman, London.

Drury, S. A. (1987). *Image interpretation in geology*. Allen and Unwin, London.

Gonzalez, R. C. and Wintz, P. (1977). *Digital image processing*. Addison-Wesley, Reading, MA.

Lillesand, T. M. and Kiefer, R. W. (1987). *Remote sensing and image interpretation*. Wiley, New York.

Peebles, C. (1988). *Guardians*. Presidio Press.

Robinson, I. S. (1985). *Satellite oceanography*. Wiley, London.

Sabins, F. F. (1986). *Remote sensing: principles and interpretation*. Freeman, San Francisco.

Schowengerdt, R. A. (1983). *Techniques for image processing and classification in remote sensing*. Academic Press, New York.

Short, N. M. (1982). *The Landsat tutorial workbook*. NASA Reference Publication 1078.

Siegal, B. S. and Gillespie, A. R. (ed.) (1980). *Remote sensing in geology*. Wiley, New York.

Stroebel, L., Todd, H., and Zakia, R. (1980). *Visual concepts for photographers*. Focal Press, London.

Image atlases

European Space Agency (1985). *Atlas of Meteosat Imagery*. ESA, Darmstadt.

Ford, J. P., Blom, R. G., Bryan, M. G., Daily, M. I., Dixon, T. H., Elachi, C., and Xenos, E. C. (1980). *Seasat views North America, the Caribbean, and Western Europe with imaging radar*. JPL Publication, 80–67.

Ford, J. P., Cimino, J. B., and Elachi, C. (1983). *Space Shuttle Columbia views the world with imaging radar—the SIR-A experiment*. JPL Publication, 82–95.

Francis, P. and Jones, P. (1984). *Images of Earth*. George Phillip, London.

Sheffield, C. (1981). *Earthwatch*. Sidgwick and Jackson, London.

Sheffield, C. (1983). *Man on Earth*. Sidgwick and Jackson, London.

Short, N. M., Lowman, P. D., Freden, S. C., and Finch, W. A. (1976). *Mission to Earth: Landsat views the world*. NASA Special Publication 360.

Short, N. M. and Blair, R. W. (1986). *Geomorphology from space: a global overview of regional landforms*. NASA Special Publication 486.

Source references

Carter, D. J. (1986). *The remote sensing sourcebook: a guide to remote sensing products, services, facilities, publications and other materials*. Kogan Page and McCarta, London.

Cracknell, A. and Hayes, L. (1987). *Remote sensing yearbook 1987*. Taylor and Francis, London. (Revised yearly.)

Specialist journals

IEEE Transactions of Geoscience and Remote Sensing

International Journal of Remote Sensing

Remote Sensing of Environment

Photogrammetric Engineering and Remote Sensing

Regular symposia

International Symposia on Remote Sensing of the Environment: organized by Environmental Research Institute of Michigan, Ann Arbor, Michigan 48107, USA.

Annual Conference of the Remote Sensing Society: organized by The Remote Sensing Society, Department of Geography, University of Reading, Reading, Berkshire RG6 2AU, UK.

William T. Pecora Memorial Remote Sensing Symposium: contact US Geological Survey, EROS Data Center, Sioux Falls, South Dakota 57116, USA.

Annual Symposium on Machine-Processing of Remotely Sensed Data: organized by the Laboratory for Applications of Remote Sensing, Purdue University, 1220 Potter Drive, West Lafayette, Indiana 47906, USA.

Annual Conference of the International Geoscience and Remote Sensing Society (IGARSS): contact IEEE Service Center, 445 Hoes Lane, Piscataway, New Jersey 08854, USA.

Biennial Conference of the International Society for Photogrammetry and Remote Sensing (ISPRS): contact International Institute for Aerospace Survey and Earth Sciences (ITC), PO Box 6, 7500A Enschede, The Netherlands.

Glossary

This Glossary defines many terms and concepts with which readers may not be familiar and which crop up in the book and in other remote-sensing literature. It should be used in conjunction with the Index, which shows where the terms are defined more fully and in context. Terms highlighted in the definitions are also contained within the glossary. It does not contain items related to applications in the environmental sciences.

absolute temperature Temperature measured on the Kelvin scale, whose base is absolute zero, i.e. −273 °C; 0 °C is expressed as 273 K.

absorptance A measure of the ability of a material to absorb EM energy at a specific wavelength.

absorption band A range of wavelengths in the EM spectrum where a material absorbs EM energy incident upon it, often resulting from various energy-matter transitions.

achromatic vision The perception by the human eye of changes in brightness, often used to describe the perception of monochrome or black and white scenes.

active remote sensing A system based on the illumination of a scene by artificial radiation and the collection of the reflected energy returned to the system. Examples are **radar** and systems using lasers.

acuity A measure of human ability to perceive spatial variations in a scene. It varies with the spatial frequency, shape, and contrast of the variations, and depends on whether the scene is coloured or monochrome.

additive primary colours The spectral colours red, green, and blue, which Thomas Young (1773–1829) discovered to be capable of reproducing all other colours when mixed by projection through filters, and each of which cannot be produced by mixtures of the other two.

albedo The fraction of the total EM energy incident on a material which is reflected in all directions.

analogue image An image where the continuous variation in the property being sensed is represented by a continuous variation in image tone. In a photograph this is achieved directly by the grains of photosensitive chemicals in the film; in an electronic scanner, the response in, say, millivolts is transformed to a display on a cathode-ray tube where it may be photographed.

artefact A feature on an image which is produced by the optics of the system or by digital image processing, and sometimes masquerades as a real feature.

atmospheric shimmer An effect produced by the movement of masses of air with different refractive indices, which is most easily seen in the twinkling of stars. Shimmer results in blurring on remotely sensed images, and is the ultimate control over the resolution of any system.

atmospheric window A range of EM wavelengths where radiation can pass through the Earth's atmosphere with relatively little attenuation.

AVHRR Advanced Very High Resolution Radiometer, a multispectral imaging system carried by the TIROS-NOAA series of meterological satellites.

azimuth In general this is the compass bearing of a line given in degrees clockwise from North. In **radar** terminology it refers to the direction at right angles to the radar propagation direction, which is parallel to the ground track in a sideways-looking system.

band In remote sensing, this is a range of wavelengths from which data are gathered by a recording device.

bin One of a series of equal intervals in a range of data, most commonly employed to describe the divisions in a **histogram**.

binary A numerical system using the base two. Examples are 0 = 0, 1 = 2^0 = 1, 10 = 2^1 = 2, 11 = $2^1 + 2^0$ = 3.

bit An abbreviation of a binary digit, which refers to an exponent of two. A bit is represented by 0 or 1 for 'on' or 'off' in a digital computer.

blackbody A perfect radiator and absorber of EM energy, where all incident energy is absorbed and the energy radiated from the body at a particular temperature is at the maximum possible rate for each wavelength, as governed by the **Stefan–Boltzmann Law**. No natural material has these ideal properties.

blind spot The point of the optic nerve to the retina where no radiation is detected by the eye.

byte A group of eight **bits** of digital data in binary form. A byte represents **digital numbers** up to 255, and is the standard adopted by most digital remote-sensing systems where the range of energies is coded from 0 to 255.

cell assemblies The linked receptors, retinal neurons, and neural cells in the visual cortex of the brain which enable interaction between perception and past experience.

charge-coupled device (CCD) A light-sensitive capacitor whose charge is proportional to the intensity of illumination. They are able to be charged and discharged very quickly, and are used in **pushbroom devices, spectroradiometers** and modern video cameras.

chromatic vision The perception by the human eye of changes in **hue**.

classification The process of assigning individual *pixels* of a multispectral image to categories, generally based on spectral characteristics of known parts of a scene.

coherent radiation Electromagnetic radiation whose waves are equal in length and are in phase, so that waves at different points in space act in unison, as in a **laser** and **synthetic aperture radar**.

cones Receptors in the retina which are sensitive to colour. There are cones sensitive to the red, green, and blue components of light.

context The known environment of a particular feature on an image.

contrast The ratio between the energy emitted or reflected by an object and its immediate surroundings.

contrast stretching Expanding a measured range of **digital numbers** in an image to a larger range, to improve the contrast of the image and its component parts.

corner reflector A cavity formed by three planar reflective surfaces intersecting at right angles, which returns radar directly back to its source.

cut off The **digital number** in the **histogram** of a **digital image** which is set to zero during **contrast stretching**. Usually this is a value below which atmospheric scattering makes a major contribution.

CZCS Coastal Zone Colour Scanner, a multispectral imaging system carried by the Nimbus series of meteorological satellites.

density slicing The process of converting the full range of data into a series of intervals or slices, each of which expresses a range in the data.

depression angle In radar usage this is the angle between the horizontal plane passing through the antenna and the line connecting the antenna to the target. It is easily confused with the **look angle**.

digital image An image where the property being measured has been converted from a continuous range of analogue values to a range expressed by a finite number of integers, usually recorded as binary codes from 0 to 255, or as one **byte**.

digital number (DN) The value of a variable recorded for each **pixel** in an image as a binary integer, usually in the range of 0–255. An alternative term used in some texts is brightness value (BV). The plural of its acronym is also DN.

directional filter A spatial-frequency filter which enhances features in an image in selected directions.

Doppler shift A change in the observed frequency of EM or other waves caused by the relative motion between source and detector. Used principally in the generation of **synthetic-aperture radar** images.

edge A boundary in an image between areas with different tones.

edge enhancement The process of increasing the **contrast** between adjacent areas with different tones on an image.

emissivity The ratio of the energy radiated by a material to that which would be radiated by a **blackbody** at the same temperature. A blackbody therefore has an emissivity of 1 and natural materials range from 0 to 1.

emittance A term for the radiant flux of energy per unit area emitted by a body. (Now obsolete.)

EOSAT Earth Observation Satellite Company, based in Lanham, Maryland, USA. A private company contracted to the US Government since September 1985 to market Landsat data and develop replacements for the Landsat system.

ERBSS Earth Radiation Budget Sensor System, carried by NOAA satellites.

EROS Earth Resources Observation System, based at the EROS Data Center, Sioux Falls, South Dakota, USA. Administered by the US Geological Survey, it forms an important source of image data from Landsat 1, 2, and 3 and for Landsat-4 and -5 data up to September 1985, as well as airborne data for the USA.

ESA European Space Agency, based in Paris. A consortium between several European states for the development of space science, including the launch of remote-sensing satellites.

false-colour image A colour image where parts of the non-visible EM spectrum are expressed as one or more of the red, green, and blue components, so that the colours produced by the Earth's surface do not correspond to normal visual experience. Also called a false-colour composite (FCC). The most commonly seen false-colour images display the **very-near infrared** as red, red as green, and green as blue.

fluorescence A property of some materials where EM energy of one wavelength is absorbed and then re-emitted at a longer wavelength.

foreshortening A distortion in **radar** images causing the lengths of slopes facing the antenna to appear shorter on the image than on the ground. It is produced when radar wavefronts are steeper than the topographic slope.

fovea The region around that point on the retina intersected by the eye's optic axis, where receptors are most densely packed. It is the most sensitive part of the retina.

frequency (ν) The number of waves that pass a reference point in unit time, usually one second.

geographic information system (GIS) A data-handling and analysis system based on sets of data distributed spatially in two dimensions. The data sets may be map oriented, when they comprise qualitative attributes of an area recorded as lines, points, and areas often in **vector format**, or image oriented, when the data are quantitative attributes referring to cells in a rectangular grid usually in **raster format**. It is also known as a geobased or geocoded information system.

geostationary orbit An orbit at 41 000 km in the direction of the Earth's rotation, which matches speed so that a satellite remains over a fixed point on the Earth's surface.

grid format The result of interpolation from values of a variable measured at irregularly distributed points, or along survey lines, to values referring to square cells in a rectangular array. It forms a step in the process of contouring data, but can also be used as the basis for a **raster format** to be displayed and analysed digitally after the values have been rescaled to the 0–255 range.

ground-control point A point in two dimensions which is common to both an image and a topographic map, and can be represented by (x, y) coordinates based on the map's cartographic projection and grid system; used in geometric correction of distorted images, and their registration to a convenient map projection.

HCMM Heat Capacity Mapping Mission, the NASA satellite launched in 1978 to observe thermal properties of rocks and soils. It remained in orbit for only a few months.

high-pass filter A spatial filter which selectively enhances contrast variations with high spatial frequencies in an image. It improves the sharpness of images and is a method of **edge enhancement**.

HIRIS High Resolution Imaging Spectrometer, possibly to be carried by the Space Shuttle.

HIRS High Resolution Infrared Spectrometer, carried by NOAA satellites.

histogram A means of expressing the frequency of occurrence of values in a data set within a series of equal ranges or **bins**, the height of each bin representing the frequency at which values in the data set fall within the chosen range. A cumulative histogram expresses the frequency of all values falling within a bin and lower in the range. A smooth curve derived mathematically from a histogram is termed the probability density function (PDF).

hue The attribute of a colour that distinguishes it from grey with the same intensity.

image dissection The breaking down of a continuous scene into discrete spatial elements, either by the receptors on the retina, or in the process of capturing the image artificially.

image striping A defect produced in **line scanner** and **pushbroom** imaging devices produced by the non-uniform response of a single detector, or amongst a bank of detectors. In a line-scan image the stripes are perpendicular to flight direction, but parallel to it in a pushbroom image.

incidence angle The angle between the surface and an incident ray of EM radiation—most usually referring to radar.

instantaneous field of view (IFOV) The solid angle through which a detector is sensitive to radiation. It varies with the intensity of the radiation, the time over which radiation is gathered, and forms one limit to the **resolution** of an imaging system.

intensity A measure of the energy reflected or emitted by a surface.

Landsat A series of remote-sensing satellites in **sun-synchronous**, **polar orbit** that began in 1972. Initially administered by **NASA**, then **NOAA**, and since 1985 by **EOSAT**. It carried **MSS**, **RBV**, and **TM** systems.

laser **L**ight **a**rtificially **s**timulated **e**lectromagnetic **r**adiation: a beam of **coherent** radiation with a single wavelength.

lidar **L**ight **i**ntensity **d**etection **a**nd **r**anging, which uses lasers to stimulate **fluorescence** in various compounds and to measure distances to reflecting surfaces.

layover A distortion in radar images when the angle of surface slope is greater than that of radar wavefronts. The base of a slope reflects radar after the top, and since radar images express the distance to the side in terms of time, the top appears closer to the platform than the base on an image, giving the impression of an overhanging slope.

LIMS Limb Infrared Monitoring of the Stratosphere experiment, carried by Nimbus-7.

line drop out The loss of data from a scan line caused by malfunction of one of the detectors in a **line scanner**.

line scanner An imaging device which uses a mirror to sweep the ground surface normal to the flight path of the platform. An image is built up as a strip comprising lines of data.

look angle The angle between the vertical plane containing a radar antenna and the direction of radar propagation. Complementary to the **depression angle**.

look direction The direction in which pulses of radar are transmitted.

look-up table (LUT) A mathematical formula used to convert one distribution of data to another, most conveniently remembered as a conversion graph.

luminance A measure of the luminous intensity of light emitted by a source in a particular direction.

Mach band An optical illusion of dark and light fringes within adjacent areas of contrasted tone. It is a psychophysiological phenomenon which aids human detection of boundaries or **edges**.

median filter A spatial filter, which substitutes the median value of DN from surrounding **pixels** for that recorded at an individual pixel. It is useful for removing random **noise**.

mid-infrared (MIR) The range of EM wavelengths from 8 to 14 micrometres dominated by emission of thermally generated radiation from materials; also known as thermal infrared.

Mie scattering The scattering of EM energy by particles in the atmosphere with comparable dimensions to the wavelength involved.

minus-blue photograph A panchromatic black and white photograph from which the blue part of the visible range has been removed using a yellow filter.

mixed pixel A **pixel** whose DN represents the average energy reflected or emitted by several types of surface present within the area that it represents on the ground; sometimes called a mixel.

MOMS Modular Optical-electronic Multispectral Scanner, a West German imaging system carried experimentally by Shuttle missions, and which is based on **pushbroom** systems.

modulation transfer function (MTF) A measure of the sensitivity of an imaging system to spatial variations in contrast.

MOS-1 Marine Observation Satellite, launched by Japan in 1987.

multispectral scanner (MSS) A **line scanner** that simultaneously records image data from a scene in several different wavebands. Most commonly applied to the four-channel system with 80 m resolution carried by the Landsat series of satellites.

nadir The point on the ground vertically beneath the centre of a remote-sensing system.

NASA National Aeronautics and Space Administration, USA.

near infrared (NIR) The shorter wavelength range of the infrared region of the EM spectrum, from 0.7 to 2.5 μm. It is often divided into the very-near infrared (VNIR) covering the range accessible to photographic emulsions (0.7 to 1.0 μm), and the short-wavelength infrared (SWIR) covering the remainder of the NIR atmospheric window from 1.0 to 2.5 μm.

NOAA National Oceanic and Atmospheric Administration, USA.

noise Random or regular **artefacts** in data which degrade their information-bearing quality and are due to defects in the recording device.

non-selective scattering The scattering of EM energy by particles in the atmosphere which are much larger than the wavelengths of the energy, and which causes all wavelengths to be scattered equally.

non-spectral hue A hue which is not present in the spectrum of colours produced by the analysis of white light by a prism or diffraction grating. Examples are brown, magenta, and pastel shades.

orthophotograph A vertical aerial photograph from which the distortions due to varying elevation, tilt, and surface topography have been removed, so that it represents every object as if viewed directly from above, as in a map.

orthophotoscope An optical-electronic device which converts a normal vertical aerial photograph to an orthophotograph.

parallax The apparent change in position of an object relative to another when it is viewed from different positions. It forms the basis of **stereopsis**.

parallax difference The difference in the distances on overlapping vertical photographs between two points, which represent two locations on the ground with different elevations.

passive microwaves Radiation in the 1 mm to 1 m range emitted naturally by all materials above absolute zero.

passive remote sensing The capture of images representing the reflection or emission of EM radiation that has a natural source.

pattern A regular assemblage of **tone** and **texture** on an image. Often refers to drainage systems.

photon A **quantum** of EM energy.

photopic vision Vision under conditions of bright illumination, when both **rods** and **cones** are employed.

pixel A single sample of data in a **digital image**, having both a spatial attribute (its position in the image and a usually rectangular dimension on the ground), and a spectral attribute (the intensity of the response in a particular waveband for that position (*DN*)). A contraction of 'picture element'.

Planck's Law An expression for the variation of **emittance** of a **blackbody** at a particular temperature as a function of wavelength.

point spread function (PSF) The image of a point source of radiation, such as a star, collected by an imaging device. A measure of the spatial fidelity of the device.

polar orbit An orbit that passes close to the poles, thereby enabling a satellite to pass over most of the surface, except the immediate vicinity of the poles themselves.

polarized radiation Electromagnetic radiation in which the electrical field vector is contained in a single plane, instead of having random orientation relative to the propagation vector. Most commonly refers to **radar** images.

principal component analysis The analysis of covariance in a multiple data set so that the data can be projected as additive combinations on to new axes, which express different kinds of correlation among the data.

principal point The centre of an aerial photograph.

probability density function (PDF) A function indicating the relative frequency with which any measurement may be expected to occur. In remote sensing it is represented by the **histogram** of **DN** in one band for a scene.

pushbroom system An imaging device consisting of a fixed linear array of many sensors which is swept across an area by the motion of the platform, thereby building up an image. It relies on sensors whose response and reading is nearly instantaneous, so that the image swathe can be segmented into **pixels** representing small dimensions on the ground.

quantum The elementary quantity of EM energy that is transmitted by a particular wavelength. According to the quantum theory, EM radiation is emitted, transmitted, and absorbed as numbers of quanta, the energy of each quantum being a simple function of the frequency of the radiation.

radar The acronym for **ra**dio **d**etection **a**nd **r**anging, which uses pulses of artificial EM radiation in the 1 mm to 1 m range to locate objects which reflect the radiation. The position of the object is a function of the time that a pulse takes to reach it and return to the antenna.

radar altimeter A non-imaging device that records the time of radar returns from vertically beneath a platform to estimate the distance to and hence the elevation of the surface; carried by **Seasat** and the **ESA ERS-1** platforms.

radar cross section A measure of the intensity of backscattered radar energy from a point target. Expressed as the area of a hypothetical surface which scatters radar equally in all directions and which would return the same energy to the antenna.

radar scattering coefficient A measure of the back-scattered radar energy from a target with a large area. Expressed as the average **radar cross section** per unit area in decibels (dB). It is the fundamental measure of the radar properties of a surface.

radar scatterometer A non-imaging device that records radar energy backscattered from terrain as a function of depression angle.

radial relief displacement The tendency of vertical objects to appear to learn radially away from the centre of a vertical aerial photograph. Caused by the conical field of view of the camera lens.

range In radar usage this is the distance in the direction of radar propagation, usually to the side of the platform in an imaging radar system. The slant range is the direct distance from the antenna to the object, whereas the distance from the ground track of the platform to the object is termed the ground range.

raster The scanned and illuminated area of a video display, produced by a modulated beam of electrons sweeping the phosphorescent screen line by line from top to bottom at a regular rate of repetition.

raster format A means of representing spatial data in the form of a grid of **DN**, each line of which can be used to modulate the lines of a video **raster**.

Rayleigh scattering Selective scattering of light in the atmosphere by particles that are small compared with the wavelength of light.

RBV The Return-Beam **Vidicon** system aboard Landsat-3 which produced panchromatic digital images with a resolution of 40 m.

real-aperture radar An imaging radar system where the azimuth resolution is determined by the physical length of the antenna, the wavelength, and the **range**; also known as brute-force radar.

redundancy Information in an image which is either not required for interpretation or cannot be seen. Redundancy may be spatial or spectral. The term also refers to multispectral data where the degree of correlation between bands is so high that one band contains virtually the same information as all the bands.

reflectance/-ivity The ratio of the EM energy reflected by a surface to that which falls upon it. It may be qualified as spectral reflectance. The suffix '-ance' implies a property of a specific surface. The suffix '-ity' implies a property for a given material.

resampling The calculation of new **DN** for pixels created during geometric correction of a digital scene, based on the values in the local area around the uncorrected pixels.

resolution A poorly defined term relating to the fidelity of an image to the spatial attributes of a scene. It involves the **IFOV** and **MTF** of the imaging system, and depends on the contrast within the image, as well as on other factors. It is usually expressed as line pairs per millimetre of the most closely spaced lines that can be distinguished, and therefore depends on human vision, scale, and viewing distance.

ringing Fringe-like **artefacts** produced at **edges** by some forms of **spatial-frequency filtering**.

rods The receptors in the retina that are sensitive to brightness variations.

SAMII Stratospheric Aerosol Measurement experiment, carried by Nimbus-7.

SAMS Stratospheric and Mesospheric Sounder, carried by Nimbus-7.

saturation In a digital image this refers to the maximum brightness that can be assigned to a pixel on a display device, and corresponds to a **DN** of 255. In colour theory it means the degree of mixture between a pure **hue** and neutral grey.

SBUV Solar Back-scatter Ultraviolet Instrument, carried by NOAA satellites.

scale In cartography this refers to the degree of reduction from reality that is represented on a map, usually expressed as a ratio (e.g. 1:250 000), a representative fraction (e.g. 1/250 000), or an equivalence (e.g. 1 cm = 2.5 km). A large-scale map represents ground dimensions by larger cartographic dimensions than a smaller-scale map. In common usage a large-scale feature has larger dimensions than a smaller-scale feature.

scattering An atmospheric effect where EM radiation, usually of short visible wavelength, is propagated in all directions by the effects of gas molecules and aerosols. See **Rayleigh**, **Mie**, and **non-selective scattering**.

scene The area on the ground recorded by a photograph or other image, including the atmospheric effects on the radiation as it passes from its source, to the ground and back to the sensor.

scotopic vision Vision under conditions of low illumination, when only the **rods** are sensitive to light. Visual acuity under these conditions is highest in the blue part of the spectrum.

Seasat Polar-orbiting satellite launched in 1978 by NASA to monitor the oceans, using imaging radar and a radar altimeter. It survived for only a few months.

signal to noise ratio (S/N) The ratio of the level of the signal carrying real information to that carrying spurious information as a result of defects in the system.

signature In remote sensing this refers to the spectral properties of a material or homogeneous area, most usually expressed as the range of **DN** in a number of spectral **bands**.

SIR Shuttle Imaging Radar, **synthetic-aperture radar** experiments carried aboard the NASA Space Shuttle in 1981 and 1984.

SMIRR Shuttle Multispectral Infrared Radiometer, a non-imaging **spectroradiometer** carried by the NASA Space Shuttle covering ten narrow wavebands in the 0.5–2.4 μm range.

SMMR Scanning Multichannel Microwave Radiometer, carried by Nimbus-7.

Space Station A planned series of three polar-orbiting, sun-synchronous satellites to be launched by NASA, the European Space Agency, and the Japanese Space Agency in the 1990s. They will carry a large range of remote-sensing devices.

spatial-frequency filtering The analysis of the spatial variations in DN of an image and the separation or suppression of selected frequency ranges.

specific heat The ratio of the heat capacity of unit mass of a material to the heat capacity of unit mass of water.

spectral hue A **hue** which is present in the spectral range of white light analysed by a prism or diffraction grating.

spectroradiometer A device which measures the energy reflected or radiated by materials in narrow EM wavebands.

SPOT **S**atellite **P**robatoire pour l'**O**bservation de la **T**erre, a French satellite carrying two **pushbroom** imaging systems, one for three wavebands in the visible and VNIR with 20 m resolution, the other producing panchromatic images with 10 m resolution. Each system comprises two devices which are pointable so that off-**nadir** images are possible, thereby allowing stereoptic viewing. It was launched in February 1986.

Stefan–Boltzmann Law A radiation law stating that the energy radiated by a **black body** is proportional to the fourth power of its absolute temperature.

stereopsis The ability for objects to be perceived in three dimensions as a result of the **parallax differences** produced by the **eye base**.

stereoscope A binocular optical instrument used to view two images with overlapping fields which contain **parallax differences**, as a means of stimulating **stereopsis**.

SSU Stratosphere Sounding Unit, carried by NOAA-series satellites.

subtractive primary colours The colours cyan, magenta, and yellow, the subtraction of which from white light in different proportions allows all colours to be created.

sun-synchronous orbit A **polar orbit** where the satellite always crosses the Equator at the same local solar time.

synthetic-aperture radar (SAR) A radar imaging system in which high resolution in the azimuth direction is achieved by using the **Doppler** shift of back-scattered waves to identify waves from ahead of and behind the platform, thereby simulating a very long antenna.

TDRS Telemetry and Data Relay Satellites, launched by the US Department of Defence into geostationary orbit, primarily for military communications but used for data transfer by Landsat.

texture The frequency of change and arrangement of tones on an image, often used to describe the aggregate appearance of different parts of the surface, but sometimes used for the spacing of drainage elements.

Thematic Mapper (TM) An imaging device carried by Landsats 4 and 5, which records scenes in seven wavebands, six in the visible and **NIR** with a resolution of 30 m, and one in the **MIR** with a resolution of 120 m.

thermal capacity The ability of a material to store heat.

thermal conductivity A measure of the rate at which heat passes through a material.

thermal inertia A measure of the response of a material to changes in its temperature. The apparent thermal inertia is calculated from the diurnal change in emitted thermal energy by a material.

THIR Temperature-Humidity Infrared Radiometer, carried by Nimbus-7.

tie-point A point on the ground which is common to two images. Several are used in the coregistration of images.

TIMS Thermal Infrared Multispectral Scanner, a device used by NASA to measure and create images of **MIR** energies emitted by the surface in six wavebands. Currently deployed on aircraft, but planned for use from the Space Shuttle.

tone Each distinguishable shade from black to white on a monochrome image, sometimes called **greytone**.

TOVS TIROS Operational Vertical Sounder.

training area A sample of the Earth's surface with known properties; the statistics of the imaged data within the area are used to determine decision boundaries in **classification**.

transmittance/-issivity The ratio of the EM energy passing through a material to that falling on its exposed surface.

transpiration The production and emission of water vapour and oxygen by plants.

tristimulus colour theory A theory of colour relating all hues to the combined effects of three **additive primary colours** corresponding to the sensitivities of the three types of **cone** on the retina.

variance A measure of the dispersion of the actual values of a variable about its mean. It is the mean of the squares of all the deviations from the mean value of a range of data.

VAS **VISSR** Atmospheric Sounder, carried by GOES satellites.

vector format The expression of points, lines, and areas on a map by digitized Cartesian coordinates, directions, and values.

vertical exaggeration The extent to which the vertical scale exceeds the horizontal scale in stereoptic viewing of two overlapping images with **parallax differences**. It is directly proportional to the base:height ratio.

vidicon An imaging device based on a sheet of transparent material whose electrical conductivity increases with the intensity of EM radiation falling on it. The variation in conductivity across the plate is measured by a sweeping electron beam and converted into a video signal. Now largely replaced by cameras employing arrays of **charge-coupled devices (CCDs).**.

vignetting A gradual change in overall **tone** of an image from the centre outwards, caused by the imaging device gathering less radiation from the periphery of its field of view than from the centre. Most usually associated with the radially increasing angle between a lens and the Earth's surface, and the corresponding decrease in the light-gathering capacity of the lens.

visual dissonance The disturbing effect of seeing a familiar object in an unfamiliar setting or in an unexpected colour.

VISSR Visible Infrared Spin-Scan Radiometer carried by the GOES satellites.

volume scattering Scattering of EM radiation, usually **radar**, in the interior of a material. It may apply to a vegetation canopy or to the subsurface of soil.

wavelength The mean distance between maxima or minima of a periodic pattern. In the case of EM radiation, it is the reciprocal of the **frequency** multiplied by the velocity of light.

Wien's Displacement Law A radiation law stating that the peak of energy emitted by a material shifts to shorter wavelengths as absolute temperature increases.

Index